D0962065

THE CODE

Also by Margaret O'Mara

Pivotal Tuesdays: Four Elections That Shaped the Twentieth Century

Cities of Knowledge: Cold War Science and the Search for the Next Silicon Valley

THE CODE

Silicon Valley and
the Remaking of America

MARGARET O'MARA

PENGUIN PRESS
New York
2019

PENGUIN PRESS

An imprint of Penguin Random House LLC
penguinrandomhouse.com

Photograph credits appear on page 485.

LIBRARY OF CONGRESS CATALOGING-IN-PUBLICATION DATA
Names: O'Mara, Margaret Pugh, 1970– author.
Title: The Code : Silicon Valley and the remaking of America / Margaret O'Mara.
Description: New York : Penguin Press, 2019. | Includes
bibliographical references and index.
Identifiers: LCCN 2019003295 (print) | LCCN 2019006563 (ebook) |
ISBN 9780399562198 (ebook) | ISBN 9780399562181 (print)
Subjects: LCSH: Santa Clara Valley (Santa Clara County, Calif.)—Economic
conditions. | Business enterprises—California—Santa Clara
Valley (Santa Clara County)
Classification: LCC HC107.C22 (ebook) | LCC HC107.C22 S33973 2019
(print) | DDC 338.709794/73—dc23
LC record available at https://lccn.loc.gov/2019003295

Printed in the United States of America
1 3 5 7 9 10 8 6 4 2

Set in Minion Pro
Designed by Cassandra Garruzzo

To Jeff

—

Want to know why I carry this tape recorder? It's to tape things. I'm an idea man, Chuck, all right? I've got ideas all day long, I can't control them, it's like, they come charging in, I can't even fight 'em off if I wanted to.

Night Shift (1982)[1]

On behalf of the future, I ask you of the past to leave us alone. You are not welcome among us. You have no sovereignty where we gather.

John Perry Barlow,
"A Declaration of the Independence of Cyberspace," 1996[2]

The machine that is everywhere hailed as the very incarnation of the new had revealed itself to be not so new after all, but a series of skins, layer on layer, winding around the messy, evolving idea of the computing machine.

Ellen Ullman,
Life in Code, 1998[3]

CONTENTS

LIST OF ABBREVIATIONS

ACM	Association for Computing Machinery
AEA	American Electronics Association
AI	Artificial intelligence
AMD	Advanced Micro Devices
ARD	American Research and Development
ARM	Advanced reduced-instruction-set microprocessor
ARPA	Advanced Research Projects Agency, Department of Defense, renamed DARPA
AWS	Amazon Web Services
BBS	Bulletin Board Services
CDA	Communications Decency Act of 1996
CPSR	Computer Professionals for Social Responsibility
CPU	Central processing unit
EDS	Electronic Data Systems
EFF	Electronic Frontier Foundation
EIT	Enterprise Integration Technologies
ENIAC	Electronic Numerical Integrator and Computer
ERISA	Employee Retirement Income Security Act of 1974
FASB	Financial Accounting Standards Board
FCC	Federal Communications Commission
FTC	Federal Trade Commission
GUI	Graphical user interface
HTML	Hypertext markup language
IC	Integrated circuit
IPO	Initial public offering
MIS	Management information systems
MITI	Ministry of International Trade and Industry (of Japan)
NACA	National Advisory Committee for Aeronautics, later superseded by NASA
NASA	National Aeronautics and Space Administration
NASD	National Association of Securities Dealers
NDEA	National Defense Education Act
NII	National Information Infrastructure

NSF	National Science Foundation
NVCA	National Venture Capital Association
OS	Operating system
OSRD	U.S. Office of Scientific Research and Development
PARC	Palo Alto Research Center, Xerox Corporation
PCC	People's Computer Company
PDP	Programmed Data Processor, a minicomputer family produced by Digital
PET	Personal Electronic Transactor, a microcomputer produced by Commodore
PFF	Progress and Freedom Foundation
R&D	Research and development
RAM	Random access memory
RMI	Regis McKenna, Inc.
ROM	Read-only memory
SAGE	Semi-Automatic Ground Environment
SBIC	Small Business Investment Company
SCI	Strategic Computing Initiative
SDI	Strategic Defense Initiative
SEC	Securities and Exchange Commission
SIA	Semiconductor Industry Association
SLAC	Stanford Linear Accelerator Center, later SLAC National Accelerator Laboratory
SRI	Stanford Research Institute, later SRI International
TCP/IP	Transmission Control Protocol / Internet Protocol
TVI	Technology Venture Investors
VC	Venture capital investor
VLSI	Very large-scale integration
WELL	Whole Earth 'Lectronic Link
WEMA	Western Electronics Manufacturers Association, later AEA

THE CODE

INTRODUCTION

The American Revolution

Three billion smartphones. Two billion social media users. Two trillion-dollar companies. San Francisco's tallest skyscraper, Seattle's biggest employer, the four most expensive corporate campuses on the planet. The richest people in the history of humanity.

The benchmarks attained by America's largest technology companies in the twilight years of the twenty-first century's second decade boggle the imagination. Added together, the valuations of tech's so-called Big Five—Apple, Amazon, Facebook, Google/Alphabet, and Microsoft—total more than the entire economy of the United Kingdom. Tech moguls are buying storied old-media brands, starting transformative philanthropies, and quite literally shooting the moon. After decades of professed diffidence toward high politics, the elegant lines of code hacked out in West Coast cubicles have seeped into the political system's every corner, sowing political division as effectively as they target online advertising.[1]

Few people had heard of "Silicon Valley" and the electronics firms that clustered there when a trade-paper journalist decided to give it that snappy nickname in early 1971. America's centers of manufacturing, of finance, of politics were three thousand miles distant on the opposite coast. Boston outranked Northern California in money raised, markets ruled, and media attention attracted.

Even ten years later, when personal computers mushroomed on office desks and boy-wonder entrepreneurs with last names like Jobs and Gates seized the public imagination, the Valley itself remained off to the side of the main action. An ochre haze of smog hung over its tidy bedroom suburbs when the wind didn't blow right, its dun-colored office buildings were impossible to tell apart, and you were out of luck if you tried to order dinner in a restaurant past 8:30 p.m. One horrified British visitor called it "the land of polyester hobbitry."[2]

Keeping the hobbits but losing some of the sleepiness, the Valley and its sister technopolis of Seattle soared to staggering heights in the dot-com 1990s—"the largest single legal creation of wealth we've witnessed on the planet," quipped venture capitalist John Doerr—only to plummet to earth as the new millennium dawned with a massive, NASDAQ-pummeling pop, leaving the carcasses of once-shining Internet companies strewn across the landscape. Magazine cover stories declared the end of the mania, grim-faced analysts switched their "buy" ratings to "sell," and Wall Street attention shifted back to the more predictable rhythms of blue-chip stalwarts. The rocketing rise of Amazon felt like a fever dream, Apple had run out of product ideas, Microsoft had been ordered to split itself in two, and Google was a garage operation whose leaders seemed more interested in going to Burning Man than turning a profit.[3]

How quickly things change. Fast-forward to the present, and Silicon Valley is no longer merely a place in Northern California. It is a global network, a business sensibility, a cultural shorthand, a political hack. Hundreds of places around the world have rebranded themselves Silicon Deserts, Forests, Roundabouts, Steppes, and Wadis as they seek to capture some of the original's magic. Its rhythms dictate how every other industry works; alter how humans communicate, learn, and collectively mobilize; upend power structures and reinforce many others. As one made-in-the-Valley billionaire, Marc Andreessen, put it a few years back, "software is eating the world."[4]

This book is about how we got to that world eaten by software. It's the seven-decade-long tale of how one verdant little valley in California cracked the code for business success, repeatedly defying premature obituaries to spawn one generation of tech after another, becoming a place that so many others around the world have tried and failed to replicate. It also is a history of modern America: of political fracture and collective action, of extraordinary opportunity and suffocating prejudice, of shuttered factories and surging trading floors, of the marble halls of Washington and the concrete canyons of Wall Street. For these, as you shall see, were among the many things that made Silicon Valley possible, and that were remade by Silicon Valley in return.

FROM THE FIRST MOMENT THAT SILICON VALLEY BURST INTO THE PUBLIC CONsciousness, it was awash in revolutionary, anti-establishment metaphors. "Start your own revolution—with a personal computer," read an ad for the new *Personal*

Computing magazine in 1978. "The personal computer represents the last chance for that relic of the American Revolution, our continent's major contribution to human civilization—the entrepreneur," proclaimed the tech industry newsletter *InfoWorld* in 1980.

Four years later, as they prepared to announce the new Macintosh computer to the public, Apple executives focused on marketing messages that emphasized "the radical, revolutionary nature of the product." One result was one of the most famous pieces of television advertising in history, the jaw-dropping spot broadcast into millions of American living rooms during the 1984 Super Bowl, when a lithe young woman ran through a droning audience, hurled a hammer at a Big Brother–like image projected on a blue screen, and shattered it.[5]

The barely veiled punch at IBM, Apple's chief rival, reflected a broader anti-establishment streak in this techie rhetoric that went beyond marketing plans and ad slogans. "Mistrust Authority—Promote Decentralization," read one plank of the "hacker ethic" journalist Steven Levy used in 1984 to describe the remarkable new subculture of hardware and software geeks who had helped make the computer personal. "Authority" meant Big Blue, big business, and big government.

It was the perfect message for the times. After more than ten years of unrelentingly dismal business news—plant shutdowns, blue-collar jobs vanishing overseas, fumbling corporate leaders, and the pummeling of American brands by foreign competitors—high-tech companies presented a bright, promising contrast. Instead of exhausted middle managers and embittered hard hats, there were flashy executives like James "Jimmy T" Treybig of Tandem Computers, who threw weekly beer parties for his staff and held alfresco press briefings beside the company swimming pool. There were CEOs like Jerry Sanders of Advanced Micro Devices (AMD), who bought a Rolls Royce one week and a top-of-the-line Mercedes the next. And of course there were Steve Jobs of Apple and Bill Gates of Microsoft, who came to exemplify a new sort of corporate leader: young, nonconformist, and astoundingly rich.

Then there was the man who gave his name to the era, Ronald Reagan, crusader against big government, defender of deregulated markets, standard-bearer of what he called "the decade of the entrepreneur." For the Great Communicator, no place or industry better exemplified American free enterprise at work than Silicon Valley, and he was particularly enthusiastic about extolling its virtues to foreign audiences.

During his historic visit to the Soviet Union in the spring of 1988—the first appearance of an American president there in fourteen years and a stunning move for a leader who had referred to the USSR as an "evil empire" only a few years earlier—Reagan stood before an audience of 600 computer science students at Moscow State University and rhapsodized about the glories of the American-made microchip. These miracles of high technology, the president told the crowd as a giant statue of Vladimir Lenin loomed behind his podium, were the finest expression of what American-style democracy made possible. Freedom of thought and information allowed the surge of innovation that produced the computer chip and the PC. No one better demonstrated the virtues of American free enterprise—particularly the low-tax, low-regulation variety beloved by Reagan—than the high-tech entrepreneurs ("no older than you," he reminded the students) who started out tinkering in suburban garages and ended up leading hugely successful computer companies.

The next revolution, Reagan explained that day in Moscow, would be technological. "Its effects are peaceful, but they will fundamentally alter our world, shatter old assumptions, and reshape our lives." And leading the way would be the young technologists who had worked up the courage "to put forth an idea, scoffed at by the experts, and watch it catch fire among the people."[6]

Many of the men and women who had been at the ground floor of what they called the "personal computing movement" were children of the sixties counterculture, whose leftist politics were as far from Reagan's conservatism as you could get. Yet here was one place where the hippies and the Gipper could agree: the computer revolution had a free-market soul.[7]

The revolutionary metaphors weren't new, of course. Since the age of Franklin and Hamilton, American inventors and their political and corporate patrons had made bold (and prophetic) claims about how new technology would change the world. From Horatio Alger to Andrew Carnegie to Henry Ford, politicians and journalists lifted up the figure of the ingenious, bootstrapping entrepreneur as an example and inspiration of what Americans could and should do. Only in America could you rise from rags to riches. Only in America could you be judged on your own merits, not your pedigree. In this telling, Silicon Valley seemed just like the latest and greatest example of the American Revolution in action.

RONALD REAGAN WAS RIGHT. THE HIGH-TECH REVOLUTION *WAS* AN ONLY-IN-America story. And he and so many others were right to laud people like Jobs and Gates and Hewlett and Packard as entrepreneurial heroes. Silicon Valley could never have come to be without the presence of visionary, audacious business leaders. Reagan and his conservative allies also were right when they argued that overly regulated markets and nationalized industries could present big hurdles to entrepreneurial innovation—many of the globe's would-be Silicon Valleys attest to that.

Yet, in its celebration of the free market, the individual entrepreneur, and the miracles of a wholly new economy, the Silicon Valley mythos left out some of the most interesting, unprecedented, and quintessentially American things about the modern tech industry. For these entrepreneurs were not lone cowboys, but very talented people whose success was made possible by the work of many other people, networks, and institutions. Those included the big-government programs that political leaders of both parties critiqued so forcefully, and that many tech leaders viewed with suspicion if not downright hostility. From the Bomb to the moon shot to the backbone of the Internet and beyond, public spending fueled an explosion of scientific and technical discovery, providing the foundation for generations of start-ups to come.

To declare that Silicon Valley owes its existence to government, however, is as much of a false binary as declaring that it is the purest expression of free markets in action. It is neither a big-government story nor a free-market one: it's both.

As we'll see in these pages, as important as the fact that the U.S. government invested in tech is *how* that money flowed—indirectly, competitively, in ways that gave the men and women of the tech world remarkable freedom to define what the future might look like, to push the boundaries of the technologically possible, and to make money in the process. Academic scientists, not politicians and bureaucrats, spurred the funding for and shaped the design of more-powerful computers, breakthroughs in artificial intelligence, and the Internet—a marvelous communication network of many nodes but no single command center.

Government largesse extended beyond the military-industrial complex, too. Deregulation and tech-friendly tax policies, lobbied for and especially benefiting computer hardware and software companies and their investors, helped the Valley grow large; ongoing public investments in research and education trained and subsidized the next generation of high-tech innovators. All the while, an increasing political distaste for grand government schemes and centralized planning kept political and military leaders largely out of the industry's way. Despite the millions in federal investment coursing through its veins, the region's tech cluster was allowed to grow organically, over time, largely off the political radar screen.

This freedom had unanticipated consequences. From the mainframe era forward, national politicians used a remarkably light hand in regulating the data-gathering behaviors of an industry whose technologies they only vaguely understood, but whose hockey-stick growth boosted the domestic economy. When the government-built Internet finally opened up to commercial activity at the start of the 1990s, both Democratic and Republican politicians agreed that regulation should be minimal, with companies largely policing themselves when it came to things like user privacy. All this ultimately permitted a marvelous explosion of content and connectivity on social media and other platforms, but the people designing the rules of the Internet didn't reckon with the ways that bad actors could exploit the system. The people building those tools had little inkling of how powerful, and exploitable, their creations would become.

Another twist on a seemingly familiar story: the high-tech revolution is the result of collective effort as well as individual brilliance, and many non-technologists played critical roles. Success came thanks to a vibrant and diverse cast of thousands, not just the marquee players who became the subjects of bestselling biographies and Hollywood films. Some were brilliant engineers; others were virtuoso marketers, lawyers, operators, and financiers. Many became rich; many more did not. Operating far away from the centers of political and financial power in a pleasant and sleepy corner of Northern California, they created an entrepreneurial Galapagos, home to new species of companies, distinctive strains of company culture, and tolerance for a certain amount of weirdness. It was a place filled with smart people who had mostly come from somewhere else—other ends of the country, other sides of the globe—and who had a willingness to leave the familiar behind and jump into the unknown. "All the losers came here," a tech veteran once told me in wonderment, "and by some miracle they pulled it off."[8]

The geographic and psychic separation between the Valley and the hubs of finance and government—not to mention the ivied halls of East Coast academia—was both its great advantage and its Achilles heel. Innovation blossomed within a small, tightly networked community where friendship and trust increased people's willingness to take professional risks and tolerate professional failures. Yet the Valley's tight circle, born in an era when the worlds of engineering and finance were all-white and all-male, programmed in sharp gender and racial imbalances—and narrowed the industry's field of vision about the products it should make and the customers it could serve.

Myopia extended further. The Valley's engineering-dominated culture rewarded singular, near-maniacal focus on building great products and growing markets, and as a consequence often paid little attention to the rest of the world. Why care too much about the way government institutions or old-line industries worked, when your purpose was to disrupt them in favor of something far better? Why care about history when you were building the future?

But there, again, revolutionary reality departs from revolutionary myth. For all its determination to push away the gatekeepers, dismantle ossified power structures, and think differently, the "new economy" of tech was deeply intertwined with the old.

Venture capital came from Rockefellers and Whitneys and union pension funds. Microprocessors powered Detroit autos and Pittsburgh steel. Amid 1970s stagflation and 1980s deindustrialization, when all of America was looking for a more hopeful economic narrative, old-line media and old-line politicians championed technology companies and turned their leaders into celebrities. The whole enterprise rested on a foundation of massive government investment during and after World War II, from space-age defense contracts to university research grants to public schools and roads and tax regimes. Silicon Valley hasn't been a sideshow to the main thrust of modern American history. It has been right at the center, all along.

The Valley's tale is one of entrepreneurship *and* government, new *and* old economies, far-thinking engineers *and* the many non-technical thousands who made their innovation possible. Even though every other industrialized nation has tried in some form to mimic its entrepreneurial alchemy, even though its companies have spread their connective tissue and disruptive power across the globe, it is an only-in-America story. And it is one born of a particularly lucky place and time: the West Coast of the United States in that remarkable quarter century after the end of the Second World War, where grand opportunities could await a young person with technical inclinations, the right connections, and a sense of adventure.

ACT ONE

START UP

We all grew up together, really, and it all turned out very well.

FRED TERMAN[1]

Arrivals

PALO ALTO, 1949

The sunshine hit his face as soon as he stepped off the train. Born in South Carolina, raised in Florida, and living amid the snowdrifts of Erie, Pennsylvania, David Morgenthaler missed this kind of warm and welcoming climate. It was January 1949. Tall and loose-limbed, with traces of the South in his speech and the restless energy of the North in his manner, Morgenthaler was thirty years old and already had a remarkable résumé. He had graduated from MIT at age twenty-one, turned down a job offer from GE at the age of twenty-two, and had wartime command of 300 soldiers by the age of twenty-four. The young officer built airstrips in North Africa for Allied bombers, became the Army's Chief Technical Officer in the Eastern Mediterranean, and was preparing to ship out to the Pacific when the United States dropped an atomic bomb on Hiroshima, one day after his twenty-sixth birthday.

Returning to civilian life, Morgenthaler joined an up-and-coming engineering firm in the busy industrial city of Erie. Now, he had moved on to a second entrepreneurial company, a maker of superheated steam boilers for electrical power generation, and his bosses had sent him out West to give a series of lectures on their cutting-edge products.

You have to think of it like a horse race, Morgenthaler would explain. That's how the high-tech game worked. The horse was the technology. The race was the market. The entrepreneur was the jockey. And the fourth and last ingredient was the owner and trainer—the high-tech investor. You could have the best jockey, but if he rode a slow horse, then you wouldn't win. Same thing if you had a fast horse, but a terrible jockey. Great technology without good people running the shop wouldn't get very far. And the race had to have good stakes. Riding a fast horse to a win at the county fair wouldn't reap many rewards, but the Kentucky Derby was another matter indeed. So it went with the market. There needed to be customers, and growth, not saturation.

This stop on the speaking tour: Stanford University's campus nestled against the coastal foothills thirty miles south of San Francisco. "What a lovely place to get a job," Morgenthaler thought. But when he started asking around, his hopes deflated. "We'd love to have you," his local contacts told him, "but we can't afford you." Palo Alto was a place for farmers and ranchers, or for people who were already rich. The war had made the region a hive of activity, but peacetime had returned it to rural sleepiness. Nearly everything happening in the electronics business was three thousand miles distant; local electronics companies were still small. Their finances were often unsteady. Plus, it was so far away from everything. A long-distance phone call to family back East would have cost 5 percent of his monthly salary.

Palo Alto didn't seem to have horse, race, jockey, or owner. David Morgenthaler reluctantly returned to the snows of Erie.[1]

NEW YORK, 1956

Before she got on the subway for her interview at IBM, Ann Hardy looked up the word "computer" in her college dictionary. She didn't find an entry. No matter. She wanted an interesting job, and work in computing sounded like it might be the ticket.

Cheerful, confident, and technically minded, the twenty-three-year-old Hardy had always hoped to be a scientist. She had been frustrated at every turn. First came her early years in Chicago's North Shore suburbs, in a family that deeply disapproved of female professional ambition. When she won admission to Stanford, her mother forbade her to attend. So she consoled herself by enrolling in Pomona, another California college where she could find sunshine, distance from family, and a major in science.

There, once again, authority figures got in the way. "Women can't major in chemistry," the chair of Pomona's department told Hardy flatly. Similar turndowns came from elsewhere in the hard sciences. The resourceful freshman looked further. She found her future at the edge of campus, in the university field house. Phys ed: an academic program, chaired by a woman, whose premed degree requirements included all the math and science classes a geek could desire.

After graduating in 1955, Hardy headed east to New York City and to Columbia University's graduate program in physical therapy. Within weeks came further disappointment. Doctors knew better than physical therapists, explained the professor to a lecture hall full of quietly attentive students. It would be unwise to disagree with a physician's diagnosis, publicly or privately. That did it. Hardy knew that she would not last long in a position that wouldn't let her speak her mind. Getting an MBA might help, but those programs didn't accept women. She dropped out of graduate school and began a furious round of professional networking. Surely someone knew someone who might help her find something exciting, perhaps even exceptional. At last: a man she knew from Chicago now worked as a computer programmer at IBM. They needed lots of programmers, he told her. No prior experience needed. All you had to do was pass an aptitude test.

IBM was well on its way to controlling 75 percent of the computer market on the day Ann Hardy walked through its front doors. Its plate-glass windows shone brightly around the clock, framing a clean, fluorescent-lit roomful of what newly installed CEO Thomas J. Watson Jr. called "well-dressed technicians" busily programming the newest IBM model. Young women got the most visible seats in the house. If passing businessmen saw the women working the computers, Hardy remembered the IBM brass telling her, the machines "will look simple and men will buy them."

All those undergraduate classes in the phys ed department paid off. IBM was a company whose success depended on a strong understanding of their customers, and it sought out premed types who, as Hardy put it, "cared about people" along with having a scientific background.

Hardy passed the aptitude test with flying colors and proceeded to hammer through IBM's six-week training program, emerging as one of the top three graduates in a cohort of fifty. While men with comparably high scores got offered jobs in sales, all she got offered was a spot as a "Systems Service Girl," a winsome assistant who'd help new customers learn how to work their machines. No, thank you. She'd stick to being a computer programmer. Downstairs she went, to the gleaming lobby showroom.[2]

PALO ALTO, 1957

One year later and three thousand miles to the west, another bright and ambitious young person started a new job, too. Burton J. McMurtry was just about the same age as Ann Hardy, old enough to remember depression and war, entering young adulthood in the prosperity and anxiety of the Cold War 1950s. The second son of a modestly middle-class Houston family, McMurtry set his sights early on going to college across town at tuition-free Rice University. There still wasn't much money around to pay for college living expenses, so he spent several summers out in the oilfields, where a young man could earn more than $1,000 a season.

The hard, physical work was completely different from anything they taught in school. Throughout the sun-blasted, 100-degree days, he learned how to handle a six-foot-long pipe wrench while high up on a refinery platform. His roustabout coworkers had little formal education, but they understood how to solve problems. "You learned a lot," he remembered, "about how to do things in a practical way."[3]

After graduating with an engineering degree from Rice in 1956, McMurtry landed in Schenectady, New York, as a summer intern at General Electric's microwave laboratory. He was hooked. Microwave technology was an "electronic war baby"—a tool that came into its own during World War II, its R&D fueled almost totally by military contracts. A decade later, it was turning into a big business. Those frequency radio waves had many applications, from beaming wireless television signals to cooking potatoes to generating vast amounts of energy through particle acceleration.

If Burt McMurtry wanted to get into the microwave business, his GE supervisors told him, he needed to head to Northern California. Stanford had the best graduate training in the field. Nearly every big electronics firm in the country was opening up microwave labs nearby. Applying there was a pretty simple matter back then, particularly for a sharp young man with the right references. In those days, the school accepted about half of the undergraduates who applied; for a Rice graduate with good connections, admission would be a breeze.

Since David Morgenthaler's 1949 visit, Stanford's engineering program had more than doubled in size. An added bonus for a newly married man with a baby on the way: Stanford had a cooperative program with local companies that allowed grad students to work full-time while earning their degree, all tuition free.[4]

It was just a matter of tracking down one of those jobs, and McMurtry found it with New York–based Sylvania, which had just expanded its microwave laboratory operations in Mountain View just to be close to Stanford's faculty expertise and talent pool. Not even a recruiting visit during an especially cold and rainy February day dampened his excitement at the prospect of going to Stanford and working in microelectronics. Six months later, the twenty-three-year-old and his wife, Deedee, were packing up their car and driving out West.[5]

BURT MCMURTRY AND HIS BRIDE JOINED A MASS MIGRATION OF SOME FIVE MILlion people who came to California in the 1950s, an exodus that included some of the best engineers in the country making their way to the twenty-square-mile patch of South Bay countryside. Most were just like him. They were in their twenties and early thirties, nearly all white and male. Many hailed from small towns and cities of the Midwest and Southwest rather than the metropolises of the East. Some were veterans who had picked up engineering skills while working on battleships and radar stations during the Korean War. They were Silent Generation squares who sported sharp crew cuts and neatly pressed shirts and slacks. They lacked pedigree, prep school ties, or Ivy League diplomas. Instead, they had energy, mobility, and engineering degrees—a most valuable currency in a Cold War world propelled by technology.

No single thing is responsible for the economic phenomenon that later became famous as "Silicon Valley." It came from a perfect storm of luck and circumstance, geopolitics and macroeconomics, talent and leadership, and young men in search of the sunshine. Many of these drivers of transformation were budding when David Morgenthaler made his 1949 visit. By the time the McMurtrys rolled into town in 1957, they were in full blossom.

For one big thing had changed in those eight years. The U.S. government got into the electronics business and became, in a sense, the Valley's first, and perhaps its greatest, venture capitalist.

CHAPTER 1

Endless Frontier

Palo Alto, California, in the mid-1950s was a tidy railroad village, filled with wood-frame Victorians, low-slung bungalows, and prim storefronts. Freshly built subdivisions of ranch houses fanned out from the city center, their winding roads dotted with scrawny, newly planted trees. Yet this was no ordinary suburb. Letters to the editor debated the merits of Mozart, classical records outsold rock-and-roll, and "appallingly brainy" high schoolers regularly scored in the "genius" level of IQ tests. At a time when only 7 percent of American adults had completed four years of college, more than a third of Palo Alto men had college degrees. Even more remarkably, so did 20 percent of the women. Unfortunately, all the educational firepower didn't bring much in terms of nightlife. "You couldn't pay me to live here," said one young bachelor to a visiting reporter. "Night falls," the correspondent noted, "with a deep yawn over much of this city."[1]

THE FARM

The university at the center of this sleepy outpost had been unusual from the very start. Opened in 1891 by Southern Pacific Railroad tycoon Leland Stanford and his wife, Jane, Stanford University was no liberal arts enclave like Harvard and Yale, founded to educate clergymen and the literate gentry. The Stanfords weren't pointy-headed intellectuals, and their charge was pragmatic: their university's purpose was "to qualify its students for personal success, and direct usefulness in life." They admitted women. Tuition was free. Stanford wasn't a place for the elite; it was for working-class strivers like Leland Stanford once had been himself. It was to be a place where anyone, no matter how humble, might one day become a tycoon too.

"We have started you both on the same equality," Jane Stanford told the men and women of the first entering class, "and we hope for the best results."[2]

What's more, the founders bequeathed the school thousands of acres of land with the proviso that the property could be leased, but never sold. A grand campus of sandstone and red tile—so different from the Gothic quads of the Ivy League—was only one piece of "the Farm," which stretched west, north, and south into open fields and grass-covered hills where horses and sheep grazed and students and professors picnicked and hiked.

Before World War II, a very similar landscape had stretched north and south from Palo Alto along the fertile Santa Clara Valley. The region had been best known for its prodigious harvest of prunes, occasionally making national headlines during its annual "prune week" (slogan: "Eat five prunes a day and keep the doctor away"). Travel magazines wrote glowing features and local writers penned sentimental poems praising "the Valley of Heart's Delight."[3]

Plenty of fruit trees remained in the middle of the 1950s, but prunes were beginning to give way to bigger things. Dotted with military installations and airplane factories, the state of California was the number-one recipient of the giant swell of federal defense spending during and after World War II. During the war, soldiers streamed through San Francisco on their way to the Pacific Theater. Thousands more civilians had migrated west for work in the state's shipyards and armaments factories. Many stayed for good, and the military installations and defense contractors webbed up and down the West Coast roared to life as the Cold War intensified. While Seattle, Los Angeles, and parts of the Bay Area specialized in building big—warplanes, battleships—the towns and cities that stretched south down the San Francisco Peninsula specialized in building small. Thanks to federal spending, the Valley of Heart's Delight was quickly becoming the valley of sophisticated electronics and instrumentation.[4]

The postwar explosion didn't come out of nowhere. A few Bay Area start-ups had long been in the business of making the delicate, high-tech components that gave larger computing and telecommunications machines their power: vacuum tubes, wireless transmitters, electromagnetic tape.

Stanford was a critical catalyst from the start. Seed capital from the university's first president, David Starr Jordan, and other faculty investors started up a wireless radio company called Federal Telegraph in a Palo Alto bungalow in 1909. Stanford grad Charles Litton worked at Federal Telegraph, then left to start his own wireless company in a Redwood City backyard in 1932. Litton's fellow ham radio enthusiasts Bill Eitel and Jack McCullough left a company founded by another Stanford alum to start a pioneering manufacturer of exquisite, expensive vacuum tubes needed to power radar systems. Faculty member William Hansen partnered with Sigurd and Russell Varian to invent the klystron—a foundational wave-frequency technology—in a Stanford laboratory in 1937. The brothers commercialized their innovation a decade later as Varian Associates. Last but hardly least, Stanford alums Bill Hewlett and David Packard scraped together $595 to start an electronic instrumentation firm in a Palo Alto garage in 1939.[5]

The military had a prewar presence as well. In 1930, the heyday of giant gas-filled dirigibles, the Santa Clara Valley beat out San Diego to become home to a major U.S. Naval blimp station. The tipping point came when a group of local boosters pooled resources to buy the thousand acres the Navy needed. It also didn't hurt that the man signing the authorizing legislation for the station, President Herbert Hoover, was a Stanford alumnus with strong local ties. The result was Moffett Field, a major hub of aviation and aerospace research sprawling along the newly constructed Bayshore Freeway, straddling the border of Mountain View and Sunnyvale. It opened in 1933; the National Advisory Committee for Aeronautics (NACA) opened a research center next door six years later.[6]

As tempting as it might be to read history backward, however, the San Francisco Bay Area was not unique. In the furiously industrious early decades of the twentieth century, you could find similar little clusters of young entrepreneurs in cities across the continent. High-tech products blossomed everywhere: automobiles in Detroit, biplanes in Dayton, cameras in Rochester, lightbulbs in Cleveland, radios in New York. Military installations dotted the landscape too. Yet Northern California quickly pulled away from, and ultimately surpassed, all the competition. The region did so because of extraordinary opportunity flowing its way in the 1950s—and because of extraordinary people who took advantage of that opportunity.

ARMY OF BRAINS

It began with the Bomb. To scientists and politicians alike, the technological mobilization of World War II—and its awesome and ominous centerpiece, the Manhattan Project—showed how much the United States could accomplish with massive government investment in high technology and in the men who made it. America's wartime investment not only marshaled physics to produce the most fearsome and powerful weapon in human history, but it also catalyzed development of sophisticated electronic communications networks and the first all-digital computer—technologies that undergirded the information age to come.

Even before peace had been declared, leading scientists had started to make the case for continued public spending on blue-sky technological discovery. As Cold War chill set in after the hot war's end, national security now depended on having the most sophisticated weapons in one's arsenal. Government spending on research and development shot up further. Thus began an extraordinary market disruption that accelerated the growth of new companies, sectors, and markets.

The human catalyst for much of this change was an engineering professor with an odd name and a talent for connecting people and ideas: MIT's Vannevar Bush, tapped by President Roosevelt to run the wartime Office of Scientific Research and Development (OSRD), an operation that mobilized thousands of PhDs and spent half a billion government dollars by war's end. Bush also was co-founder of one of the century's earliest high-tech start-ups, Raytheon, founded in 1922 to market gas rectifier tubes that provided an inexpensive and efficient power source for home radios.[7]

As the architects of the atom bomb labored in secrecy, Bush became the most prominent public face of the government's research efforts. A 1944 *Time* cover story dubbed him the "General of Physics." Yet the MIT man was more than a mere political operator or bureaucrat. He was an audacious, prescient technical thinker. In 1945, he published a long essay in *The Atlantic* that proposed a mechanized system for organizing and accessing knowledge. Bush called it the "memex," a machine he cheerfully described as being designed for "the task of establishing useful trails through the enormous mass of the common record." Generations since have hailed the memex as inspiration for the hypertextual universe of the World Wide Web.[8]

Bush's more immediate impact, however, came from another of his networked creations: the OSRD's great army of scientific men and university research laboratories, all mobilized at breakneck speed to provide the computational power needed to win the war. Making modern weaponry—from the millions of conventional weapons disgorged from the bellies of B-52s to the singular atomic bomb itself—was at its core a *math* problem, demanding thousands of lightning-fast calculations to determine a missile's arc, a radar system's nodes, a mushroom cloud's spread. "More than a hundred thousand trained brains working as one," was how *The New York Times* described Bush's scientists at the time, and elite brains they were: having to mobilize a team quickly, the man from MIT drew on people and institutions he knew the best.[9]

One of those people entering this close and clubby world of scientific men was a genial but intense Californian who had been Bush's very first doctoral student: Stanford's Frederick Emmons Terman. Like so many who would later make their mark on the tech industry, Fred Terman was a faculty brat, the son of famed Stanford psychologist and IQ-testing pioneer Lewis Terman. (The elder Terman's research into intelligence testing was only a shade removed from the stridently racist work of Stanford's founding president David Starr Jordan, whose investigations of human "fitness" made the university's now-notorious early reputation as a hub of eugenics research.) Fred Terman forged a wholly different path, heading east in the early 1920s to study electrical engineering at a place that at the time was still known as "Boston Tech."[10]

After finishing his MIT degree in two years, Fred Terman returned home to become the hardest-working man on the Stanford faculty, working seven days a week and relishing every moment. He spent his brief moments of leisure playing competitive bridge. When once asked why he never took vacations, Terman replied, "Why bother when your work is so much fun?" At one point, he was the main advisor for half of the graduate students in his department. Terman also embarked on a lifelong project of encouraging the best of them to strike out on their own as entrepreneurs. Nine months before Hitler's invasion of Poland, it was Terman who had persuaded two of his most beloved Stanford protégés, Hewlett and Packard, to start up their eponymous company across town.[11]

Although deeply loyal to his university and hometown, when Terman received Bush's call to come back to Boston and join the grand project of war work, he didn't hesitate. Destination: Harvard, where Terman headed up a lab devoted to "radar counter-measures." The furious pace of war work was a perfect fit. "Everything is just ducky here," wrote his wife, Sybil, to her sister, but "Fred is so busy, I don't see how he lives."[12]

Terman's wartime experience was typical. Needing to move fast and produce immediately applicable results, Bush's research operation operated largely via outsourcing, sending contracts for basic and applied research to university laboratories, and pulling in a team of experts from all over—even if it meant asking people to move across the country for the duration of the war. The prominent role of government was a significant culture shift for American universities, as was the scale of spending. Where before a department counted itself lucky to receive a private donation of a few thousand dollars for an industrial research project or a modest amount of foundation work, now it regularly received government grants many orders of magnitude larger.[13]

The bounty wasn't evenly spread, however. Massachusetts institutions alone received one-third of all the money spent by OSRD. Most of that third went to one place: MIT's Radiation Laboratory, or "Rad Lab," tasked with developing a top secret system of radar technology to win the war. (The Rad Lab was offense, and Fred Terman's lab was defense, developing technologies to jam enemy radar.) Thanks to the fat contracts going to the Manhattan Project architects affiliated with the University of California, Berkeley, California came in second. New York State, home to most of the nation's largest electronics companies, followed closely on its heels. Everywhere else lagged far behind.[14]

BIG SCIENCE

A few months after Bush's *Time* cover splashed over the nation's newsstands, Franklin Roosevelt wrote an open letter to his "General of Physics" asking for formal recommendations of how the government might encourage research on a permanent basis. The language of the president's request was as audacious as Bush's emergent idea of the memex, and likely took editorial cues from the ebullient science advisor himself. "New frontiers of the mind are before us," Roosevelt wrote Bush, "and if they are pioneered with the same vision, boldness, and drive with which we have waged this war we can create a fuller and more fruitful employment and a fuller and more fruitful life."[15]

An already ailing FDR did not live to see the results of his pitch. In July 1945,

Bush delivered his report to a new president, Harry Truman. Titled *Science, the Endless Frontier*, the brief took up Roosevelt's high-flying, politically evocative language, and ran with it. "The pioneer spirit is still vigorous within this nation," wrote Bush. "It is in keeping with the American tradition—one which has made the United States great—that new frontiers shall be made accessible for development by all American citizens." Scientific discovery could be America's manifest destiny for the twentieth century, just as westward conquest had been to the nineteenth. The way to do it: a new agency run by scientific experts, a "National Research Foundation." The next month, the detonation of atomic weapons over Hiroshima and Nagasaki gave Bush's argument about technological capacity a grim and forceful proof point.[16]

From the very beginning, then, there was a cognitive dissonance in the way America's postwar politicians and technologists talked about the world-changing upsides of high-tech investment—expanding the frontiers of knowledge, pushing out into the unknown, bettering society, furthering democracy—and the far more bellicose and disquieting reasons that this investment happened in the first place. Vannevar Bush talked about an "endless frontier," but political leaders agreed to such lavish public spending on science and tech in order to fight endless war. Bush's idea of a public agency devoted to funding basic research (now renamed the National Science Foundation, or NSF) became reality after Congress received the alarming news in 1949 that the Soviets had managed to build a bomb of their own, showing American leaders that the USSR clearly had a scientific capacity far greater than previously imagined. One year later, as the U.S.-Soviet conflict heated up with war in Korea, the Truman White House issued NSC-68, authorizing a spike in military spending—and in this age of physics-driven weaponry, this meant more money for science and technology. By the end of 1951, the U.S. had put more than $45 billion into military procurement.[17]

The "New Look" military strategy adopted by Dwight Eisenhower and his Secretary of State John Foster Dulles beginning in 1953 accelerated the shift toward advanced electronics, moving defense spending away from ground troops and conventional arms and toward ever-more-sophisticated weapons and the calculating machines that helped design them. Military planners reckoned that they'd need the electronics industry to increase its production levels five times over to accommodate national security demands.[18]

The significance of the push wasn't just the considerable amount of money America's political leaders agreed to start spending on science in the decade following the war—it was *how* they spent it. For the era also was the height of McCarthyite witch hunts, when bold, centralized government planning efforts smacked of socialism and authoritarianism. Thus, the National Science Foundation followed the precedent of Vannevar Bush's OSRD: it did not conduct basic research itself, but allocated grants to university researchers through a highly competitive selection process. "Every idea," wrote NSF officials in their first annual report, "must compete against all other ideas in the market place of science."[19]

The same thing happened on the "D" side of R&D: the Army and Navy outsourced the job of designing and building high-tech weapons to private electronics and aerospace companies, reanimating industries that had boomed during wartime and slumped after V-J Day. Defense Department officials persuaded Congress to authorize bigger tax breaks for electronics plant construction, and bought companies the expensive machines needed to build military-grade equipment.

New technologies saturated every branch and activity of the Cold War military machine. From walkie-talkies soldiers carried in their pockets to radar systems strung across continents, electronic communications equipment powered nearly every aspect of the modern military. A single bomber now carried twenty different pieces of electronics, each costing as much as the whole plane had merely one decade before. Supersonic planes demanded sophisticated electronics to help their human pilots, because "the airplanes simply blast through space faster than the human mind can think," as one aerospace executive put it. By 1955, thanks to the money being shoveled in the electronics industry's direction, it had revenue of $8 billion a year—the third largest in the United States, behind only autos and steel.[20]

THE YOUNG AND TECHNICAL

The military-industrial complex ran on people power too. Ramping up R&D required thousands of physicists, engineers, mathematicians, and chemists with cutting-edge skills and a Terman-esque work ethic. Need far outpaced supply. Altogether, the nation's universities had produced only 416 physicists and 378 mathematicians between 1946 and 1948. It was a classic chicken-and-egg problem. The military needed the best scientists. At the time, nearly all of them worked in universities. Lure these people away to work on defense projects, and frontier-busting basic research would suffer—as would the capacity of universities to produce more scientists. Pentagon spokesman Eric Walker observed dourly, "we are adding to a strategic national resource at a slow rate indeed." By 1952, the NSF estimated that the U.S. was close to 100,000 scientists short. But military planners put a positive spin on it. "In a sense," reckoned Truman's defense mobilization chief, the former GE executive "Electric Charlie" Wilson, "shortages are a symbol of our progress."[21]

For the young and technical, the world was their oyster. A New Yorker paging through the Sunday *Times* on any given weekend in the mid-1950s would have found its executive employment section stuffed with pitches aimed at the technical and ambitious. "Wanted: Scientific Frontiersmen," proclaimed Connecticut defense contractor Avco. "The scope of your future can be as LIMITLESS as the ATOM'S," exulted Boston-based Tracerlab. "Are you an engineer or draftsman who must see growth?" asked Pennsylvania's Westinghouse. Men who had worked with military technology during wartime were particularly valuable prospects ("veterans preferred," noted Sperry Gyroscope), but even more prized were the few possessing PhDs in electrical engineering, physics, or math. "If you have more to offer," IBM promised, the company would return the favor. "At IBM men find the kind of facilities, associates and climate which stimulate achievement."[22]

All these ads appeared in the "employment wanted—men" section in those days of gender-segregated classifieds, of course, and the tidal wave of technical talent hired on by defense contractors in these early years usually looked like the science and engineering classrooms in American colleges: nearly all male, all white, and under forty. "Modern science," *The New York Times* reminded its readers, "is a young man's business."[23]

Dig a little deeper, though, and you could find engineers that didn't fit that stereotype. An integrated wartime military had produced a good number of technically trained black veterans, too, and the fierce demand and shortage of engineering talent opened up professional opportunities rarely seen in a racially segregated and deeply unequal America. The nation's black newspapers held up these men's accomplishments as a credit to their race. A reader could find glowing stories about men like Raymond Hall, a graduate of Purdue University's top-ranked engineering program who worked at RCA. Or about an ex-preacher and physicist named Edward W.

Jones who supervised junior engineers and ran top secret tests at Westinghouse. "We are going to be physicists like Daddy," proclaimed his four children (three boys, one girl) to a reporter for the black daily *The Pittsburgh Courier*.[24]

The same went for women. War work had produced scores of female computer programmers, given training and opportunity not simply because so many men were at war, but also because the hardware designers believed that programming was a rote and non-technical job, comparable to a telephone operator or stenographer. The work became known as "coding" precisely because it was considered to be little more than transcription or translation, rather than the creation of original content. In wartime and after, rhapsodic press coverage celebrated the mathematical prowess of the mainframe machines—the "electronic brains"—but paid little attention to the mostly female operational labor force that made those feats possible. In the 1950s, despite the mounting evidence that programming was a creative profession that required a great deal of skill and tacit knowledge, it retained its clerical reputation—giving young women an opportunity to get in the door, and learn by doing. If those women had taken a decent amount of college science and math, like Ann Hardy, they were eligible for supervisory roles.[25]

It wasn't easy to rise up the ranks, though. Hardy managed to do it in her six years at IBM due to her programming ability, a powerful work ethic, and an unwillingness to acquiesce to the sexist nonsense surrounding her. As a result, she had the opportunity to work on groundbreaking computing projects. She became an early member of IBM's Stretch project (also known as the IBM 7030), an audacious effort to build a scientific supercomputer for the government's nuclear research at its Los Alamos National Laboratory. Like so many government-funded projects, the Stretch literally stretched the boundaries of what was possible in computing. It sold for close to $7 million; only nine were built. The feds were a demanding client. Los Alamos staff reminded IBM in early planning meetings that "high reliability [was] required" and "compact" size desired. The result was a machine that, at least for a few years, reigned as the world's fastest computer. Ann Hardy was one of the few people who knew how to program it.[26]

By the time the Stretch was done, however, Hardy had become tired of fighting the old boys' network at Big Blue. She had loyally moved "up and down the Hudson" as she transitioned from one unit of the company to another, bouncing from

New York to Ossining to Poughkeepsie as IBM demanded. Her standout programming skills had gotten her promoted to middle management, but she couldn't advance any further without an MBA—ideally, her supervisors suggested, from Harvard. Except Harvard didn't take women. "Everything I looked at, I couldn't do it," she remembered in exasperation. "There was some roadblock there." The last straw was discovering that the men she supervised all made more than she did. After she confronted the higher-ups, she got a massive raise—but one that still left her making less than the lowest-paid man on her team.

That was it. Hardy took a leave from IBM and left the East Coast altogether. If she couldn't get a Harvard MBA, she would go back to school at one of the best places in the world to learn science and engineering: the University of California, Berkeley. One year later, she joined the federal laboratory in Livermore, a top secret facility at the far reaches of the Bay Area's sunbaked eastern hills. She was no longer the only technical woman in the room. "I actually shared an office with a woman," she marveled. Corporate America was utterly unused to having women managers in the ranks; America's great military-run technical machine, in contrast, had had scores of technical women present at its wartime creation, and the egalitarian spirit lingered on. In the no-nonsense confines of what was then the heart of the U.S. military's nuclear research operation, "they didn't completely discount women like you didn't matter."[27]

Ann Hardy wasn't alone in making the westward exodus. In an era when so many Americans were on the move, and when so many companies were falling over themselves to employ the young and technical, a great many were picking up and heading toward the sunshine.

GO WEST

From his wartime office on Cambridge's brick-lined quads, Fred Terman had seen the outlines of America's high-tech future coming into sharp relief, and had become determined that his dusty and quiet little slice of Northern California would be part of it. This was Stanford's critical moment, he wrote a colleague frankly in 1943. "I believe that we will either consolidate our potential strength, and create a foundation for a position in the West somewhat analogous to that of Harvard in the East, or we will drop to a level somewhat similar to that of Dartmouth, a well thought of institution having about 2 per cent as much influence on national life as Harvard." Now that high-tech research was a national priority, becoming a powerhouse university was not only a point of academic pride. It could unlock a wave of new economic growth for an entire region. Upon returning to California, Terman set about persuading Stanford administrators to take advantage of the "wonderful opportunity" presented by the coming surge of government contracts—even if it meant reorganizing the university to do it.[28]

This would not be an easy task. Even after the war, the action in electronic computing remained overwhelmingly on the East Coast, home to companies large and small, banks and financiers, and most of the private-sector customers. Philadelphia had UNIVAC, the first digital mainframe maker, which had been commercialized from ENIAC, the legendary all-digital machine built at the University of Pennsylvania during the war years. ("UNIVAC" became shorthand for early mainframe computers in the way that "Kleenex" and "Google" later stood in as proxies for whole product or service categories.) New York had IBM, the company that triumphantly positioned itself as "a business whose business was how other businesses do business" and whose sales and marketing prowess swiftly turned it into the dominant mainframe producer.

MIT and Harvard weren't just the biggest players in the growing federal research complex—the leaders of those universities were the *creators* of that research complex. Their dominance turned Boston into the postwar era's first start-up hub, home to companies spinning out of university labs and to the first high-tech venture capital fund. When it came to the marvelous and fearsome world of the giant, blinking, beeping mainframe, the makers and the market were almost entirely restricted to the 500-mile strip of the Northeast Corridor.

Yet Terman could see what others could not. The great surge of military spending was remapping the nation's high-tech geography, knocking the East Coast off its perch as the capital of advanced electronics, and creating an extraordinary opportunity for entrepreneurial Westerners to rush in. The nuclear age had given new industrial purpose to the arid zone beyond the Hundredth Meridian, whose vast high deserts had provided the remoteness, the openness, the tiny population base needed to conduct nuclear research and testing in secrecy.

From the Grand Coulee to the Hoover to every river and cataract in between, the massive dam-building projects begun amid farm drought and Great Depression in the 1930s came online to provide cheap hydropower for the electricity-guzzling demands of the postwar aerospace industry. Up and down the Pacific Coast, cities whose military bases and shipyards had powered the fight against Japan now had factories running at full throttle, producing airplanes and missiles and all kinds of armaments to fight the many fronts of the global struggle between American capitalist democracy and Soviet communism. The Pacific tilt made aerospace companies headquartered on the West Coast—from Boeing to Lockheed to Hughes Aircraft—some of the biggest manufacturers in the nation.

For the region's universities, science was, truly, an endless frontier. MIT and Harvard and other Ivies still rested atop the list of federal grantees, but the overall pool of research spending had become so large that institutions in other parts of the country now won considerable chunks of that bounty. Research money gushed toward the universities of the Pacific West, from the evergreen-limned quads of the University of Washington in Seattle to the flower-fragrant plazas of Cal Tech in Pasadena. Buoyed by streams of public money and expanding student populations, they transformed into what University of California Chancellor Clark Kerr famously called "multiversities" of immense economic and political impact.[29]

To observers gazing at the faraway strangeness of California from the comforts of the Eastern Seaboard, it made sense that Berkeley, home to the physicists who built the atomic bomb, became one of the most important parts of the postwar research machine. Outsiders were a touch more surprised by Stanford, once written off as the sentimental folly of a nineteenth-century robber baron and his wife, a school best known for its pretty scenery, rugged football team, and Herbert Hoover—who had retired there after his electoral pummeling in 1932. Who could have imagined that this university would become a hub of cutting-edge electronics research? Who would have thought a small Northern California college town would become the capital of the high-tech world?

Fred Terman never doubted it for a minute.

CHAPTER 2

Golden State

Terman returned home ready to get to work. In his quest, he found an invaluable ally in J. E. Wallace Sterling, a historian who took over the Stanford president's office in 1949. Built like a linebacker and brimming with charm, Wally Sterling was a specialist in foreign relations who had come to Stanford from Cal Tech, giving him a strong appreciation for both the nuances of the Cold War struggle and for the growing importance of research universities within it. Sterling promoted Terman from Dean of Engineering to Provost, and gave his blessing to the reorganization of the university around what Terman called "steeples of excellence," building up programs like physics, materials science, and electrical engineering.[1]

Such a radical reshuffling could never have been possible at Harvard or MIT. But Leland and Jane Stanford had given their university an applied, engineering-first approach from the very start, and had laid down no hard-and-fast rules about how the university must organize itself. The university was also quite young—in business for barely sixty years by the start of the Korean War—and had relatively few traditions or entrenched practices that might resist the machinations of an engineer-administrator determined to turn Stanford into the perfect laboratory for the military-industrial complex.

Such license allowed Terman to not only build up basic research capacity but also move his university into even more applied work, bringing together star faculty and lab resources into the new Stanford Electronics Laboratories. The facility quickly became one of the military's most important hubs of reconnaissance and radar R&D. At a moment when America lived in fear of missiles and bombs raining out of the sky, Stanford researchers made the signal jammers and wave tubes that kept that from happening. Humanities professors howled at the shifting of resources away from disciplines that didn't have much relevance to the Cold War research enterprise. Yet the strategy proved remarkably effective. Within a few years Stanford became one of the largest recipients of federal research dollars and had vaulted up the ranks in prestige.

Terman and his colleagues also exploited Stanford's other great, unique asset: its massive real estate holdings. The nearly nine thousand acres bequeathed to the school in the Stanfords' founding grant had been something of a white elephant for years; in the prune capital of the world, the only parties interested in leasing it were farmers and ranchers. But during the postwar boom, in which not only military riches but new suburban residents poured into the San Francisco Peninsula, Stanford's acreage went from being a drag on the balance sheet to a money-making engine.

Going against the advice of consultants who encouraged the university to take advantage of the postwar suburban exodus and cover its hillsides with ranch homes and cul-de-sacs, Sterling and Terman in 1952 began developing a 350-acre chunk of open space into a research park for advanced industry. This was something few universities had ever done before (and scores would do afterward, in hopeful imitation of Stanford's model). Lucky tenants got special access to Stanford students and faculty as well as Class A commercial buildings just a short bike ride away from campus. Designed to blend in seamlessly with the homes and gardens around it, the park's buildings looked more like suburban middle schools than industrial facilities, sitting in solidary splendor amid extensive landscaping.[2]

The big bet paid off, handsomely. At Terman's urging, homegrown start-ups Hewlett-Packard and Varian Associates became the anchor tenants of the Industrial Park when it opened. And blue-chip East Coast companies like GE and Kodak agreed to join them, paying top dollar to be near university "brains" of both the human and digital variety. Firms across the region signed up to Stanford's partnership programs. Other electronics behemoths like Litton and Sylvania established microwave-tube research labs nearby. "Obviously, this is not a coincidence," Terman preened. "There was a real technical fall-out from the Stanford activity." David Packard agreed. "These people have come to Palo Alto for one reason and one reason only," he noted two years after the Park's opening. "They want to be close to Stanford University because [it] is a great source of ideas of the electronic industry and a source of well-trained engineers."[3]

The Stanford Industrial Park brought high-tech companies literally next door to campus and made them "Industrial Affiliates" who (for a fee) could obtain special access to Stanford faculty and graduate students. Professors and grad students bounced between college classrooms and high-tech start-ups, often getting in on the lucrative ground floor of companies that later became high-tech behemoths. Wherever Terman and his engineering faculty found the newest thing in industrial technology, they adjusted academic programs accordingly, ensuring Stanford produced the kinds of graduates these companies needed. As Terman once declared, the West Coast electronics industry was about "selling the products of the intellect."[4]

The new academic enterprise grew remarkably swiftly. By the time Burt McMurtry showed up on campus in the fall of 1957, he found a research park crammed with tenants and a campus brimming with star faculty and sharp graduate students. "Stanford was a candy store," he remembered. "There was this openness that was very different. Academia usually thought of itself as on a higher plane than industry, but Terman had—by force of will—insisted that Stanford was going to be outward-looking." Faculty were encouraged to spend time in industry, and were welcomed when they returned. Many students, like McMurtry, worked at electronics firms around town.[5]

The flow wasn't about transfer of technology, it was about talent—about people who moved back and forth from the labs of Stanford to the offices of its research park to the ramshackle warehouses and prefab office buildings that began stretching southward down El Camino Real. Everywhere else in the 1950s, academia was a true ivory tower, surrounded by impregnable walls between town and gown, between "pure" research and business enterprise. At Stanford, those walls dissolved.

BILL AND DAVE

By the middle of the 1950s, the firm founded by two of Fred Terman's favorite graduate students had become a potent example of how new industries could bloom in the Santa Clara Valley. Buoyed by both defense contracts and by growing commercial demand for their sophisticated electronic test and measurement devices, HP had become a powerhouse, with more than a thousand employees and

$30 million in net revenue. "A top-drawer Palo Alto electronics outfit," proclaimed *BusinessWeek*. The company also earned a local and national reputation for its distinctive business culture, one far removed from the gray-suited ranks of organization men and executive hierarchies of midcentury American capitalism. Creatures of the engineering lab instead of the Fortune 500, the two founders scoffed at modern theories of how to run a corporation—"I've never been much on management experts," Packard once observed—and they made a hands-on, nonhierarchical approach central from the start. The two founders liked to call it "management by wandering." It was the start of a broader corporate culture and organizational mission dubbed "The HP Way."[6]

At the company's headquarters at the heart of the Stanford Research Park, employees and shirtsleeved executives ate lunch together on the building's sun-drenched patio. They played volleyball and horseshoes on the lawn out back, and spent off-work hours socializing together with their wives. Although this was a high-water mark of worker unionization in the U.S., Hewlett and Packard had about as much patience for card-carrying union members as they did for three-piece suits and corner offices. Instead, to build loyalty and camaraderie, HP gave out stock.

After HP went public in November 1957, fortunes rose along with its share price. Yet from the start, the two founders consciously presented their firm as a business concerned with higher and better things. "I think many people assume, wrongly, that a company exists simply to make money," Packard once told HP managers. "While this is an important result of a company's existence, we have to go deeper to find the real reasons for our being." Nonhierarchical, friendly, a change-the-world ethos paired with an unflagging focus on market growth and the bottom line—HP created the blueprint for generations of Silicon Valley companies to come.[7]

HP's founders really were about more than just business. As his company grew, Dave Packard became one of the area's most important civic leaders, including a stint leading Stanford's Board of Trustees. A staunch Republican and champion of free markets and entrepreneurial public policy, Packard had sharp political instincts to match his ideological convictions. A friend to one generation of politicians and mentor and donor to the next, he leveraged his business reputation to become a de facto Silicon Valley brand ambassador in the world of politics. And

Packard may have been a free-enterprise man, but he was keenly attuned to how the tech industry's relationship with Washington shaped its fortunes.[8]

Like his mentor Fred Terman, Dave Packard's political education began during World War II, when he had rallied fellow executives to form the West Coast Electronics Manufacturers Association, known as WEMA, which successfully lobbied for a chunk of wartime military contracts. As electronics companies flocked to the Valley in the 1950s, Packard became the region's booster-in-chief, traversing a speaking circuit that ranged from national conferences to local Rotary breakfasts. In any venue, he made sure to point out what set the West Coast industry apart from the rest. It was fine for the Eastern electronics companies to have the lock on radio and television production, he told one audience. "The West Coast attraction has been . . . the technical and scientific aspects of the business," and that was where true growth and innovation came from.[9]

Packard also used the rubber-chicken circuit to talk politics. By the early 1960s, Packard had become an outspoken critic of the activist, expansive liberalism that was having its heyday in Kennedy's Washington. "Socialists in our government—and among our people—would place the importance of the state above the importance of the individual," Packard warned an audience in his hometown of Pueblo, Colorado, in April 1963. "They would direct our lives from Washington! They would take our wealth and distribute it as they see fit!" The way to protect individual liberties, in Packard's telling, was to unfetter markets from burdensome taxes and regulation. Lyndon Johnson's Great Society made things even worse. "Once social welfare becomes a government monopoly—as it is rapidly becoming—it is only a matter of time before we see it requires only another series of steps to put business under government monopoly," he lamented to a Palo Alto audience in 1965.[10]

Packard was not alone in articulating these views. Other postwar businessmen, especially those from the booming precincts of the Sunbelt, were becoming particularly vocal about their free-market beliefs and starting to donate generously to conservative candidates and causes. An ideological godfather to this movement—and someone that Packard had come to know very well—was ex-president and Stanford alumnus Herbert Hoover, who made the campus his home for the final three decades of his life. Hoover, too, was a fierce critic of the liberalism afoot in

Washington, the original anti–New Dealer whose conservative views had solidified into strong anticommunism as the Cold War commenced. In 1959, he turned his namesake Hoover Institution from a modest foreign-policy research institute into a powerful, sharply opinionated think tank. Packard chaired Stanford's Board of Trustees at the time, bestowing the institutional seal of approval permitting this transformation.[11]

Hoover's new chapter opened with a bang. "The purpose of this institution," the former president wrote in the institution's newly revised charter, "must be, by its research and publications, to demonstrate the evils of the doctrines of Karl Marx." To keep "left-wingers" from taking control, Hoover personally selected the institution's new director, a thirty-five-year-old economist named W. Glenn Campbell.

Protests from the faculty ensued almost immediately—Fred Terman was entering into unholy alliances with industry, and now Herbert Hoover was playing politics?—but Hoover, with Packard's assist, prevailed. The institute would remain independent from the rest of Stanford, and Campbell would report directly to President Sterling. Campbell proved to be exactly the man the institution needed. Over a close to thirty-year tenure as its leader, he turned the Hoover Institution into a powerhouse with a massive endowment and a reputation as the nation's premier conservative think tank.[12]

Thus, at the same time that Fred Terman was turning the Farm into the nation's most entrepreneurial technical university, Glenn Campbell was making the campus home to an all-star roster of conservative thinkers and politicians—undeterred by the periodic angst they stirred up in the more liberal campus precincts surrounding Hoover Tower. In the decades to come, Packard became one of the Hoover Institution's most faithful and generous donors.

Meanwhile, Dave Packard continued to nurture WEMA and its growing member base of young California-based electronics companies. The industry was so new that more seasoned folk didn't have qualms about helping newcomers. "Executives felt free to reveal inside information about getting started, or how to grow a company," recalled one entrepreneur. While still ferocious competitors, the Westerners knew they were underdogs who'd fare better if they stuck together rather than going it alone.[13]

LIFTOFF

Despite Northern California's later legendary reputation for growing start-ups, locally born companies like HP and Varian were a small piece of the region's story in the 1950s. Company founders and entrepreneurs had not reached the mythical, kingly status they would one day attain in the Valley. Those that went out on their own were outliers.

Burt McMurtry saw this firsthand. "If you were starting a company," he observed, "it probably meant that you were odd and couldn't work for people." The most visible, interesting, and prestigious places to work were large national firms. They had the money and clout to recruit top-notch staff, build out top-end facilities, and secure the largest defense contracts. The majority of the engineers in the Valley worked for companies that already comfortably rested in the Fortune 500.[14]

The two biggest of these players—and shapers of what the Valley would become—were Lockheed and IBM. Big Blue opened a Northern California electronics research laboratory in 1952. Father and son CEOs Thomas Watson Sr. and Jr. were beginning their major push into digital computing and were having little luck persuading California-based engineers to return to upstate New York snowdrifts. After five years in a modest building in downtown San Jose, IBM built a lavish new manufacturing facility that won one trade magazine's 1957 "Factory of the Year" award for its "carefully cultivated 'campus' atmosphere." *San Jose Mercury* columnist Frank Freeman raved that the IBM plant was "so different from the usual pell-mell workaday experiences that it's like being in another world, a sort of Buck Rogers' world . . . peopled by young brains, no long hairs."[15]

However, the real Buck Rogers action happened at the local branch of another national company, a facility closed off to casual visitors because of the top secret nature of most of its operations. Right along the freeway, next door to the looming government airship hangar at Moffett Field, sprawled a giant plant that was the Valley's largest high-tech employer for decades: Lockheed Missiles and Space Company. Here's where the Cold War economy of the region had its beating heart.

Lockheed had come up from Southern California to open up shop in Sunnyvale in 1954, drawn by a desire to be close to Stanford's electronics experts and the high-speed aerodynamics research going on at NACA's Ames Aeronautical Laboratory.

Lockheed soon became one of the most stalwart and eager of Fred Terman's industrial affiliates. Another reason for the move: security. With a nation on high alert at the possibility of Soviet bombs raining down from the sky, the Defense Department encouraged its contractors to locate in "dispersed" areas away from major population centers that, presumably, would be ground zero for attack. For Lockheed, it also made sense not to have all of its many sensitive military operations located at its Southern California headquarters. The Missiles and Space division moved north. Thus properly dispersed, the company soon landed the prime contract from the Air Force for a new missile-launched satellite.[16]

The next year, the company won another huge contract, for the Polaris, one of five giant, long-range ballistic missiles that made up an extraordinary federal push to find ways to zoom ever-more-powerful nuclear weapons halfway across the planet. They were among the highest of high tech at the time, marrying two technologies of World War II—the atomic bomb and the German V-2 rocket—into a deadly and powerful delivery system. By the spring of 1957, the missile program was entering its first flight-test phase, and was already estimated to cost two times the price of the entire Manhattan Project. One Stanford researcher confidently predicted that the missile program would result in "the greatest explosion of scientific devices" in history. He wasn't that far off the mark.[17]

In addition to bringing big-ticket government contracts to the Santa Clara Valley, Lockheed brought people, people, and more people. By the end of the 1950s, tens of thousands of engineers streamed daily into Lockheed's doors in their white shirts and narrow ties, working on cutting-edge technologies so top secret that they couldn't tell their families over the dinner table what they did at work that day. Nearby suburbs filled with Lockheed men and their wives and children, further skewing the Valley's demographics toward the white, the middle class, and the college educated.

Plenty of blue-collar workers found jobs in Sunnyvale as well; like the other electronics firms of the era, Lockheed had assembly lines alongside its research labs. But the lack of diversity carried through even in jobs that didn't require an engineering degree. In an era before affirmative action, Lockheed and the other major electronics firms of the Valley had no pressure to recruit minority or female employees. Even after the enactment of federal minority hiring laws that required contractors like Lockheed to meet certain hiring targets, the percentage of Latino,

Asian, and black workers at its Sunnyvale facility only reached 10 percent. More than 85 percent of its workforce was male.[18]

THE MISSILE MAKER, THE ENTREPRENEURIAL UNIVERSITY, THE DISTINCTIVE business sensibility, the professional networks, the government money, the elite (and homogeneous) workforce: many of the key ingredients were coming together in Palo Alto by the middle of the 1950s. Some people were also starting to make considerable amounts of money, as the Valley's first generation of tech stars—Varian, HP, Ampex—had splashy public debuts on Wall Street.

But HP and other early electronics firms were not computer companies. The Valley made instrumentation and communication components—oscillators, high-frequency radar, magnetic tape—not digital hardware and software. Its biggest tech employers had corporate headquarters elsewhere. Over time, Lockheed would subcontract out a good deal of its work, driving business to other, smaller Valley companies, but in the 1950s it was a mostly closed world.[19]

In many respects, mid-1950s Santa Clara Valley was merely a smaller-scale Los Angeles, home to aerospace companies, light manufacturing, and some smart academic scientists. While California was gaining a reputation for being an excellent place to hire an electrical engineer, it didn't offer much in terms of capital and operational support for those who wanted to strike out on their own and start a new company. To do that, you really needed to return east, to the place most tightly connected into the Cold War research machine: postwar America's first start-up capital, Boston.

For Boston had MIT, which continued to rake in more federal money than any other university, and which had strong ties to Pentagon brass and the largest East Coast electronics incumbents. MIT's famous Rad Lab had disbanded after the war, but spectacularly innovative military-funded laboratories had risen in its stead: MIT's Lincoln Laboratory, home to high-powered digital computing, and Instrumentation Lab, which designed guidance systems for rockets and missiles; the Air Force Cambridge Research Laboratories, devoted to radar defense. Down the way was Harvard, the second-largest recipient of federal dollars, boasting an ivied intellectual pedigree and a large stable of faculty stars, including the imperious and mesmerizing father of high-tech venture capital, Georges Doriot.

Lockheed Missiles and Space might rule the Valley, but Boston had Vannevar Bush's Raytheon, now one of the nation's most important defense and aerospace contractors, and plenty of others. In the New England countryside encircling Boston's metropolitan fringe ran "America's Technology Highway," Route 128, and a string of modern research parks and corporate campuses that lured electronics firms out of grubby Cambridge and into the sprawling suburbs.

Yet even for all its power and might, Boston, too, was missing a key ingredient, something that David Morgenthaler later characterized as the ultimate "enabler," a cheap and potent "fuel" to propel electronics in the same way that refined oil propelled the auto industry. It was the magic something that could make already sophisticated electronics even faster and smaller, something that could scale up the Cold War's "products of the intellect" to market-disrupting proportions.

That something was the silicon transistor.

SHOCKLEY & CO.

In the same busy summer of 1957 that Burt McMurtry prepped for his cross-country move, Stanford's chair of electrical engineering dispatched another young Texan to a one-year-old start-up in a refurbished Quonset hut next to a suburban Sears store in neighboring Mountain View. The company was Shockley Semiconductor Laboratory, and the man detailed there was a brand-new professor named James Gibbons. Born and raised in Texarkana (he and future computer magnate and presidential candidate H. Ross Perot were high school classmates), and attending MIT before going on to graduate study at Stanford, Jim Gibbons had recently returned to his graduate alma mater after a prestigious research fellowship in England. His job offer came with one condition: that he spend the first six months embedded at Shockley, the hottest new firm in town. Gibbons's marching orders were to learn about semiconductor manufacturing from the company so that Stanford could "transistorize the curriculum" and build a solid-state electronics lab of its own.[20]

The young assistant professor arrived to find a firm filled with some of the brightest engineers in town. But the shop was already becoming a toxic workplace, thanks to its brilliant and mercurial founder, Palo Alto native and co-inventor of the transistor, William Shockley.

Shockley had departed the nation's premier industrial research facility, Bell Laboratories, after becoming convinced that germanium (the material used for the first generation of transistor technology) was too weak and unreliable to power the tiny switches at the heart of electronic circuitry. Instead, he set out to commercialize transistors made with purer, stronger silicon. Unmarried and untethered, Shockley considered Southern California as a possible location, but ultimately decided to come home. His much-beloved elderly mother was unwell. Plus, his fellow hometown boy Fred Terman was making a pretty persuasive case that Palo Alto—and Stanford—would be a fitting destination.[21]

While brilliant, Shockley had an eccentricity that was sometimes charming (he loved playing magic tricks with his staff) and sometimes difficult (he made all prospective hires take a battery of IQ tests and brain teasers). Tellingly, he had been unable to lure anyone who'd worked with him at Bell Labs to join him in the sunshine. So he recruited a group of young stars from other industrial labs and universities, a group mostly from modest backgrounds, whose greatest pedigree was their technical talent. One, Robert Noyce, was the son of an Iowa clergyman. Another, Jay Last, came from a family of Pennsylvania schoolteachers. A third, Eugene Kleiner, arrived in the U.S. as a teenage refugee from war-torn Europe. Only one was actually from Northern California, the shy and detail-obsessed Gordon Moore, who had grown up in a modest clapboard cottage in nearby Menlo Park.[22]

The young recruits quickly concluded that Shockley was going about building his semiconductors in a wrong-headed way. He was committed to an expensive and laborious process called the four-layer diode, and refused to be persuaded that cheaper, simpler silicon chips were the way to go. Jim Gibbons showed up at the storefront just weeks before these Shockley lieutenants—Noyce, Last, Moore, Kleiner, plus four others—quit to start a company of their own called Fairchild Semiconductor, which quickly surpassed and outlasted Shockley's operation. Before they went out the door, they invited the young professor to join them. He said no. "It was probably the most expensive decision I ever made," Gibbons said later.[23]

The exit became Valley legend and inspired hundreds of high-tech ventures to come. Tiring of a cantankerous boss who refused to hear their ideas, the men of Shockley Semiconductor wrote a letter to Wall Street investment bank Hayden, Stone, & Company where Gene Kleiner's father had some business dealings. It was the ultimate shot in the dark: eight scientists who'd never run anything more than a few laboratory experiments, writing someone they barely knew, asking for help finding someone who'd hire them all as a group, to start a new venture making high-tech devices that nearly no one outside the scientific community had heard of. And they wanted to stay in California to do it. The banker who received the letter didn't know a thing about semiconductors, so he passed it along to the firm's junior high-tech analyst: Arthur Rock.

Rock wasn't your typical white-shoe banker. He was from Rochester, New York, the son of Jewish immigrants who owned a candy store. From there, he had made it to Harvard Business School and then to Wall Street, taking a short detour in between to run the New York Young Republicans' campaign for Eisenhower-Nixon '52. Rock wasn't an engineer by training, so his investment instincts were people-driven. "Good ideas and good products are a dime a dozen," he later explained. "Good execution and good management—in a word, good *people*—are rare."

Rock looked at the Shockley Eight and saw people much like him: family men in their early thirties, strivers with first-rate credentials, willing to buck the system just a little bit to get things done. In short, good people. Enough for Rock to branch out from his day job and go on the hunt for an angel investor—someone comfortably rich, but quirky enough to take a willing gamble on a new technology and a pack of unknowns. He found Sherman Fairchild, an eclectic high-tech enthusiast who had become a multimillionaire thanks to his inheritance of a massive amount of IBM stock. The eight scientists of Shockley became the founding employees—and shareholders—of a new company called Fairchild Semiconductor.[24]

Modern Silicon Valley started with Fairchild and the "Traitorous Eight" who founded it. Financed by an eccentric trust-funder in a deal brokered by an East Coast financier, the firm's origins underscored how tightly wedded the Valley was to outside, old-economy interests from the very start. But this was a new take on the relationship—not a company that was merely a Californian outpost of an Eastern electronics giant, but a wholly new enterprise founded by the engineers themselves. Noyce and Moore went on to cofound Intel, bringing a Fairchild colleague named Andy Grove along. (Grove, like Gene Kleiner, had been a teenage refugee, escaping war-torn Hungary for a new start in opportunity-rich postwar America.) Kleiner became the founder of one of tech's most influential venture capital firms, funding and shaping generation-defining companies from the PC era to the age of social media. Rock moved West in 1961, partnering with a young money manager named Tommy Davis to form Davis & Rock, leading to later investments in Intel and Apple. Other Fairchild founders and early employees founded other semiconductor companies that made billions and replaced the mechanical innards of nearly every consumer product with tiny microchips.

The company founded with IBM money didn't make computers, but it was the spark for the technologies that ultimately upended the mainframe market Big Blue had dominated for so long. Most important, the men of Fairchild established a blueprint that thousands followed in the decades to come: find outside investors willing to put in capital, give employees stock ownership, disrupt existing markets, and create new ones.[25]

The Fairchild founders took a huge gamble in abandoning a Nobel Prize–winning legend to set out on their own. But it turned out that their timing could not have been better. A mere three days after the Traitorous Eight officially incorporated their company, the Soviet Union launched the Sputnik satellite.

CHAPTER 3

Shoot the Moon

Eighteen thousand miles per hour. That was the orbiting speed of the beach ball–sized hunk of metal the Soviets shot into space that Friday evening in early October 1957. By the time news of the launch hit America's Saturday morning papers, the satellite was on its tenth orbit, soaring more than five hundred miles above the Hudson River Valley as Ann Hardy was having breakfast at her kitchen table in Poughkeepsie. Seconds later, it flashed over Boston as Vannevar Bush puffed on his pipe in his book-lined study. By its eleventh loop, it was well past Cleveland, where thirty-seven-year-old David Morgenthaler was raking early autumn leaves and thinking about the big new job he would start in a few days—after a decade moving up the management chain, he was a company president at last, running the American branch of a British chemical operation. The satellite sailed southward over the military bases of Alabama and Texas, past the sun-bleached expanse of the nuclear test sites of New Mexico and Nevada, above California's and Washington's vast airplane factories and the great dams whose hydropower fueled them.

The morning dew was still burning off the Bay Area hills when the metal orb passed overhead on its twelfth trip around the earth, far above Palo Alto's oak-lined streets and the modest apartment where Burt and Deedee McMurtry had only just cleared out the cardboard boxes left over from their cross-country move. McMurtry was only two weeks into his new life as a Stanford graduate student. He'd been at Sylvania barely longer than that. As Northern California's young newcomers like the McMurtrys sipped morning coffee on their back patios, squinting up into the sky in wonderment at the tiny machine soaring up beyond their sight, they had little inkling of how much Sputnik was going to change everything.[1]

SPACE RACE

The Sputnik I launch came as an utter surprise to the U.S. political establishment and upended contented assumptions about American scientific supremacy. "The Russians could no longer be regarded as 'backward' and 'beaten,'" remembered then President Dwight D. Eisenhower some years later. "There was no point in trying to minimize the accomplishment or the warning it gave." Washington, D.C., had barely recovered from the shock when another gut punch arrived: the early November launch of a second Soviet "moon," this time an 1,100-pound air-conditioned spaceship bearing a stray dog from the streets of Moscow named Laika. (America's horrified dog lovers protested the cruelty of sending an animal into orbit with no hope of returning alive. Reporters, desperate to inject some humor into the grim situation, dubbed the pooch "Muttnik.")[2]

Then, landing with a thunk on the president's desk a matter of days later, came the report of a civilian panel commissioned by Eisenhower earlier in the year to assess the nation's ability to withstand a Soviet nuclear attack. The committee, chaired by San Francisco attorney and RAND Corporation co-founder H. Rowan Gaither, had very bad news.

Despite the billions spent by the Americans on ballistic missile development since the start of the 1950s, despite the money poured into Eisenhower's "New Look" military, they'd been outpaced. Despite having a far smaller economy, the Soviets had spent about as much. They had trained far more scientists, developed technology much more quickly, and were on a fast track to go even further. America's Cold War enemy now had rockets powerful enough to put a dog in space, which meant they had the capacity to send nuclear warheads thousands of miles across the ocean—and could do so with such speed and force that U.S. civil defense systems would be utterly overwhelmed. All the schoolroom duck-and-cover drills and backyard fallout shelters would be for naught; the Soviets now possessed the ability to annihilate whole cities. The only way to counter this threat, this committee made up of defense industry executives and advisors informed the president, was to close "the missile gap." The panel projected that would require a boost to defense R&D outlays of more than $40 billion.[3]

In a matter of weeks, the top secret Gaither Report leaked to the press, and Sputnik's autumn of anxiety ballooned into a full-blown political panic. Ike Eisenhower deeply disliked untrammeled government spending, but the combined pressure from his own advisors and a Democratic-led Congress was too strong to resist. The defense contracting spigot opened into a firehose. Dollars flowed out of D.C. to propel ever-more-powerful missiles up into the heights of the atmosphere and down into the depths of the ocean. Further billions poured into the Strategic Air Command, the radar- and transistor-driven communications network on which American military survival now depended.

By the autumn of 1958, the nation had a new, considerably beefed-up space agency, the National Aeronautics and Space Administration, or NASA. And Congress agreed to fund a wholly new agency within the Department of Defense devoted to state-of-the-art space and satellite research, a place where researchers could, in the words of Secretary of Defense Neil McElroy, "follow these various will-of-the-wisps . . . and carry them through to a point where there can at least be a determination of their feasibility and what their probable cost might be." The little operation became known as the Advanced Research Projects Agency, or ARPA.[4]

Another huge beneficiary of the post-Sputnik boom: higher education, as the president's reliance on the academics he called "my scientists" intensified. Responding to pressure from Congress, Eisenhower named MIT President James R. Killian Jr. the first presidential science advisor. While the different branches of the military fought over leadership of newly enlarged research programs, Killian and the other scientists advising Eisenhower pointed out that universities were a logical place to nest much of this activity. They already were important partners in the Cold War fight, and the race into space required basic research that only universities could perform. Plus, outsourcing to universities would keep an already big government from getting even bigger.[5]

By the end of the following summer, the boost to universities spread far further with passage of the National Defense Education Act (NDEA) of 1958, which shoveled millions in public funds toward building classrooms and labs, hiring faculty and research staff, and boosting scholarships so that America could produce as many scientists and mathematicians as the Soviets. It was a new era. "With all their irritating faults," the science advisors told Eisenhower, "universities are essential agencies of our national hopes, and they must be treated accordingly."[6]

The surge of new spending on missiles, strategic defense systems, and the brains who made them dwarfed all that had come before, including the Manhattan Project itself. Dwight Eisenhower was never entirely comfortable with the world that had grown on his watch, warning as he left office in January 1961 of the creeping influence of what he dubbed "the military-industrial complex" on American life. But he had set a seemingly unstoppable flywheel in motion.[7]

Its speed accelerated further when Eisenhower's successor, John F. Kennedy, proclaimed within months of taking office that American astronauts would reach the moon by the end of the 1960s. Bruised by the Bay of Pigs crisis and needing to burnish his foreign policy credentials, Kennedy saw the moon shot as the way to definitively prove American scientific prowess to the world—and to the American people, most of whom still worried that Russia would reach the moon first. "We choose to go to the moon in this decade and do the other things, not because they are easy, but because they are hard," Kennedy declared in September 1962 in Houston, as the foreign policy stakes of the gambit shot higher in the wake of the Cuban Missile Crisis.

Kennedy spoke these words on a blazingly hot South Texas day, before an enthusiastic crowd at Burt McMurtry's alma mater, Rice University. NASA's newly established Mission Control Center was just across town. "Those who came before us made certain that this country rode the first waves of the industrial revolution, the first waves of modern invention, and the first wave of nuclear power, and this generation does not intend to founder in the backwash of the coming age of space." The goal was audacious, the costs stratospheric, but the challenge was "one which we intend to win."[8]

Propelled by missiles and the moon shot, R&D made up more than 10 percent of the entire U.S. federal budget for the first half of the 1960s, and further skewed the high-tech world's emerging Sunbelt tilt. In the far West, already busy aerospace centers in Southern California and the Pacific Northwest revved up even further. *The New York Times* enthused in early 1963 that Los Angeles had become "the Detroit of the space industry." While ex-President Eisenhower grumbled about the Kennedy Administration's "fiscal recklessness" when it came to the space program, NASA transformed all the places it landed.[9]

The Santa Clara Valley was no exception. Eisenhower's missile program and Kennedy's moon shot boosted demand for precisely the kinds of things the region's electronics labs already developed and sold: microwaves and radar to track satellite trajectories, oscillators and transistors to provide lightweight and potent energy sources, and networks to communicate with spacecraft hurtling beyond the outer atmosphere. Stanford's quads of sandstone and tile sprawled outward amid the bounty of post-Sputnik higher education spending; tenants paid top dollar to lease space in its research park.

For nearly all of the 1950s, this dusty, blossom-bedecked slice of California was a place of industrious young engineers who mainly worked for very large electronics companies. In the post-1957 space age, start-ups began to grow—not only as a result of the technological breakthroughs born in the Valley during this period, but also because of the ways in which military and aerospace contracting changed—more money flowing, and flowing in ways that opened up significant commercial opportunity for new entrants. When it comes to explaining why Northern California eventually came to have such an outsized role in the high-tech universe, the missiles and rockets of America's race into space shoot right through the center of the story.

THE EYES OF THE WORLD

By the time Sputnik burned out in orbit three months after its launch, the Valley already was feeling the change in atmosphere. At Lockheed Missiles and Space in Sunnyvale, which had been conducting space and satellite work since its inception, "suddenly 'hurry up' became 'hurrier up,'" one employee remembered. Within a matter of months, the division became Lockheed's biggest and most lucrative unit and the Valley's largest and richest employer, its 300-acre campus along U.S. 101 employing 19,000 people and raking in close to $400 million in sales in 1959 alone.[10]

It was Lockheed's Sunnyvale command center that guided the first American satellites to make the round trip into space, beginning with the *Discoverer* series that took a capsule into orbit and brought it back safely late in the summer of 1959. The payload of that first successful mission wasn't dog or man, but an American flag that traveled half a million miles during its 26½-hour journey. The capsule toured the country that fall, making stops in the U.S. Senate and at aerospace conventions before ending up on display back in Sunnyvale. For a facility built on secrets—nearly everything that happened at Lockheed was classified research, never to be shared with those without the right security clearances—it was a rare opportunity for employees' families and curious neighbors to come visit the mysterious campus and see just a little bit of what it was about.[11]

Down the road at Stanford, the combined boon of more science spending and a busier Lockheed meant that Fred Terman's big bets on physics and applied technology paid off even more handsomely. A beefed-up Strategic Air Command and satellite surveillance program increased demand for the traveling-wave tubes and signal jammers of the Stanford Electronics Lab. Money pouring into Sunnyvale for aerospace work spilled over into the engineering labs on the Farm. Then, in 1959, the Eisenhower Administration endorsed Stanford's bid for a federally sponsored particle accelerator for high-energy physics research. Built at a cost of more than $100 million, the giant Stanford Linear Accelerator soon sprawled out across rolling grasslands in the countryside a few miles west of campus.[12]

With all this happening, the world began to take notice. At the 1958 World's Fair in Brussels, a crisp diorama of Stanford Industrial Park was a standout feature of America's industrial pavilion, a symbol of the clean and modern knowledge work being made possible by the United States' commitment to applied technology. "Of the nine parks featured in the display," the *Stanford University Bulletin* proudly informed its readers, "the co-sponsors considered the Stanford Park the most photogenic."[13]

Then, in the fall of 1959, none other than Soviet Premier Nikita Khrushchev came to visit IBM San Jose as he made a swing through California on the heels of his "kitchen debate" in Moscow with Vice President Richard Nixon. IBM executives were apprehensive about the visit. Khrushchev had caused a stink in Los Angeles a day earlier after not being allowed to visit Disneyland. He showed up in San Jose wearing a longshoreman's cap given to him by the firebrand San Francisco labor organizer Harry Bridges earlier that morning. Yet the premier genially allowed himself to be ushered around the facility by CEO Tom Watson Jr. He had a roaring good time at lunch in the employee cafeteria. While Khrushchev "was effusive in his praise and flattery" for America's technical achievements, one reporter noted, he "promptly added that it would not last long."[14]

Six months later, another presidential visit to the Valley's orchards and cul-de-sacs: Charles de Gaulle of France, on a visit to San Francisco, asked to see the marvelous research facilities down by Stanford in its industrial park before he left. The buzz from the Brussels exhibition had given the park a global reputation, and de Gaulle wanted to see what all the fuss was about. His motorcade trawled down Skyline Drive to Palo Alto, *le General* resplendent in a convertible limousine, the French tricolor flapping on its hood in the spring sunshine. The town's quiet neighborhoods came alive with spectators. On tree-lined Waverly Street, down the way from the original garage of Hewlett and Packard, a teenage prankster dressed up as Napoleon and lay in the gutter as the presidential procession went by. Decades later, lifelong residents remembered the strange and exciting sight of de Gaulle's convertible as it swept through the Town & Country Village shopping center, his familiar Gallic profile framed by the incongruous backdrop of Stickney's Hick'ry House and the Village Sudsette.[15]

The ultimate destination of this scenic driving tour was HP, where Bill Hewlett showed him the company's assembly line and some of its latest space-age models. De Gaulle nodded pleasantly, thinking all the while about the ways France might build similarly gleaming California-style factories and laboratories. Undersecretary of State Douglas Dillon, the president's official host for his California visit, declared that the president had been "much impressed." Five years later, increasingly alarmed at U.S. computer makers' aggressive forays into French markets, de Gaulle announced "Le Plan Calcul." The massive, multiyear effort toward building a homegrown French computer industry featured expansive education programs, corporate partnerships, and—yes—research parks.

De Gaulle was among the first, but hardly the last, to look to the Santa Clara Valley as an alluring model well worth imitating. Worshipful delegations of curious officials began to make pilgrimages in the French leader's wake: parliamentarians from Japan, university administrators from Canada, economic development functionaries from Scotland. Before the 1960s were out, Fred Terman had started traveling the globe as an economic development consultant, explaining the secrets of his industrious little valley to eager foreign audiences.[16]

By that time, even more of the world's eyes were watching, for the region had become known not just for HP and Lockheed and the perpetual motion machine of Fred Terman, but for a new breed of tech firms. These were operations for whom defense contracting wasn't the primary market, and whose products were making all kinds of machines—from computers to cars to contraptions running assembly lines—faster and lighter and more powerful. They were the companies that put the silicon in Silicon Valley, and the space race helped launch them into the stratosphere.

FUNNY LITTLE COMPANIES

This was what happened to Fairchild Semiconductor. Three months into their start-up's existence, without having yet made a single chip, the Traitorous Eight landed a contract to manufacture 100 silicon transistors for an onboard computer for "the manned missile," a new long-range bomber. Wisely, Bob Noyce and Gordon Moore were adamant that Fairchild conduct its own research, rather than depending on government contracts that would not let them own resulting patents. "'Government funding of R&D has a deadening effect upon the incentives of the people," declared Noyce. "This is not the way to get creative, innovative work done." But having the government as a customer? That was less of a problem. In 1958, 80 percent of Fairchild's book of business came from government contracts. That was only a preview of an even bigger windfall.[17]

In those Fairchild labs in early 1959, Jean Hoerni discovered a way to place multiple transistors on a single silicon wafer by protecting them with a coating of chemical oxide. Hoerni's "planar process" allowed his colleague Bob Noyce to experiment with linking the transistors together, creating an integrated circuit, or IC, more powerful than any device before it. Another advantage: the material. Back in Dallas, Jack Kilby of Texas Instruments came up with the same idea nearly simultaneously, fabricating his IC using germanium instead of silicon. Noyce was the first to file a patent, in 1961, spurring a fierce patent fight between the two companies. Both men eventually were credited with the invention, but Noyce's silicon device proved easier to manufacture at volume, giving the silicon-powered California firms an edge over the germanium-powered producers elsewhere.[18]

These elegant and tiny devices were no longer transistors—they were "chips" that would launch an entire industry and, ultimately, power a personal computing revolution. But that was all in the future, and the IC was painfully expensive to make and market. The first chips produced by Fairchild cost about $1,000 apiece—far more than ordinary individual transistors. Why would a corporate customer need such an expensive device? The ICs were so technical, so bleeding-edge, that people outside the small electrical engineering fraternity really didn't understand their capabilities to store enormous amounts of information in a sliver of silicon no bigger than the head of a pin.

NASA understood. The space agency began using ICs in the guidance system of the Apollo rockets, then to improve the guidance system for Minuteman intercontinental ballistic missiles. Fairchild received hefty contracts for both. By 1963, demand created by the Apollo and Minuteman programs had driven the price of silicon chips down from $1,000 to $25—now putting them at a price point where a much broader market might buy in. The feds remained a stalwart customer, however. As the Apollo tests continued, the demand to go tinier, lighter, faster intensified. "Make those babies smaller and smaller," the military brass barked at the chipmakers. "Give us more bang for the buck."[19]

The demand for building more cheaply came from the top. Eisenhower had always been anxious about the influence and expense of the defense complex, and he had ordered DOD to trim its budgets toward the end of his term. His Democratic successors accelerated the cost-cutting—less out of nervousness about the power of the generals, and more because they had so much else they wanted to do. Kennedy and Johnson had bold ambitions to wage war on poverty, deliver new social programs, go to the moon, and cut taxes at the same time. And then there was that increasingly expensive little war in Vietnam. Deficit spending could only go so far; more fiscal reshuffling needed to happen, particularly in the defense and aerospace budget, where uncapped, sole-source contracts had led to ballooning project costs that the military wasn't well-equipped to control.

With a push from the White House and Congress, defense agencies opened contracts to competitive bidding and started to adopt fixed-price contracts as the standard. As Johnson put it in an address to Congress in the first grieving weeks after Kennedy's November 1963 assassination, the government needed to get "a dollar's value for a dollar spent." To do that, Defense Secretary Robert McNamara added pointedly, the military-industrial complex needed to bring in new players. "Subcontracts should be placed competitively," he said, "to ensure the full play of the free enterprise system." The pugnacious McNamara, former president of Ford Motor Company, had come into office pledging to run government more like a business. He was determined to deliver.[20]

The moves triggered a shakeout in the Northern California electronics industry. Large East Coast firms that had set up transistor and tube operations in the Valley cut back their operations, or got out of them altogether. Some firms merged, others were acquired. Eitel-McCullough and Varian, pioneering local firms that had dominated the microwave business, were left reeling. Engineers started calling it "the McNamara depression." In contrast, companies that were in the IC business flourished. A swelling number of young men spun out new semiconductor firms of their own. Many had worked at Fairchild, and their start-ups became known, naturally, as "the Fairchildren."[21]

Contributing to the blossoming of an industry was the extraordinary power of its marquee product. Inventors had been searching for faster and cheaper power sources since the dawn of the age of steam. The silicon chip delivered. With every year, semiconductor firms managed to multiply the interconnected logic transistors on each chip, making the computers surrounding them faster, smaller, and more powerful. Gordon Moore predicted in 1965 that the number of components per chip would double every year. An astounding prediction, but within a decade, "Moore's Law" proved true. Once set in motion, silicon technology defied all the rules of economics, becoming the propulsive force behind a second industrial revolution.[22]

The case of the Valley chipmakers underscores that public investment mattered greatly, but that *how* that money was spent mattered even more. The decentralized, privatized, fast-moving public contracting environment encouraged entrepreneurship. The Valley already had a tightly fraternal Valley culture, where people were used to sharing ideas rather than clinging to them. As the space age reached its crescendo, this characteristic was reinforced by the fact that no single company could have a monopoly on the windfall. As the semiconductor market grew bigger and richer, the geographic concentration of the industry intensified. From the founding of Fairchild until man walked on the moon, nearly 90 percent of all the chip-making firms in America were located in the Santa Clara Valley.[23]

CHAPTER 4

Networked

The sunshine and silicon of Northern California might have drawn the attention of curious foreign leaders and dazzled reporters. Yet Boston—the city its nineteenth-century boosters crowned "the hub of the universe"—remained the Valley's more popular and more serious older brother throughout that decade of missiles and moon shots. MIT continued to reign supreme in the computer-research world; spinoff companies from the university labs of Cambridge turned Boston into a start-up hub well before California earned that reputation. The big business of mainframe computing—one so dominated by IBM that its other market competitors were referred to as "the Seven Dwarfs"—remained rooted in the eastern half of the continent. It also was back East that other seminal technological developments of the space-age Sixties emerged. First came the minicomputer, which shrank down digital machines to a more manageable and relatively affordable size. Then, aided in part by the spread of the minicomputer, came the earliest networks that turned computers into tools of communication as well as calculation.

Although minis and networks weren't born in Northern California, their widespread adaptation over the course of the decade vastly enlarged and diversified the world of computing. In doing so, they further altered the geography of the tech business. No longer was computer power restricted to very large and deep-pocketed institutions and a small priesthood of technical specialists. No longer was the game only one of selling electronic hardware (both whole computers and the components for them). It was about building and selling and distributing operating platforms and software services, as well as networks and devices to support communication.

The electronics business truly became an *information* business in the 1960s, creating new markets, engaging new users. And those first years—while decades ahead of the eventual explosion of online commercial platforms and software—set the stage for the Santa Clara Valley to one day become the most networked place of them all.

THINKING SMALL

But first: back to Boston, where, six months before Sputnik, two MIT researchers got an entrepreneurial itch to strike out on their own. Ken Olsen was one of those lucky young men who'd been able to be on the ground floor of some of the most exciting developments in digital computing, thanks to the public money coming into university labs during and after World War II. The glacial pace of academia, however, wasn't a good fit for Olsen, an inveterate tinkerer who'd spent his boyhood summers working in a machine shop and fixing radios in his basement. Partnering up in early 1957 with a fellow researcher, Harlan Anderson, Olsen secured $70,000 from Harvard Business School professor Georges Doriot, who had established a tech-focused investment fund he called "venture capital" to support young and untested entrepreneurs. Olsen and Anderson departed MIT and moved into a shuttered textile mill in the old industrial town of Maynard. Their new company, Digital Equipment Corporation, was open for business.

Digital sold a new breed of computer, transistorized and programmable, a fraction of the size and price of mainframes. Its power source was Noyce and Kilby's integrated circuit, made newly affordable by the scale and scope of the space program. Olsen called it the "programmed data processor," or PDP. Retailing for under $20,000 at a time when mainframes could cost over $1 million, the mini officially ushered in the next generation of digital computing.[1]

The refrigerator-sized machine was still a far cry from the personal, desktop computers to come, but it brought its users in from the remote and chilly world of punch cards and batch processing, allowing them to program and perform computer operations in real time. By the decade's end, the IC-powered model Digital introduced in 1965, the PDP-8, had become a familiar sight in computer labs across the country, and a gateway to the digital world for an entire generation of tinkerers and homebrewers and future Ken Olsens: college kids and grad students playing computer games after hours, young coders writing their first programs, schoolchildren who became test subjects for early educational software.

Just as Fairchild gave birth to the chip business in the Valley, Digital turned Boston into the capital of minicomputing, an industry that would employ hundreds of thousands and generate billions in revenue for over two decades. In 1968, a key engineer on the PDP-8 project departed Digital to found another minicomputer company, Data General. Four years later came Prime Computer, half an hour's drive down the road in Natick, another aging industrial town. In crumbling brick factories that once had pumped out boots and blankets for Civil War soldiers, the mini makers of Massachusetts built machines that ushered in a second, transistorized generation of digital computing, hacking away at IBM's dominance of the market more than most of its mainframe competitors ever could.

Into the stuffy confines of the Boston business world, the minicomputer companies imported some of the renegade, improvisational spirit of the postwar electronics lab—the same spirit that was the hallmark of California companies from HP to Fairchild and its children. Ken Olsen hadn't worked anywhere but academia before he founded his company. He had limited patience for management experts and blue-suited sales types. A devout churchgoer with little interest in the trappings of great wealth, he liked to spend his off hours paddling quietly on New England ponds in his favorite canoe. Olsen considered himself "a scientist first and foremost."

While this too often made Olsen tone-deaf to the nuances of the business and consumer markets, it made his company remarkably effective at building the products that scientists and engineers needed. By the early 1970s, Digital had become the third-largest computer maker in the world. Doriot's $70,000 investment—for which he obtained a 70 percent stake in the company—was worth $350 million. It was one of the most lucrative deals in high-tech history.[2]

SHARING TIME

About the same time that Ken Olsen set up shop in his Maynard mill, John McCarthy began musing about a better way to distribute computer power.

The Boston-born and Los Angeles–raised son of a labor organizer, McCarthy had the soul of a radical and the mind of a scientist. He had graduated high school two years ahead of schedule and earned degrees from Cal Tech and Princeton before joining the faculty of Dartmouth. Of the many scholars entranced by the computer's potential to imitate and complement the human brain, McCarthy was the one who in 1955 put a name to the phenomenon and the field of research that rose around it: "artificial intelligence."

By 1958, McCarthy had been scooped up by MIT, where he promptly established an Artificial Intelligence Laboratory with another rising faculty star, Marvin Minsky. The two men were close friends from graduate school and had been born within a month of each other thirty years before. Along with a shared generational sensibility, they were both convinced that computers could and should be far more than mere adding machines. But in order to accomplish that, electronic brains needed to change the way they interfaced with their human users. This meant that they had to stop making people wait their turn in line.[3]

For that was the reality of life in the computer world of the late 1950s. Mainframes were immensely powerful, but they could only process one batch of data at a time. Impatient researchers had to line up with punch cards in hand, then spend hours or even days waiting for an answer to their query. And, in those very buggy days of early programming, they could only run the program in the first place after they'd spent considerable time refining their instructions—another stage that took even more time and further batches of cards. There had to be a better way.

The fix, argued McCarthy, was to adapt a mainframe so that multiple users could work with the machine at once, creating a hub-and-spoke system with a mainframe at the center and user terminals at the periphery, networked by coaxial cable. Instead of waiting, users could receive results in seconds, and immediately try again if there were errors in the data. Instead of submitting punch cards, people would type instructions on terminals, in real time. "I think the proposal points to the way all computers will be operated in the future," McCarthy wrote to the head of MIT's computing lab in early 1959, "and we have a chance to pioneer a big step forward in the way computers are used."[4]

John McCarthy was hardly the only person in Boston who was thinking about how to improve the human-computer interface. A ferment of conversation had been brewing around the issue ever since the late 1940s, when the father of cybernetics, Norbert Wiener, had led a legendary series of weekly seminars in Cambridge to debate questions of man and machine. The notion of getting computers to talk to one another wasn't impossible to imagine: digital networking had been around nearly as long as the digital computer itself via another born-in-Boston system, the Semi-Automatic Ground Environment, or SAGE. Sponsored by the Air Force and designed at MIT's Lincoln Laboratory, the project linked military computers to radar, creating a digital command-and-control system for air defense. First beta-tested on Cape Cod in 1953, this very first online network launched nationally in that fateful post-Sputnik summer of 1958.[5]

And just on the other side of town McCarthy could find a psychologist named Joseph C. R. Licklider—known to all as "Lick"—who had become consumed with questions of how the radically different timescales of computer thinking (lightning-fast) and human reaction (not so fast) might be overcome by better design. In 1960, he mused on this problem in a paper entitled "Man-Computer Symbiosis," a short treatise that became one of the most influential and enduring documents in high-tech history. In seven pithy pages, Licklider laid out a new world order, one where men (the masculine being proxy for all genders) did the creative thinking: "set the goals, formulate the hypotheses, determine the criteria, and perform the evaluations." Computers would do the routinized tasks of data gathering and calculation that, according to Lick's estimates, sucked up 85 percent of human researchers' time. But such symbiosis would require new tools—including computer time-sharing.[6]

The development of Digital's first computer provided both Licklider and McCarthy with the opportunity they needed. In 1961, the two joined forces to adapt a brand-new Digital mini so that it could work as a time-sharing machine. The same year, the MIT computing center took up McCarthy's challenge to reconfigure a newly delivered IBM so that it, too, could accommodate multiple users at once. Time-sharing had moved from wild idea to reality.

THEN, TIME-SHARING BLEW UP TO IMMENSE SCALE. FOR IN 1962, J. C. R. Licklider went to Washington, appointed to run a new computing initiative within that little Defense Department agency created in late 1957 to chase those post-Sputnik "will-o-the-wisps": ARPA. Leveraging his considerable reputation and the full force of his personality, Licklider secured a remarkably generous budget and considerable administrative freedom to spend it where he wished—and he proceeded to direct a considerable chunk of these resources to MIT and to Stanford, where John McCarthy had just joined the faculty.

As money flowed from ARPA to academic computing projects dedicated to advancing "man-computer symbiosis," the Sputnik-inspired research agency effectively created an entirely new discipline at America's leading universities: no longer merely a variant of physics or electrical engineering, but the stand-alone intellectual pursuit of "computer science." Stanford got its computer science department in early 1965, the last new program established by Fred Terman as Provost. John McCarthy became one of six founding faculty members of the small-but-mighty unit.[7]

With ARPA dollars as the spark, time-sharing spread rapidly. Now, operations that couldn't afford their own mainframe (or even their own minicomputer) could tap in to the thinking power of an electronic brain in another room, another building, or across town. Policymakers and journalists soon were rhapsodizing about the networks' ability to bring computer access to the masses, becoming a "computer utility" just like electricity or telephone service. "Barring unforeseen obstacles," cheered MIT management professor Martin Greenberger in the *Atlantic Monthly* in 1964, "an on-line interactive computer service, provided commercially by an information utility, may be as commonplace by A.D. 2000 as the telephone service is today."[8]

THE TIME-SHARING BOOM

Although time-sharing services were born to serve academic people performing academic work, there now was money to be made in providing them for users hungry to access computing power from their offices and homes. Thus, the second half of the 1960s saw a surge of start-up companies trying to hack into the business of sharing computer time. In Texas, there were operations like H. Ross Perot's Electronic Data Systems and Sam Wyly's University Computing Company, which distributed computer power and provided software services to mainframe users. But companies like those were really not *networking* in the way that McCarthy and Minsky and Licklider imagined it; human programmers, not computers, fed data into the machines and determined each run's order and length. It was batch processing by telephone wire.

True commercial time-sharing came from a smaller, more technologically sophisticated group of start-ups, who programmed powerful scientific computers to determine the order and run time of jobs, and who built out computer-communication networks atop the phone lines to provide faster, cheaper, and more powerful processing services to their customers. It was in a company that would eventually become a leader in that second category, a rangy little Palo Alto start-up called Tymshare, that Ann Hardy found herself at the beginning of 1966.

Tymshare was a California creation, through and through. It was the idea of two engineers working at the Santa Clara Valley outpost of General Electric, Tom O'Rourke and Dave Schmidt, who looked at the region's electronics companies and saw a robust customer base for time-sharing just waiting to happen. Tymshare's first demos happened at Lockheed Missiles and Space, where engineers had been waiting twenty-four hours for their FORTRAN code to process and jumped at the chance to speed things up. It used a powerful scientific computer manufactured by Southern California–based Scientific Data Systems, or SDS, which had been modified into a time-sharing machine by an outfit based in Berkeley.[9]

Ann Hardy had been living and working at Livermore Lab for five years, but her husband's job moved over to Palo Alto, and it simply wasn't feasible to do the cross-Bay commute (and unimaginable that the wife's work would take precedence). New in town, jobless and aimless, Hardy decided that she needed a home time-sharing terminal to keep her skills sharp, and began to look around for a service to which she could subscribe.

Hearing about Tymshare, she cold-called Dave Schmidt, who was so bewildered by the idea of a local housewife who wanted to "play around" on a time-shared computer that he confessed that Tymshare's machine wasn't yet operable. Accustomed to working on a time-sharing machine at GE that already was able to accommodate multiple users, O'Rourke and Schmidt hadn't realized that the machine they ordered up from Berkeley wouldn't have a similarly capacious operating system (OS) on it. The machine had been tested to accommodate up to two users; a commercial operation like Tymshare needed one that could share time among at least twenty. They needed to upgrade the system. "I told them they needed to hire me," remembered Hardy. Schmidt agreed, little realizing how fundamental Hardy's programming would be to Tymshare's eventual success. "If I had known what the system meant to this company," he later admitted to her, "I never would have hired a woman to work on it."[10]

Tymshare quickly became a hit. Aerospace engineers at Lockheed and Philco-Ford thirsted for access to computing power, and now at last came a service that let them tap into all the computer power they needed and wanted. But with a user base that was almost entirely engineers, the Tymshare computer quickly began to get hacked. They probably didn't mean to stir up trouble, Hardy reasoned. "As soon as you let an engineer on a computer, they try everything."[11]

The tech world was still decades away from encryption standards, but the Tymshare team almost immediately started to encrypt their passwords to stabilize the system. Hardy buckled down to start building other checks and balances into the OS so that it was harder for users to precipitate a crash. While big computer companies were still figuring out shared protocols to allow different machines to talk to one another, Tymshare had leapt ahead to find ways to make these communications secure.

Then came the network itself. Tymshare's founders had started their company with only a local market in mind, but the success of the service soon had them pushing out to new locations. But since the only feasible way for its customers to send data over the wires was via local telephone lines, going national meant creating full-bore computer centers in every city they wanted to serve—an expensive and hardly scalable proposition.

As Tymshare's executives tussled with the dilemma of building out a national data network in an era when no infrastructure for the thing existed, Hardy reached out to a programmer she had known at Livermore Labs whose name (LaRoy Tymes) and expertise (computer networking) made him seem fated to work at Tymshare. When first hired at Livermore, Tymes had been a twenty-year-old without a college degree; he had intended to become an apprentice electrician. Suddenly, thanks to the government's great technical hiring spree, he was operating computers in one of the military's most important, top secret facilities. "It was way beyond anything I'd ever seen in my life or even heard of," he remembered. "I was . . . very impressed by the fact that they would entrust these multimillion-dollar machines to somebody like me."[12]

Like Ann Hardy, LaRoy Tymes had stumbled into programming, and it proved to be his great calling. He came to Tymshare in early 1968 and set about building a brilliant work-around to the networking problem: a national network of minicomputers that could effectively act as information middlemen between the customer and the all-powerful SDS 940, tied together with multiplexed lines that allowed many messages to go across the wires at the same time. Tymes's work-around—branded Tymnet—turned out to be a pace-setting innovation of its own. From its

1971 launch until the advent of the commercial Internet in the early 1990s, there was no service for business customers that was close to comparable to it.[13]

Tymshare boomed, but it still felt like a start-up: small, convivial, and technically exciting. After finishing a seventy-hour workweek, staff would retreat to a steakhouse around the corner for beers and shop talk. Hardy had small children, but she stayed on, adapting her family life around the demands of work. Perhaps it was the suburban bedroom-community culture, or the overwhelmingly male demographics of the engineering profession, but it was rare to find a working mother like her in the Valley's tech ranks, much less one who was a technical executive. The feminist revolution of the 1960s and 1970s was in full flame, but among Palo Alto's playgrounds and cul-de-sacs it still felt like the Eisenhower era.

As the market grew, Tymshare expanded to become a leading performer in an industry coming into commercial maturity right as Wall Street entered several consecutive years of a bull market. In September 1970, Tymshare went public, outpacing its issue price in its first day of trading and creating a windfall for its founders and early employees. Ann Hardy, however, wasn't one of them. "They didn't give me stock options like they gave all the men," she said matter-of-factly. No early retirement for her, and she didn't want to stop working anyway. There were far more interesting things ahead.[14]

DEREGULATED

Although commercial time-sharing was taking off in the late 1960s, the future of the networked computer still depended on politics and policy. Across the Atlantic, Britain and France had kept telecoms government-run operations. In America, the networks were privately owned, but the government exercised a great deal of control through regulation designed to keep service universal and prices low. In return, telecommunication companies like AT&T and Western Union had the market to themselves. AT&T's control of the telephone wires was one reason it was so difficult for companies like Tymshare to build out nationally, and it extended to control of the machines that connected into them. Clunky Bell telephone handsets, leased monthly from the phone company, became a near-universal feature in mid-century American homes.

In the 1960s and early 1970s, however, several things happened to ensure that America's regulatory environment for computing would be very different from that of telephony. Little noticed at the time, these differences ultimately proved hugely consequential to the development of the tech economy to come.

First, there was the case of the Carterfone. Texan Thomas F. Carter was a man of the prairie and the oilfield, who in 1959 patented a device that hooked a regular phone line into a two-way radio, giving ranchers and oilfield workers a way to communicate across distances that a regular walkie-talkie couldn't muster. It didn't take very long for AT&T to pounce. Only Ma Bell's equipment could be used on the Bell network, the telecommunications giant informed Carter. The Carterfone would have to go. Thomas Carter's company may have been tiny, but his outrage was Texas-sized. He decided to fight. In 1965, Carter filed a private antitrust suit against all of the Goliaths of the industry: AT&T, its twenty-two regional subsidiaries, and the number two telephone service provider, General Telephone & Electronics (or GTE).

Three years later, in July 1968, the FCC ended the case with a decisive ruling: AT&T did not have the right to ban third-party equipment from being used on its lines. "We pinned their ears back," Carter crowed. "We made fools of them." It was, one Washington lawyer declared, "the most important communications case in the last decade." The *Carterfone* decision helped accelerate the entry of new manufacturers into the phone market as well as forcing AT&T to develop more-appealing consumer products. A generation of American children would grow up with Mickey Mouse phones on their bedside tables. More important for the online future ahead, a market opened for a new set of high-tech entrants to build handsets, couplers, routers, and network infrastructure that allowed computer communication.[15]

A second David-vs.-Goliath battle pushed the door open further. This time the setting wasn't Texas oilfields, but the pinstriped canyons of Wall Street.

If the New York Stock Exchange was the Cadillac of the American stock market, the over-the-counter, or OTC, market was its slightly dented Model T. These were companies too tiny for a listing on the big board or even regional exchanges, their buying and selling controlled since the 1930s by a self-regulating group for brokers called the National Securities Dealers (NASD). The market was opaque and archaic in its methods; its printed stock tables were highly unreliable, and buyers could usually only learn a stock's price by calling a broker. As the market began to surge, the broker-dealers of the NASD realized they needed to move into the twentieth century, not to mention smarten up their slightly seedy reputation. Computerizing the whole system would be the quickest and splashiest way to do it.

In 1968, the dealers put out the job of "automating" their system for bid, and the firm that won was a four-year-old Southern California start-up named Bunker Ramo. The firm had defense industry roots: founded by Martin Marietta president George Bunker and TRW vice president Simon Ramo, the firm was dedicated to what the two founders termed "a national need in the application of electronics to information handling." An early client was NASA, for which Bunker Ramo built one of the world's first computerized information retrieval systems, using the networked computer to classify and categorize large data sets *a la* Vannevar Bush's memex.[16]

At first, the system Bunker Ramo designed for the dealers was simply another digital database that put paper stock tables on line. But when the firm added a feature that allowed brokers to buy and sell over the network, AT&T again cried foul. This wasn't time-sharing anymore, AT&T lawyers argued; it was two-way telecommunication, and Ma Bell would no longer lease Bunker Ramo its lines.[17]

Just like Thomas Carter, Bunker and Ramo pushed back. The complaint the firm filed with the FCC culminated in another landmark ruling by the commissioners, this time in 1971. Called "Computer I," the FCC decision declared a compromise: a new "hybrid services" category of communication into which most two-way communication via computer would fall. It didn't end the debate; "Computer I" would be joined by "Computer II" and "Computer III" before the 1970s were out. By the time the FCC got around to properly muddling through some rules for how to categorize communication by computer, the pace of technological change had galloped far ahead.

The FCC's indecision essentially settled the question. Gone was the strong regulatory approach that had characterized the previous four decades of rulemaking and enabled the all-powerful Ma Bell universe. Instead, a range of commercial networked services inhabited the wires, all delivering more and more content and, over time, enabling e-mail and chat and all kinds of two-way communication: CompuServe, founded in Ohio in 1969; The Source, coming out of Virginia in 1979; Prodigy, a joint venture between CBS, IBM, and Sears, in 1984. LaRoy Tymes's Tymnet never became a familiar brand name like these others, but the first true time-sharing network was the most durable of them all, running without a crash from 1971 to 2003. "He never had a bug in his network code, ever," marveled Ann Hardy. "Amazing work."[18]

Carterfone and Bunker Ramo, and the FCC rulings that came in their wake, meant that digital communications portals became the realm of *many* private companies, not just one or two. And, while hugely popular, none of these services became a publicly governed "computer utility," a computerized analogue to AT&T. Instead, Congress and the FCC actively blocked AT&T from delivering online content and deregulated the entirety of the system, forcing Ma Bell's breakup in the early 1980s.

In that vacuum, the only network powerful and widespread enough to eventually become the backbone for the world's electronic commerce and communication was something quite different—a nonhierarchical, noncommercial, wild-and-woolly network also made possible by America's race to space. It was called the ARPANET.[19]

INTERGALACTIC

By the time the time-sharing business began to take off, J. C. R. Licklider already had moved on to other, far bigger ideas. Nearly from the moment he landed at the Pentagon, Lick started to issue a flurry of memos that took time-sharing networks to the next level—proposing to interconnect time-shared computers to create a powerful national web of interactive supercomputing. With a flamboyant and intentionally humorous flourish that underscored the wildness of the idea, he called it the "Intergalactic Computer Network."[20]

Academics embraced the concept, despite the wicked technical problem presented by trying to get different makes and models of mainframes to communicate with one another. Teams of computer scientists on different campuses were pursuing similar research questions, but they worked on entirely separate systems that couldn't share data or findings. It was nearly as inefficient as waiting thirty-six hours to process your punch cards. Licklider's eventual successor at ARPA, a pipe-smoking Texan named Bob Taylor, labored over the remainder of the 1960s to fix this problem and make a "Resource Sharing Computer Network" possible.[21]

The Pentagon had by then entered the age of Robert McNamara's businesslike budget-cutting, and the brass proved receptive to Taylor's argument that networked computing would eliminate inefficiencies and free up contractors to obtain the best kind of computer at the lowest price. A second, more apocalyptic argument for a national computer network came out of RAND, whose researchers pointed out that this decentralized network could serve as a military communication lifeline in case nuclear attack wiped out regular telephone service and the command centers of the Pentagon.

The bosses said yes, but they didn't give Taylor a very large budget. For help designing the thing, he turned back to the academic researchers who had been grappling with questions of man-computer symbiosis for the better part of a decade. The resulting network, the ARPANET, was made by and for academics, and it reflected their priorities: ease of communication, nonhierarchical collaboration, with no particular hardware or software platform favored over another.

Time-sharing rapidly had become a profit-hungry marketplace, with different computer companies battling so fiercely to become the industry standard that the whole enterprise collapsed under its own weight. Other blue-sky electronic technologies—from radar and microwave to the transistor and IC—had become militarized, a fact that caused increasing distress among the academic community as opposition to the Vietnam War grew. The ARPANET was different: a child of the spit-and-polish space age, nurtured and shaped by a bunch of temperamentally anarchic professors and long-haired graduate students at the tail end of the tumultuous 1960s. It was a product of the military-industrial complex that had a countercultural soul.

RICHARD NIXON WAS SEVEN MONTHS INTO HIS PRESIDENCY WHEN NEIL ARMstrong's boot made its imprint on the lunar surface in July 1969, achieving the grand goal John Kennedy had set eight years before. The "missile gap" about which the Gaither Report had warned so darkly turned out to have been a mirage, based on bad intelligence. But the ensuing flurry had produced real results. The American push to catch up with the Soviets had accelerated all sorts of electronics innovation, further enlarged the computer industry, and turned the U.S. higher education system into the wealthiest and foremost in the world.

Three months and nine days after the moon landing, the ARPANET went live.

The network was a tiny blip on the expansive and expensive landscape of America's space program. Of the $24 billion spent on the space race between Kennedy's declaration and Neil Armstrong's "giant step for mankind," only about $1 million went to build the ARPANET. Only a few academic scientists and government bureaucrats knew of its existence. Its launch didn't earn a mention in any of the national newspapers. Yet looking backward from a half century on, the network the ARPANET became—the Internet—had a more consequential effect on the globe's political and economic evolution than all the rocket launches and orbiting satellites and moon landings put together.

CHAPTER 5

The Money Men

The history of high technology often plays out as a sharp-elbowed Boston v. Silicon Valley rivalry. A story of winners and losers, of all or nothing. Yet the way things unfurled in those first postwar decades shows just how intertwined the two regions' fates were from the start, linked by personal affection, professional envy, and the glue of federal contracting. The linkage—not just the rivalry—was part of what made each so powerful.

Fred Terman crafted his "community of technical scholars" with Vannevar Bush's MIT in mind, a project of admiration as much as of competition. Raytheon got the NASA contract for the Apollo guidance system, and then subcontracted out the ICs to Fairchild. Time-sharing started at MIT and migrated west on a wave of the Pentagon's money. The ARPANET mattered because it allowed Palo Alto and Berkeley and Cambridge and Washington and everywhere else to communicate and collaborate. And it was all about the people. The Valley teemed with MIT-trained engineers, Harvard MBAs, and people who'd once worked somewhere along Route 128. (The Valley enjoyed an advantage in that perpetual shuffle between the coasts, for few who experienced Northern California's Mediterranean climate wanted to return to New England winters.)

Yet there were distinct differences between the two regions—ones that shaped what they ultimately became. Geography *was* destiny. As AnnaLee Saxenian observed in her definitive 1994 study comparing the two regions, the Santa Clara Valley's remoteness from the centers of political and financial power turned out to be a tremendous advantage. It forced improvisation, nurtured new sorts of companies, and ultimately meant less dependence on the fickle bounty of the military-industrial complex. "In attempting to imitate" the model Bush and others had established in Boston, she writes, "they unwittingly transformed it."[1]

THE SANTA CLARA VALLEY WASN'T A BIG CITY, BUT AN ISOLATED CLUSTER OF small bedroom suburbs. "When I moved out to the Valley, it felt like an oasis in the desert," recalled Marty Tenenbaum, an engineer fresh out of MIT who came to Lockheed in 1966. He stocked up on a year's worth of books in Harvard Square before he left, just in case he couldn't find a decent bookstore. (He did find one, but not much else.) The people of the tech business were all friends and neighbors. With little to distract them, they spent their off hours talking shop on the sidelines of Little League games or debating semiconductor design in the leatherette booths of the Valley's handful of pubs.[2]

In other contexts, the situation might lead to dead-end provincialism. Tied into the rivers of money and scores of people flowing westward into the military-industrial complex and into the classrooms of Berkeley and Stanford, it sparked a disruptive and highly effective new industrial model.

For Harvard and MIT may have been undisputed academic powerhouses, but they operated far differently than the entrepreneurial, opportunistic, ever-hustling Stanford. And Boston may have had bankers and Brahmins, but Northern California had young financiers and lawyers with a ground-eye view of the technology scene and the opportunities it presented to make a good deal of money. The Santa Clara Valley became a particularly brainy and well-resourced Galapagos, developing peculiar new species of business specialists who focused exclusively on high technology. No other place—not even Boston—had what the Valley had. It was this ecosystem that ultimately gave Northern California its competitive edge.

Even after the ascension of the chipmakers and the mini companies, go-it-alone entrepreneurship was a shaky proposition, requiring market knowledge and resources that no federal contract could provide. By and large, electronics entrepreneurs were young men from modest backgrounds, book smart but not Wall Street smart. These would-be Bob Noyces and Ken Olsens needed management advice. Guidance on marketing, sales, advertising. Legal help with writing contracts and filing patents and allocating stock options. And they needed money. Traditional banks—or, for that matter, most investors—were unwilling to give it to young men who'd never worked outside an academic lab, and who often were building products for which there wasn't yet a market. Start-up founders required investors who understood the technology they were building, but also knew how to navigate the business world. And they needed someone who was a little bit of a gambler.

They needed a venture capitalist.

THE GENERAL

The morning fog hadn't yet burned off the plum orchards as the rented blue Pontiac rumbled down the country road a few miles south of Palo Alto. Bill Draper was behind the wheel. His friend and business partner Pitch Johnson was doing the same thing, in another orchard, in an identical blue Pontiac. They did this every morning in those heady early days of Kennedy's New Frontier: wake up early, hop in the rental cars, go on the hunt for deals.

Draper slowed down as he spotted a wooden barn looming up among the trees. A farmer had built it years before to dry the harvest, but as he got closer Draper could tell it wasn't full of prunes any longer. Nailed up over the door was a sign with a company name on it—syllables of "techs" and "trons" that signaled a fledgling electronics operation inside. Rolling to a stop, Draper hopped out of the Pontiac and headed to the door.

A brisk knock, and the door opened. Another young man in shirtsleeves squinted out into the morning light. Draper introduced himself and got right to the point. "I'm a venture capitalist. We buy minority interest in companies like yours, and turn over money to you to grow your business." That was all a cash-poor entrepreneur needed to hear. "Come on in," the young man replied. Draper had just snagged another one. Morning after morning, company after company, Bill Draper and Pitch Johnson became the go-to guys if you were a young and hungry electronics company.[3]

Handsome and confident, with the stride of a natural athlete, William Henry Draper III was a different sort of California migrant than heartland boys like Burt McMurtry and Bob Noyce. Born on New Year's Day 1928, he was the son of a distinguished banker and diplomat, raised in the affluence of Westchester County, a veteran of Korea, a member of the exclusive and ultra-secret Skull and Bones society at Yale. After graduating in 1950—one year behind fellow Bonesman and future president George H. W. Bush—Draper headed to Harvard Business School.

He landed in the lecture hall of the "inspirational" Georges F. Doriot. The Frenchman, known as "General Doriot"—he had become a U.S. citizen to join the Army during World War II, and had reached the rank of brigadier general—was five years into his tech-investment career as president of American Research and Development Corporation (ARD), but his Digital investment was still years in the future. Then, Doriot wasn't yet a venture capital legend, but simply one of the MBA program's most popular faculty members.

The professor's yearlong course had a deceptively bland title, "Manufacturing," but was an energizing journey of lectures and real-world experience that "taught us what it was really like to be a businessman," said Draper. "I am building men and companies," the professor proclaimed. (As HBS didn't admit women, Doriot offered a simultaneous class for students' wives that offered helpful tips on how to be a supportive executive spouse, like cutting out relevant *BusinessWeek* articles for their husbands and greeting them at home with a favorite drink after a long day at the office. Draper's wife, Phyllis, didn't like it one bit.)[4]

While venture capitalists—VCs—later got a popular reputation as convention-bucking cowboys, the early breed were tradition-bound products of the business establishment. Doriot was no exception. The fastidious Frenchman had left the Army but kept to a schedule of military precision: rising at seven each morning to walk to work, keeping long hours broken only by a ten-minute lunch in the office cafeteria, and regularly awakening at 2:00 a.m. to mull over a thorny business question. "Always remember," he warned his students with a stern glare, "that someone, somewhere, is making a product that will make your product obsolete."[5]

The venture fund Doriot helmed was the brainchild of Ralph Flanders, a three-term Republican senator from Vermont who had previously directed the Boston Federal Reserve. (The flintily independent Flanders gained fame as the first fellow Republican to censure red-baiting Senator Joseph McCarthy in 1954.) The other original financiers were similarly blue of blood: John Hancock Life Insurance head Paul Clark, Massachusetts Investors Trust's Merrill Griswold, and Karl Compton, president of MIT.[6]

Doriot may have popularized the "venture capital" label, but investment in new and risky ventures wasn't a new thing—far from it. From Queen Elizabeth I's court to the Florence of the Medici to the financiers who gave seed investments to Henry Ford, wealthy people and interests long had been willing to make bets on new and relatively untested ideas and companies. Making "adventure capital" investments in new industries became something of a hobby among the sons and grandsons of Gilded Age millionaires. By the middle of the twentieth century, enough of these bets had paid off that family investment funds—like those of the Rockefellers and Whitneys—had turned into reliable sources of financing for early-stage companies in a range of industries. Insiders called these sorts of deals "private placements."[7]

Old-money titans didn't know a thing about how electronics worked, but they could smell the money in it. Doriot's contribution was to play matchmaker, using his deep familiarity with Boston's research scene to sniff out the tech, find the entrepreneurs, and connect them to the moneyed elite. By the time Bill Draper strolled into his Harvard lecture hall, the General had raised millions of dollars from both individual and institutional investors that seeded investments in quite a few companies. Most were in Massachusetts. Draper sensed an opportunity.

So did his father. The senior William Draper had been Truman's Undersecretary of the Army and then headed over the Atlantic to direct European operations of the great American postwar rebuilding project, the Marshall Plan. His deputy there, Fred Anderson, was another distinguished military man and diplomat. The two became fast friends. Returning back to the States, the two men joined up with a third partner: none other than H. Rowan Gaither, the Paul Revere of the missile gap.

Two years after Gaither's fateful report helped set the tech-spending flywheel in motion, the three men blended their Wall Street expertise and Washington connections to bring Doriot's venture model to the West Coast. The structure, though, was a little different. The General ran ARD as a closed-end, publicly traded mutual fund; the trio opted to make their operation a limited partnership, in which the partners made most of their money on the "carry"—a hefty chunk of any profits made by the fund.

The limited-partnership model, which became the prevailing one for Valley VCs in later decades, intertwined the money man's financial fate even more tightly with those of his portfolio companies. It encouraged him to choose bets carefully and maintain an intensely hands-on managerial relationship with his entrepreneurs. By securing a large percentage of a company as payment for his seed capital, the VC became the entrepreneur's most important business partner—and possibly the one enjoying the biggest returns. The hit rate was only one in ten. But, with the money to be made in electronics, you only needed one big hit.

After raising $6 million from New York financiers, including a cool $2 million from the Rockefeller family, Draper, Gaither, and Anderson opened for business as Palo Alto's first venture capital firm in 1959. Young Bill Draper came west to join them shortly afterward. Funnily enough, the firm that pioneered the limited-partnership structure wasn't all that successful, largely because the three founders were too busy and too grand to put all else aside, roll up their sleeves, and enwrap themselves in the day-to-day management of their portfolio companies. You needed young men to do that, men who hadn't yet made their fortunes and their reputations, men who were willing to rumble out in the blue Pontiacs and chase down the deals.[8]

THE GOVERNMENT

Venture capital might have remained something of a boutique business, the province of connected Ivy Leaguers and dabbling trust-funders, if Lyndon B. Johnson hadn't wanted to be president.

In the summer of 1958, Johnson was in his fourth year as Senate majority leader, wrapping up a truly spectacular season of legislative accomplishment, and thinking about how to win the 1960 Democratic presidential nomination. To get there, he needed the support of the Democratic Party's northern liberal wing, who saw the garrulous Texan as just another Southern obstructionist when it came to civil rights. Then, once nominated, he needed to work hard to appeal to the business constituencies inclined toward his likely Republican challenger, Vice President Richard Nixon.

One issue that hit both of these targets: small-business investment, which was widely perceived to be in short supply, stifling economic growth by discouraging would-be entrepreneurs from hanging out a shingle. It was just the kind of legislative challenge Johnson relished. Small business was also good politics when it came to the man from Abilene, President Dwight Eisenhower, a longtime champion of small entrepreneurs—operations "vitally important to the soundness and vigor of our system of free competitive enterprise," as he put it.

Democrats in Congress had been trying to pass something along these lines since the 1930s, to no end, partly because their approach involved setting up a large government bank, whose rationale and cost had little appeal for Republicans.

Johnson knew that he'd need something with more of a private-sector focus. For help, he reached out to a Harvard professor known as an expert on this sort of thing: Georges Doriot.[9]

Wrangling his fellow lawmakers as the summer recess approached, Johnson won successful passage of a "Small Business Investment Act" that set up an eye-poppingly generous set of tax breaks and federal loan guarantees for small enterprises and their investors. Set up shop as a Small Business Investment Company, or SBIC, and for every dollar you raised of your own capital, the feds would guarantee three dollars more in long-term loans. There were tax incentives for early-stage investment and special breaks for "management consulting services." It was Doriot's model, with the U.S. government subbing for the Rockefellers as deep-pocketed angel. Loosely regulated and thrown wide open to nearly anyone who wanted to create one, the SBICs became what one later critic scowled was a "license to steal."[10]

Although Johnson's measure became background noise amid all the news-making legislation of the year—NASA! NDEA! civil rights!—the new program was an immediate hit with investors. By 1959, more than 100 SBICs were in business; by 1961, the number soared to 500. The program brought a surge of new people and firms into the business, allowing ambitious young men to become venture capitalists even if they weren't already rich. The program kicked off a wave of new investment funds across a range of light manufacturing and white-collar service sectors, but especially in electronics, where the post-Sputnik surge had created a large and hungry military market.[11]

The synergy of SBIC and defense spending crackled in Palo Alto. A restless Bill Draper spun off from his father's firm and went into business with an SBIC of his own, recruiting his friend Pitch Johnson to join him. They each scraped up $75,000. Johnson had to go to his father-in-law for most of his share. The government loaned the pair $450,000.[12]

THE GROUP

By the time Draper and Johnson opened for business in 1962, they had company. The Bay Area was home to many great family fortunes, from timber and agriculture to shipbuilding and coffee, and several of these old-money enterprises jumped into electronics investing. Several San Francisco–based banks and insurance companies also opened SBICs on the side. The fund managers tended to be like Draper and Johnson, well-credentialed strivers in their twenties and thirties, not yet rich themselves, but able to be ambassadors between their generation of young electronics entrepreneurs and their fathers' generation of bankers and financiers. Over at Bank of America, there was George Quist, who later co-founded one of the tech industry's most important investment banks. At Fireman's Fund, there was Reid Dennis.

Although about the same age as his venture peers, Dennis became something of an elder statesman to the burgeoning Bay Area venture community. In 1952, six months after graduating from Stanford with the powerhouse combination of a BS in electrical engineering and an MBA, the San Francisco native had plunged every penny he had left over from his college fund—$15,000—into an investment in local magnetic-tape manufacturer Ampex. He'd seen a demonstration of the technology in an on-campus seminar and thought it looked interesting. It turned out to be far more than that; the investment eventually netted him close to $1 million.

Bow-tied, genial, and meticulously organized (he maintained color-coded day planners throughout his career, archiving them carefully for future reference), Dennis had enough money to quit his day job, but he didn't. Instead, he began convening regular lunches with other finance folk like him with an appetite for making investments in odd little tech companies down on the Peninsula. Some of the new venturers were scions of wealthy California families; others had been hired to manage the money. They had come of age amid the scientific flurry of the Eisenhower years, making them more attuned than their elders to the shiny new business opportunities in microelectronics, and they ran together in a pack.

Amid the wood paneling and the free-flowing martinis of San Francisco's clubbiest lunch spots, nervous start-up founders would make their entrepreneurial pitch before the group. Afterward, the investors would send the entrepreneur outside to cool his heels on the sidewalk while they debated whether to opt into the deal. "Sometimes," remembered Dennis, "we would get $100,000 committed over coffee." They had a remarkable hit rate. Of the first 25 investments made by the lunch bunch, 20 were successful. They only lost money on 3. "There was no competition" for what they were doing in the Valley, remembered Dennis. He and his lunchtime comrades called themselves The Group. Others called them The Boys Club.[13]

While growing to enormous size in later decades, the Silicon Valley venture industry budged very little from the all-male, all-white, upper-middle-class demographics that it had at its start. That this homogeneity was its starting point is little surprise. "The West Coast venture group," reported Pitch Johnson to an inquiring politician a couple of decades later, "is pretty much an engineers and MBAs combination." Women couldn't even enter most programs in those two fields in the 1950s and 1960s. Less than three percent of African American men had completed college or graduate school in 1960, in any field; only a handful of nonwhite students could be found in the nation's top MBA programs.[14]

The women "computers" who'd programmed the earliest mainframes had left the building, often to raise young families. When and if they wanted to get back into the work after a few years, they'd find that their skills were already two or three tech generations out of date. Hardware and software moved too fast to accommodate an intermission for parenthood.

The rare female technical experts who'd stuck with it, like Ann Hardy, were pretty much invisible to the investors who scouted labs and companies looking for someone who'd make a good CEO. As if to underscore the point, some of the places The Group met for lunch didn't allow women into their dining rooms.[15]

Why did these patterns persist, even as women and minorities made significant inroads into other professional domains? The explanation lies with the characteristic of the Valley VC community that set it apart from other regions, and that made it so good at finding and nurturing one generation of entrepreneurs after another: its personal, tightly networked nature.

Just like the broader Valley tech community, the region's first VC generation consisted of men who were alike in age, education, and temperament. They were friends as well as colleagues and competitors; they chose their investments based on gut instinct as well as more traditional metrics. Much later, one of the region's most successful and influential VCs, John Doerr, got in hot water after admitting that a major factor guiding his decisions was "pattern recognition." The most successful entrepreneurs, he found, "all seem to be white, male, nerds who've dropped out of Harvard or Stanford and they absolutely have no social life. So when I see that pattern coming in," he concluded, "it was very easy to decide to invest."[16]

The entrepreneurs of the 1960s might not have been college dropouts in jeans and hoodies, but they, too, fit a pattern: engineering degrees from certain programs, some service in the military, conservative in their bearing and their politics, and utterly consumed and fascinated by technical challenges. They were, in fact, much like the venture capitalists who financed them. The mind-meld between entrepreneur and VC was both Silicon Valley's great advantage and its greatest weakness.

THE ENGINEERS JUMP IN

The SBIC program may have provided the money and fresh talent the venture business needed to scale up, but it proved in the long run to be a bad fit for the kind of venture investing that the Valley needed. Alarmed by the suck on the U.S. Treasury—unsurprisingly, the SBIC never got back all the money it so generously loaned out—Congress loaded down the program with additional regulations and fewer giveaways, cramping the style of freewheeling fund managers. As time went on and the venture community grew, a number of investors who began with SBICs moved to the limited partnership model. Some never bothered with SBICs in the first place.

That was the case with Fairchild's original power broker, Arthur Rock. Sensing that the future lay in West Coast electronics, he departed Wall Street for San Francisco in 1961, and soon persuaded his friend Thomas J. Davis to partner up with him. Barely forty, Tommy Davis was a seasoned investor who'd already had a remarkable and varied career: law school, wartime service as an OSS spy in South China jungles, then out West to manage money for a California oil-and-cattle empire.[17]

Dissatisfied with his bosses' conservative approach to investing, Davis jumped at the chance to partner with Rock. The two had the same people-driven priorities—"back the right people" remained Rock's mantra—and very soon found themselves financing one of the biggest hits of the 1960s, founded by a distinctly quirky "right person": Max Palevsky and his scientific computer company, SDS. The deal was a fantastic one for Davis & Rock, as a cash-hungry Palevsky gave away most of his company to his investors in exchange for seed funding. When SDS was acquired by copy-machine giant Xerox in 1968, the partners got a more than $60 million return on their $250,000 investment.[18]

As the deals got richer, more people got into the venture capital business. This included big players beyond California, notably an affable and stupendously hardworking Chicagoan named Ned Heizer, who raised over $80 million to start a venture fund as Wall Street boomed, putting stakes in everything from supermarkets to semiconductors. Big funds like Heizer's were unheard-of in the Valley, however. The first generation of electronics pioneers were only starting to see their returns, and the vast majority of America's wealth clustered in the Northeast Corridor.[19]

Although the Valley wouldn't fully reap the rewards for another decade, the relatively modest scale of its venture operations kept VCs close to their companies, their entrepreneurs, and emerging technological trends in ways that were impossible for the Ned Heizers of the world. This only increased when a new wave joined the scene in the late 1960s, drawn from that great 1950s surge of migrants who had filled the labs of Lockheed and Sylvania and the classrooms of Stanford. They were operators *and* technologists, bringing a tacit understanding of the electronics business that was immensely valuable in a fast-changing and intensely competitive business environment. One of them was Burt McMurtry.

In twelve years in the Valley, the low-key Texan had risen from junior engineer to doctorate-holding director of a lab that employed 500 people. He was as loyal as ever to his Houston roots, making an annual pilgrimage to Rice each year to recruit its engineering graduates. The "Rice Mafia" he helped create in the Valley was the biggest concentration of alumni anywhere outside Texas. But California was now home, its sunshine and energy matching McMurtry's innate optimism and enthusiasm for new high-tech challenges. As venture funds proliferated and the industry diversified, the idea of going to a "funny little company" no longer seemed as far-fetched.[20]

One of those doing something different was another Valley veteran, Jack Melchor. After a long career at two successful start-ups, the second of which was acquired by HP, Melchor had started to worry about his health. Dave Packard, his mentor and boss, was such a hard worker that he'd developed a twitch; Melchor, too, had worn himself out so thoroughly that he ended up hyperventilating one day in the office and had to be wheeled out under oxygen. That's all it took for him to depart HP for the slightly less taxing world of investing. "Why don't you come join me?" he asked McMurtry. The two didn't know each other well, but Melchor had heard good things. McMurtry reckoned that this "was too good an opportunity to pass up."[21]

Incorporating as the Palo Alto Investment Company, the two-man operation set up shop in a second-floor apartment above a pizza parlor. Every night at 6:00 p.m., a giant silent-movie-theater organ installed in the restaurant roared into operation, sending him scurrying out of the office in search of peace and quiet.

"Deedee says it's the earliest I ever came home, before or since." In four years, he and Melchor invested in sixteen start-ups, never putting more than $300,000 into any of them. Compared to some of the big funds brewing back East, it was small stuff. But it turned out to be the start of one of the Valley's most storied VC careers.[22]

GALAPAGOS

The venture capital industry was a critical part of the distinctive, tech-supporting Valley ecosystem that coalesced over the 1960s—but only one part.

Another piece was law. While Draper and Johnson were trolling the orchards and Dennis was presiding over lunchtime pitch meetings, a four-man Palo Alto law office began to offer up services to young high-tech entrepreneurs and venture capitalists. McCloskey, Wilson, & Mosher was a partnership formed by three men who—like so many of their clients—had tired of the corporate scene and were ready to "take some risks," as co-founder John Wilson later put it. Wilson, the Yale-trained son of an Akron rubber-industry executive, had first come to California as a navy pilot during World War II. Stationed at Moffett Field in Sunnyvale, he'd been smitten by "the golden hills and the majestic oaks" and returned as soon as he could. His partner Paul "Pete" McCloskey was a Golden State native, a Stanford alum, and a Korean War veteran who had led six bayonet charges and earned two Purple Hearts, the Silver Star, and the Navy Cross. Both men were a little overqualified to become small-town lawyers, but Palo Alto was no longer an ordinary small town.[23]

Within a few years of its 1961 founding, the firm had developed a reputation for the close support it provided to clients, and for its growing expertise in the very particular needs of small electronics companies, from incorporation to patents to personnel. They also stepped in as high-profile advocates in fights that mattered to the community at large, as when McCloskey represented the bucolic and horsey town of Woodside—where he and many electronics-industry executives lived—in its fight to prevent the Atomic Energy Commission from stringing large overhead electrical lines across town. Ironically, the wires' destination was a facility that had helped make the Valley's high-tech reputation: the power-hungry Stanford Linear Accelerator.

By 1967, McCloskey would use that elevated public profile to run successfully for Congress. The crowded field that McCloskey vanquished in the Republican primary in this special election included Bill Draper, who'd decided to indulge a long-brewing political urge, as well as the GOP's favored candidate, Shirley Temple Black, the child movie superstar turned suburban homemaker. The race ultimately came down to Black versus McCloskey, and the lawyer's surprising victory became known as "the torpedoing of the Good Ship Lollipop."[24]

McCloskey's departure for Capitol Hill left a partnership spot at the firm, soon filled by a young Berkeley law graduate named Larry Sonsini. With Sonsini's arrival, the small-town firm began to sharpen its distinctive, only-in-Northern-California model for high-tech law practice. The newly renamed Wilson Sonsini Goodrich & Rosati was a law firm with the hands-on, multitasking style of a Valley venture capitalist. The partners worked closely with VCs from a company's start, delivering advice tailored to the needs of an enterprise with little cash on hand. They layered on new sorts of specialists—not just lawyers who understood different aspects of corporate and securities law, but also science PhDs who understood technology and its commercialization. Over time, Sonsini became increasingly interested in helping clients raise money once they outgrew the early stage: tapping into institutional investors, helping companies go public.[25]

The model endured over Silicon Valley's subsequent tech generations. "Lawyers prided themselves on matching the culture," reflected attorney Roberta Katz, who came to the Valley much later as general counsel for Web-era superstar Netscape. "There was a sense of needing to be nimble and quick and not burdened by too much bureaucratic overhead." Move fast, and do what it takes to get deals done: law became another place of increasing divergence between Boston and California. Among other things, Massachusetts placed strict limits on lawyers entering into business transactions with clients—such as the Valley practice of taking stock options instead of cash payment for legal fees—dampening the capacity of Boston start-ups to obtain high-priced legal counsel. (Decades later, some Valley insiders credited Sonsini as the original progenitor of an oft-repeated saying about the way the place worked: "no conflict, no interest.") As tech offerings became particularly hot property in the 1960s stock market, the Bay Area became home to a new cadre of investment banks that blurred the lines between venture, law, and brokerage in similar ways. But it started with the lawyers.[26]

A second, only-in-the-Valley species: high-tech real estate developers. Suburban Boston was a crowded landscape of colonial villages and nineteenth-century mill towns, with increasingly strict rules about where new homes and office parks might be built. In contrast, the Santa Clara Valley had vast tracts of land at the ready for large-scale industrial development—and a group of landowners who were willing to sell at the right price. During Spanish and Mexican rule, Northern California's countryside had been divvied up into very large land grants, or *ranchos*. Many of these big holdings remained intact or only moderately subdivided into the twentieth century, making it remarkably easy to develop large swaths of acreage at once. Land had been dirt cheap in the Valley for decades, and few farmers could resist the lure of cashing out to developers as the electronics market heated up.

More and more flowering orchards gave way to bulldozers, as developers rolled in and promptly erected "tilt-ups"—low-rise industrial buildings made of slabs of prefabricated concrete. No one was winning any architectural awards, but the Valley's ability to quickly meet market demand was yet another reason it was able to grow so quickly.

Kings of the bulldozers were Richard Peery and John Arrillaga, who had the great foresight to head out into the orchards at the same time as Draper and Johnson. Instead of funding companies, their goal was to buy up land and turn it into office parks. Arrillaga grew up poor in Los Angeles and attended Stanford on a basketball scholarship, playing professionally for a bit before returning to Palo Alto. There he met Peery, a hometown boy with a fierce entrepreneurial streak, and the two pooled $2,000 of their own money to start buying up orchards in the cheap flatlands around San Jose. The two weren't alone. Mother-and-son pair Ann and John Sobrato began similar speculative development with the proceeds gained from selling a family restaurant; by the end of the 1960s they were building hundreds of tilt-ups and were—like Arrillaga and Peery—on their way to becoming some of the wealthiest people in the Valley.[27]

Lightning-quick real estate development benefitted not only from the legacy of the *ranchos*, but also from a host of other special regional circumstances. There was the extensive water and sewage infrastructure built by San Jose under the watch of an exceedingly ambitious city manager who didn't worry too much about obtaining permits before building anything, and that allowed the city to build out quickly as a bedroom community for all the engineers flocking to the Valley to work. There was cheap electricity, negotiated during and just after World War II by local governments as a play to build up the industrial base. It fueled not only the mass manufacture of sand into silicon glass but also the fabrication of the chips themselves. Then there were the highways—public investment in the 1950s and 1960s that widened U.S. 101 on the Bay side of the Valley and built a new interstate, I-280, along the coastal hills.[28]

Last, but hardly least, there was the California factor. The men leading the state during this explosive period of growth—Republican Governor Earl Warren, Democratic Governor Pat Brown—led the way in building not only highways for the cars and sewers for the split-level homes, but also a public education system of unparalleled excellence. Well-resourced and rapidly expanding, California's public schools educated the children of white-collar Lockheed engineers and blue-collar assembly line workers alike, sending an increasing portion of them on to California's growing higher education system, which added on three new campuses in the 1960s. Tuition for in-state residents—including at the University of California, Berkeley, one of the very best universities in the world—cost next to nothing.[29]

UC Chancellor Clark Kerr was determined to match or exceed the Ivy League in research excellence, although he gently disdained the raw entrepreneurialism of Fred Terman. "The individual faculty member is a genuine entrepreneur; the real producer of the intellectual product," he observed. "Out of the free action of each the public gets the best product as business competes with business and mind with mind." Although Kerr's desire to preserve and celebrate the ivory tower might have muted the ability of Berkeley's faculty and students to capitalize on their inventions (and helps explain why the East Bay did not become a high-tech hub), the broad-based commitment to public education provided the state with an abundant, highly educated talent pool.[30]

There also was a dash of serendipity at work when it came to the Valley's California location. For example, the state's civil code prohibited enforcement of noncompete clauses in employment contracts, a prohibition that didn't have a thing to do with the intellectual property or trade secrets of the electronics industry, but instead came into being in 1870, as the state's early lawmakers attempted to reconcile the chaotic jumble of legal regimes—Spanish, Mexican, Anglo-American—that had ruled the state. But the result of this provision helped facilitate the job-hopping that became a hallmark of the Valley's tech community.

If an engineer left his job and jumped to a direct competitor, his old employer couldn't do anything about it, even though the employee's tacit knowledge could be a tech company's most valuable asset. Massachusetts, in contrast, enforced these clauses. As did Washington, Oregon, Illinois, Texas, New York, New Jersey, and more. Every other place in America that had a tech sector practiced enforcement of non-competes; but California did not. As non-compete clauses grew along with the technology industry, and spread to other sectors as well, the freedom to move kept the Valley filled with funny little companies, and kept knowledge spilling over from one technical generation to another.[31]

IT WOULD STILL BE TWO DECADES MORE BEFORE THE VALLEY DECISIVELY pulled out ahead of Boston, but the ingredients were there early. Boston may have had MIT and Harvard and leading companies of the postwar electronics world, but it didn't have that cheap land and abundant infrastructure and local people willing to capitalize upon it to such an unfettered degree. New York and Philadelphia may have had the capital and the big electronics makers and some of the universities as well, but these places didn't have the relentless focus on nurturing start-ups. Nowhere else but the Valley had the entrepreneurial and opportunistic Stanford, the thrusting bulldozers and hustling law firms, and the young money men opening up shop along Sand Hill Road. Nowhere else had the people. The California Gold Rush had been over for a century, but the Golden State remained a destination for the adventurous young from elsewhere, arriving with little to lose and an appetite for reinvention.

Arrivals

NEW YORK HARBOR, 1965

"The bill we sign today is not a revolutionary bill," President Lyndon B. Johnson said. "It does not affect the lives of millions. It will not reshape the structure of our daily lives, or really add importantly to either our wealth or power." The toned-down rhetoric wasn't what usually came out of these kinds of presidential bill-signings—particularly one staged in the grand setting of Liberty Island, the backdrop of the Manhattan skyline gleaming in the crisp autumn air. But Johnson was signing into law a sweeping reform of the nation's immigration system, one that swept aside quotas that had been in place since the anxious, nativist days of the early 1920s, and opened the door to new streams of migration from all over the world.[1]

Now, immigration law would prioritize special skills and connections to family who already lived in the United States. Gone were laws that had excluded Asian immigrants for so long; gone, too, were the limits on new arrivals from Latin America and Africa. Instead of restricting entry by country of origin, the new system would operate on "the principle that values and rewards each man on the basis of his merit as a man," said the President. Here was the next, necessary step in America's commitment to civil rights and racial equity, correcting "a cruel and enduring wrong in the conduct of the American Nation."

For the bill's opponents, the loudest of whom were Johnson's fellow Southern Democrats, immigration reform was a dangerous opening of the floodgates. What would happen to the nation's heritage, its citizens' connection to their Enlightenment roots? "I don't know of any contributions that Ethiopia has made to the making of America," huffed North Carolina's Sam Ervin. The special-skills provision was "just sanctimonious propaganda," he said, allowing immigrants to come by the tens of thousands "to compete with Americans for available jobs."[2]

In the end, the effects of the bill signed that day in New York Harbor went far beyond what Johnson or Ervin imagined. The new waves of immigration that followed in its wake brought new ethnic, racial, and religious diversity and redefined who and what was "American." Immigration from India was three times what the Johnson Administration had predicted. Nearly six million new immigrants came to the U.S. from Asia between 1966 and 1993 alone.

Few places in America were more transformed—and economically and intellectually invigorated—by these new arrivals than the hubs of the technology industry: Boston, Texas, Seattle, and, especially, Silicon Valley. Skills requirements were not "sanctimonious propaganda" in the world of high tech; immigrants from Taiwan and Hong Kong, then China, India, and the former Soviet Union became the engineering backbone of hundreds of start-ups and large tech companies. Many of them ended up founding companies themselves. In the 1980s, immigrants from India and China were at the helm of nearly one-fourth of Silicon Valley companies. By the Internet era, the number of immigrant founders in the Valley stood at 40 percent. Nationwide, more than 25 percent of high-tech companies had a foreign-born founder.[3]

The new wave included people just as critical to the scaling-up of the tech phenomenon, even though their faces never appeared on the covers of *Fortune* or *Forbes*: assembly-line workers from Mexico and Southeast Asia, who soldered the semiconductors, built the desktops, and fabricated the routers. By the end of the 1980s, over half of Silicon Valley's blue-collar workforce was Latino or Asian.[4]

"We can all believe that the lamp of this old grand lady is brighter today," concluded Johnson that long-ago October, "and the golden door that she guards gleams more brilliantly in the light of an increased liberty for the people from all the countries of the globe." Little did he know what he had started.

CHAPTER 6

Boom and Bust

"Dave, do you mind if I ask you a question?" With blinking red eye and the calm voice of a newscaster, thus spake the supercomputer at the heart of Stanley Kubrick's fearsome parable of AI run amok, *2001: A Space Odyssey.* Opening in early April 1968, the film was unlike anything moviegoers had seen before: a visual spectacular of precisely modeled spaceships and trippy psychedelics, with a nonlinear plot, plodding pace, and few recognizable actors. The closest thing the film had to a star was the omniscient and ominous HAL 9000, who seized control of a deep-space mission from its human astronauts.

2001 was a box-office dud at first, but word of mouth among the college crowd turned it into a phenomenon. Other pop culture landmarks cascaded out in 1968 as well: the Beatles' *White Album*, the Broadway musical *Hair*, and Joan Didion's lyrical and searing portrait of countercultural San Francisco, *Slouching Towards Bethlehem.* It was a year of fractured politics, of rage against the war machine, of slain heroes and violence on the streets, of trust shattered and authority questioned. At the Summer Olympics in Mexico City, San Jose State University athletes and track medalists Tommie Smith and John Carlos raised their fists in a Black Power salute on the podium, becoming one of the most enduring images of the decade.

Forty miles down the road from Didion's out-of-their-mind hippies in the Haight, and next door to the student activists of San Jose State, the electronics makers of the Santa Clara Valley could have been living in another universe. To be sure, that year's political wildfires swept through Stanford's campus and up into the hills where poets and activists and early back-to-the-landers were living in creaky old farmhouses and tumbledown cottages. Some of these freethinkers soon would start disrupting the business of high technology just as they were disrupting everything else. But not quite yet.

From California to Massachusetts and all points in between, the tech industry of 1968 remained the domain of the squares. It was a business of mainframes and minis, of specialized engineering and specialized investing, of selling to enterprises rather than to consumers. If you took a poll up and down the Valley's tilt-ups, you'd find plenty of people planning to vote for Nixon that fall. You'd also find plenty who planned not to vote at all.

American society might be fracturing, but the technology industry happily surged forward. The reason was *money*. Vietnam-era pressures had slowed the surge of research money coming from the Pentagon; the Apollo program soon would wind down. No matter. Wall Street was bullish for all things electronic. In addition to the public markets, technology companies now had a growing base of private-sector customers, new product niches, expanding overseas markets, new investment pools. The boom years turned out to be relatively brief, but they left a long shadow.

THE GO-GO YEARS

Wall Street's tech boom started in the summer of 1966, when Digital held its first public offering. The mini maker's 350,000 shares of common stock, offered at $22 a pop, sold out almost immediately. Ken Olsen instantly became a multimillionaire. High net worth didn't markedly change the company's low-key business culture, however. Olsen's biggest splurge after the Digital IPO was to buy a second canoe.[1]

Another Boston-based pacesetter was Wang Laboratories, which had its IPO in the late summer of 1967, exactly a year after Digital's debut. "I had a banker call me and ask for 100,000 shares," said one broker. "He said he had no idea what the company did but he heard it was wonderful." Wang's valuation the day before the IPO was $1 million; the day after, it was $70 million. After realizing what her 100 shares of the company now were worth, the CEO's secretary shouted, "I'm rich! I'm rich!"[2]

It was only the start. Certainly, the 1950s had minted high-tech millionaires like Hewlett and Packard, and the early years of the 1960s had seen an excited uptick in the value of what analysts called "space-age stocks." But the second half of the decade saw a dizzying explosion of electronic companies' ability to earn prodigious amounts of cash. Everyone wanted a piece of the action. As Silicon Valley journalist and entrepreneur Adam Osborne later observed, "In the late 1960s, all you had to do was walk into the middle of Wall Street and shout 'Minicomputer!' Or 'Software!' and you would be buried up to your neck in money."[3]

It wasn't just stock pickers who drove the boom. The general corporate prosperity of the middle of the 1960s created new investment pools and expanded the ranks of venture capitalists. David Morgenthaler, who had made that brief, yearning visit to Palo Alto two decades before, was a case in point. By 1968, Morgenthaler had achieved success in business that he couldn't have imagined as a girl-chasing teenager in Depression-wracked South Carolina. The company presidency he'd taken just after the Sputnik launch proved to be a stepping stone to an even larger job, as the company he led morphed into a multinational conglomerate. He'd earned enough to retire many times over, or at least to take his foot off the gas.

Morgenthaler couldn't stop thinking about Palo Alto, however. Cleveland, his adopted home, was a place where richly rewarding corporate careers could be made. But its days of entrepreneurial glory were at least five decades gone by then. The town had become a place where you worked for someone else, and "I was sort of tired of having a boss." The future, he sensed, lay in the small electronics companies popping up in California and in Boston. They were commercializing, growing, pushing out new markets even as the space race slowed and defense spending declined. He wanted in. At the age of forty-eight, David Morgenthaler went into business as a venture capitalist, seeding his first fund entirely with his own money.

He was quite different from the Valley VCs, older and a little more cautious. "I wanted to be at the edge of the new," he explained, but "I didn't want to be so new that you had too high a risk of failure." Yet he knew how to spot good technology and good talent, as his previous job had involved acquiring one small firm after another—fifty-seven in all. And in temperament and outlook, the Clevelander had much in common with the Californians. He was a businessman with the expertise of an MIT engineer and the excitement of a boy building a science project. He liked to be hands-on, opinionated, and ready to replace management if necessary. He was a people person: gregarious, curious, a connector. Although he would remain in Cleveland for another decade-plus, he began to bind himself closer to the people and products and next-generation ambitions of California.[4]

Plenty of ambition was now on display out West. Buoyed by the bull market, tech's money men and start-up pioneers got inspired to start up once more. In the span of one year, a host of Valley firms were born that would set the pace for the region for decades to come. Arthur Rock and Tommy Davis went their separate ways. Davis started a new VC firm, the Mayfield Fund, with Stanford computer science professor William Miller. George Quist left Bank of America's SBIC and joined another young San Francisco money man, William Hambrecht, to start a new breed of boutique investment bank, specializing in high-tech start-ups. And Bob Noyce and Gordon Moore departed Fairchild Semiconductor and its micro-managing investors to found a completely venture-financed company that didn't have to pay fealty to an East Coast boss. They called it Intel.

AS THE NEW TECH COMPANIES SURGED, THE FREE WORLD TRIED ENVIOUSLY TO emulate America's new-economy example. Emblematic of the nation's envied position was a somewhat unlikely volume that climbed European bestseller lists in 1967. Titled *Le Défi Américain* (*The American Challenge*), it was neither grand romance nor spy novel, but a book about the technological and economic supremacy of the United States—and Europe's failure to keep up. Written by prominent journalist and publisher Jean-Jacques Servan-Schreiber (who was so famous in France that, like Brigitte Bardot, he was known simply by his initials, JJSS), the book argued that Europe needed to become more like America by investing heavily in scientific research and development and adapting American management and marketing techniques. The message hit a nerve; *Le Défi Américain* sold over a million copies in Europe and ultimately was translated into fifteen languages.[5]

Servan-Schreiber's climb up the bestseller list was not an isolated incident. The book tapped into a growing worry on both sides of the Atlantic in the late 1960s that the United States had become so economically unassailable that an emerging "technological gap" was draining away Europe's talent, weakening its institutions, and potentially endangering fragile international alliances. The same year, European ministers convened to discuss the possibility of a "technological 'Marshall Plan,'" and the Johnson Administration had established an interagency committee to assess how the U.S. might apply its technological might to rectify this imbalance. "Unless we're careful, our concept of the Atlantic partnership can be eroded by the fear and concern about the power of American capital and technology," Vice President Hubert Humphrey remarked in an article that appeared on the front page of *The New York Times*. Defense Secretary Robert McNamara noted more astringently that Europeans' "complaint is that we are so surpassing them in industrial development that we will eventually create a technological imperialism." American innovation was riding high, and the world was scrambling to catch up.[6]

TURN OUT THE LIGHTS

How quickly things changed. As the 1960s ended and the '70s began, *Le Défi Américain* remained on the bestseller lists, but the challenge was quite different. A newly energized Europe had big, government-subsidized projects around data processing and computer networking well underway. Japan was becoming an economic powerhouse and a technological rival, rising to "the American challenge" not by imitating the United States but by developing different, more agile ways to produce and sell its manufactured goods. Elsewhere in East Asia, ambitious nation-states like Singapore lured American manufacturing operations with trade incentives, cheap skilled labor, and sparkling new industrial parks.

The nation's international economic position shifted. By 1971, Nixon Commerce Secretary Maurice Stans was warning Congress that the United States might have its first trade deficit since 1893. Wall Street's bull run came to a shuddering halt. University graduates that year entered into a weak job market and were bewildered by what they found. "Before, the kids who dropped out, at least they had a choice," one member of the University of Chicago Class of '71 lamented to a reporter. "Now it's got no meaning 'cause it looks like you had no choice. Like you dropped out because you *couldn't* get a job."[7]

As the political and economic scene changed, so did the military-industrial complex. The Vietnam War had placed intense strain on the U.S. military budget. The successful moon landing had ended the bonanza of space-age contracts; NASA's 1971 budget was half of what it was in 1966. Budget anxieties and environmental concerns had contributed to Congressional "no" votes on big-ticket projects like the Supersonic Transport (or SST), America's answer to the Atlantic-hopping Concorde jet produced by an Anglo-French consortium in 1969.

The cancellation of the SST in early 1971 threw its designated contractor, Boeing, into an economic tailspin, and set off a deep, years-long recession in the aerospace giant's hometown of Seattle. The only people in town who seemed happy about the decision, *The Wall Street Journal* reported with a scowl, were members of Seattle's radical Left. "It means more disgruntled people out of work," remarked one organizer cheerfully as he sat amid a bustling socialist food bank, a poster of Che Guevara glaring down from above. By 1971, Boeing-related job losses had so crushed the regional manufacturing economy that two commercial real estate agents erected a sardonic billboard along a major highway reading, "Will the last person leaving Seattle turn out the lights?"[8]

The political and economic changes slammed into a Santa Clara Valley that had enjoyed unchecked growth through the boom times of the military-industrial complex and the bull market of the 1960s. Defense cuts contributed to a loss of ten thousand manufacturing jobs in the San Jose metropolitan area between 1969 and 1971. And it wasn't just assembly line workers: as big contracts for missiles and satellites wound down and weren't replaced, the Valley's largest employers laid off scores of engineers.[9]

On top of this, Wall Street deals slowed to a trickle. The mania for tech IPOs that had seized the market in 1968 was completely over by 1971. It turned out that Burt McMurtry had become a venture capitalist just in time for the VC industry's lean years to begin. The first fund that he and Jack Melchor operated out of their offices above the pizzeria looked like a bust (although it later became wildly successful). "The most common exit strategy," he remembered ruefully, "was that we lost all our money."[10]

COMPUTER INTELLIGENCE

Yet change also brought new opportunity. As contracts for military R&D slowed down, new ones rose up. As Richard Nixon declared war on cancer, Stanford and Berkeley became leaders in the growing fields of medical and biotechnological research. Stanford's computer science department started its own industrial affiliate program, where researchers and local computer companies could do regular "show and tells." It was a great way to raise money for Stanford too. Soon, the "Computer Forum" was pulling in a million dollars a year in corporate support.

If electrical engineering defined Stanford's Terman years, computer science increasingly defined its new era. Federal funding for computer science once had been minuscule compared to other fields; now it was rising fast, funding graduate fellowships and faculty research activities to the tune of $250 million per year nationally by the middle of the 1970s. On the Farm, faculty member Bill Miller, an originator of the affiliate program and a co-founder of that new Stanford-focused VC, the Mayfield Fund, became the world's first "Associate Provost for Computing" and embarked on digitizing the university's administrative operations. In 1971, he became Stanford's provost.[11]

The Valley also was catching up with MIT in the fields of robotics and artificial intelligence, thanks to the AI lab helmed by John McCarthy and the work happening down the road at SRI International, a think tank recently spun off from Stanford in response to student protests about the classified military research conducted there. SRI still did plenty of military work, but it now was grabbing national attention for what it was doing to make machines think. In November 1970, *Life* magazine invited its millions of readers to "Meet Shaky," a robot that rolled the linoleum halls of SRI, capable of point-to-point navigation, "seeing" objects in its way, and rudimentary speech recognition. The mainframe-powered Shaky was a smartphone and a driverless car, fifty years before its time.[12]

Down the hall at SRI lay another future-tense lab, the Augmentation Research Center, led by a soft-spoken engineer in his forties named Douglas C. Engelbart. As academics and policymakers worried over the question of automation—the replacement of human workers with robotic machines, and human brains with artificially intelligent computers—Engelbart was one of a small and growing group of researchers interested in *augmentation* of human effort via technology. Semiconductor companies might have been laboring to crowd more and more technology on smaller and smaller chips, but Engelbart wasn't terribly interested in shrinking the size of computers. The real potential lay in the computer networks that could allow human beings to connect and communicate with one another.

In December 1968, five months after Intel's incorporation, two months after Smith and Carlos raised their fists on the Olympic podium, and one month after Richard Nixon's first presidential victory, Engelbart and his team had delivered a comparably future-altering presentation at a San Francisco computer conference. Quietly sitting down at a keyboard onstage, Engelbart proceeded to use the devices before him to send commands back to a computer at his lab thirty miles away. The results flashed up on the screens behind him in the San Francisco auditorium. And the words appearing overhead weren't inscrutable computer language, either. He typed simple commands. He edited a grocery list. He jumped his cursor from place to place on the screen by moving a square wooden box that fit under his palm, with wheels on its bottom and a cord trailing from its rear. Engelbart called it a "mouse."

The presentation went down in Silicon Valley history as "the mother of all demos." The inventions unveiled by Engelbart were a preview of a world still two and three decades in the future: the mouse, interactive computing, hyperlinks, networked video and audio. But for all its envelope-pushing vision, the demo also was important in showing how these futuristic devices could work in an ordinary office or household. It made the fearsome, HAL-like computer into an ordinary, accessible, even rather friendly kind of machine. Engelbart's demo resonated deeply with a new generation of technologists who were just emerging from the classrooms and labs of Berkeley and Stanford and MIT during the Vietnam era and its agonizing aftermath. They wanted nothing to do with the military-industrial complex. They didn't want to build the real-life HAL 9000. Instead, they wanted to make technology personal, and set information free.

SURE, SOME PEOPLE WERE DECLARING THAT THE DEFENSE DRAWDOWN MEANT tech's boom times were finished. Maybe it looked that way if you saw the lights flickering out in the Cold War boomtown of Seattle. You certainly could feel the economic pain if you were sitting inside Lockheed's sprawling campus on the east side of 101.

But travel just a few miles up and down that highway, and you'd find other companies—ones that were younger, that were smaller, that might have started out doing some federal business but didn't do much of it anymore. You'd also find lots of interesting opportunities in the academic labs of Stanford or at research institutes around town. You'd find a new generation of the young and technical, ready to redefine what the computer might be. Maybe you could be part of these labs. Or jump into those venture-backed companies.

Better yet, you could start one yourself.

ACT TWO

PRODUCT LAUNCH

Silicon Valley created the Rust Belt. All the things they did became obsolete almost overnight.

Floyd Kvamme[1]

Arrivals

PALO ALTO, 1969

Being a teenage figure skating champion had taught Ed Zschau the value of hard work and careful, repetitive practice. It also had gotten him comfortable with performing in front of a crowd, attuned to their desires, eager to meet expectations. Perhaps it was all that time calibrating the rasp of ice along metal blade, the arc of a perfect loop jump, that had made him interested in physics, too—its technical particulars, and the way it worked in the world.

Or maybe it was simply a sign of the times: born and raised in Omaha, he had set off for Princeton the same fall of 1957 that Sputnik soared into space. Science was on everyone's mind; technology became an American preoccupation. Four years later, a degree in the philosophy of science in hand, Zschau joined the exodus west, to Stanford's business school. He complemented his PhD in business administration with an MS in statistics, completing both so swiftly that he was offered a Stanford teaching position by the time he hit his mid-twenties.

Ed Zschau loved being in front of a classroom, and the students adored him back. He'd ham it up, sing corny folk songs, and leaven the dreariness of management science with philosophy and humor. He bought a motorcycle; he got married. He heard hot tips about new industry trends while playing pinball down at The Oasis on El Camino. It was the late 1960s, and the friendly young professor in blue jeans was a breath of fresh air in an academic world of tweed and pipe smoke. But places like Stanford didn't care whether someone was a good teacher. Professional success meant publishing papers, and Zschau didn't do enough of that to merit tenure. So before he hit thirty, the extroverted professor embarked on the first of several bouts of "self-renewal" and switched gears—into high-tech entrepreneurship.

The timing was perfect. Now it was 1969, and new start-ups were growing all over the Santa Clara Valley, helped along by new venture capital funds and specialist law firms and marketing operations, many of them started by young guys like him. With his charm and connections, Zschau knew he could start something up. But what was the something? He'd mull the possibilities during his morning shower. Finally, he found it.

The raft of new environmental measures just signed into law by Richard Nixon—the Clean Water Act, Clean Air Act, the creation of the EPA—created a big demand for pollution-measuring instruments. And now there were these minicomputers, selling by the thousands. Why not build a computer system that used a mini to automate and analyze all this environmental data?

It turned out to be not just a million-dollar idea, but ultimately a $100 million one—a company called System Industries that started off measuring water quality and the composition of moon rocks and then graduated to providing data storage. System Industries wasn't very large, nor did it ever have a breakthrough product, but the former figure skating champ was brilliant at making friends and navigating the cozy professional networks of the local tech community. It all came naturally to Ed Zschau, as did his overflowing optimism about where life might lead. "Do what you enjoy doing," he'd tell advice-seeking students. "Do it the best you know how, and good things will happen."[1]

PALO ALTO, 1970

When the engineer brought a sheep to work, Regis McKenna knew he wasn't in Pittsburgh anymore.

McKenna had come of age in the city of steel and smoke at a time when its mills were grinding to a halt and there were few jobs to be had. Nearly the only job offer he got was for a publishing company that wrote about the electronics industry, work that eventually brought him out to the Santa Clara Valley in 1963. The surroundings were wildly different from anything he'd known—the acres of orchards, the squat tilt-ups dotting the terrain between El Camino and Highway 101, the sweeping views of the mountains to the east and west—and the things those companies were doing were fascinating. Instrumentation and tube companies still ruled the roost when he first arrived, and he spent as much time making sales calls on aerospace companies up in Seattle as he did down in the Valley. But, soon enough, all these little semiconductor outfits started sprouting up, most of them started by guys coming out of Fairchild.

Before too long, McKenna had joined up with one of them, a small outfit called General Micro-electronics, which was notable for being the first to use the metal oxide semiconductor, or MOS, technology that later would be state of the art. That job got him in the door at National Semiconductor. The company was so new, and management was hands-on: CEO Charlie Sporck came in himself with a hammer and two-by-fours to build out a secure storeroom for newly fabricated chips. Working there also wasn't for the faint of heart. A few days in, the man who'd hired him, Don Valentine, called McKenna into his office. He waved a computer punch card with names handwritten on it in green ink. "These are the fourteen people I considered hiring before I hired you," Valentine declared.

"So why did you hire me?" McKenna wondered.

"Because you," his boss smiled craftily, "were the only one I knew I couldn't intimidate."

The ethos of National wasn't just about thick skin. It was about a conscious repudiation of the stuffy and the hierarchical, and about rewarding people for what they accomplished, not for their fancy degrees or Nobel Prizes. If you had technical brilliance, it didn't matter where you came from or how you behaved. Arrogant jerks got a pass, as long as they built and delivered greatness.

That's how you got to the engineer and the sheep. The engineer was Bob Widlar, National's lead designer, the genius mind behind the elegant linearity of the Fairchild IC. A cult hero not only because of his creative output but also because of his extreme eccentricity, he drank to excess, kept an axe in his office, and flipped the bird at nearly any camera that dared take his picture. He'd get on a plane and disappear for weeks at a time. He didn't believe anyone over thirty could design anything worthwhile. And when the thrifty Sporck decided to cut costs by not mowing the lawn in front of National's tilt-up, Widlar brought in a sheep to do the job instead.

The low-key, scrupulously organized McKenna came off like a straight arrow amid the wild men around him, but it was an environment where he thrived. His lack of an engineering degree became an advantage as he acted as translator and mediator, explaining the tight little world of young chip companies to the broader universe beyond it. You needed to educate customers, building share in existing markets and discovering entirely new ones. You needed to tell compelling stories about the tech coming out of the tilt-ups, and about the personalities who made it.

Seven years after he first arrived in the Valley, Regis McKenna left the sharp-elbowed, middle-fingered life at National and went into business on his own, as a full-service marketing consultant. There were a number of these firms blooming in the Valley those days—those engineers needed all the spin doctors they could get—but only one with so many connections and insights into the microchip world. McKenna now understood that marketing tech products required a different approach than in any other industry. For years to come, his first slide in any pitch read, "The ad is the last thing we do."

McKenna also had absorbed the chipmakers' dislike of hierarchy. As he hung out his shingle in 1970, he felt that it would have been pretentious to assume the title "President" of a one-man operation with zero clients and a mere $500 in start-up capital. Instead, he printed business cards reading "Regis McKenna—Himself." He never changed them.[2]

CHAPTER 7

The Olympics of Capitalism

Silicon Valley got its name because Don Hoefler needed a headline. A regular contributor to the industry trade paper *Electronic News*, Hoefler had just written up a multipart feature story on Northern California's booming computer-chip industry. It was the dawn of the 1970s, and despite the growing economic gloom, Moore's Law ruled the Valley. Its tiny and lightning-fast silicon semiconductors of complex circuitry continued to get smaller, cheaper, and more powerful by the month. Anything that had a spring, a transistor, a memory core, or a vacuum tube now could be powered by a silicon chip—from factory machinery to mainframe computers to wristwatches. Macro computer power was going micro. It was a revolution.

Northern California had plenty of competition—down in Dallas, electronics giants Texas Instruments and Motorola manufactured chips by the thousands—but the technological innovations emerging from the Valley were powering the largest and most sophisticated machines. IBM, king of the mainframes, made its own chips, but the rest of the American computer industry trekked out West.

"The guys back in New York and D.C. call this place 'Silicon Valley,'" a couple of visiting sales managers informed Hoefler over lunch one day. Short, memorable, and a little jokey—silicon was built on sand, after all—the name was exactly what the reporter needed to describe this laid-back, entrepreneurial slice of Northern California to his readers. "Silicon Valley, U.S.A.," blazed Hoefler's header on the cover of *Electronic News*' January 11, 1971, issue. The name turned out to be a keeper.[1]

AT THE WAGON WHEEL

The durability of the name had a lot to do with the durability of Don Hoefler, a journalistic army of one at a moment when national reporters only rarely looked away from the Northeast Corridor to consider this nerdy stretch of California suburbia. He gathered much of his intelligence from what he called his "field office" on a barstool at Walker's Wagon Wheel, one of the semiconductor industry's bars of choice in an area decidedly short on nightlife. As after-work happy hours devolved into boozy gossip sessions in the tavern's overstuffed booths, Hoefler gathered scoops. He covered it all, from new product releases and big hires to recent weddings and raucous company parties. Hoefler knew which company was about to launch a new line, and which CEO just bought a flashy new sports car. "If it didn't appear in *Electronic News*," one industry insider declared, "it didn't happen."[2]

Having gotten his start in the industry as an in-house corporate publicist, Don Hoefler understood the technology as well as having a nose for colorful, personality-driven stories. Outsiders saw bafflingly complex technologies and bland ranks of shirtsleeved engineers. Hoefler saw the machines of the future, built by cowboy entrepreneurs and genius oddballs.

At a moment when wars, protests, and moon shots dominated the national headlines, Hoefler's coverage of the strange little technological Galapagos of Silicon Valley set the tone for reporters following in his wake. The charismatic personalities and intensely competitive culture of the semiconductor industry made for great copy, and America certainly needed some new heroes.

The Wagon Wheel was a good place to find them. As big contractors and aerospace companies up and down the West Coast went into their post-Vietnam tailspin, the chipmakers boomed. The Valley still had plenty of suit-and-tie outposts of Eastern electronics giants, but its semiconductor companies were the rising stars. They still were young and agile enterprises with low fixed capital costs—a stark contrast to the increasingly sclerotic old-line manufacturers who were scrambling to stay competitive. And they no longer needed defense contracts to survive.

The king of the Fairchildren was Intel, founded when Bob Noyce and Gordon Moore decamped from Fairchild Semiconductor in 1968 after years of chafing under the micromanagement of its East Coast parent company. In contrast, Intel was entirely venture-funded by local firms. Then there was National's CEO Charlie Sporck, the cost-conscious son of a taxi driver, the guy who didn't like to pay to mow the lawn and who, while at Fairchild, had pioneered the idea of offshoring chip assembly to East Asia. Down the Valley was Advanced Micro Devices (AMD), founded by Jerry Sanders, who had grown up a street-fighting kid from the South Side of Chicago and morphed into a sales executive fond of loud suits, sporty cars, and Gucci loafers. Sanders poached twelve other Fairchild employees to come along with him.[3]

The late-'60s hot market for mergers and acquisitions added to the diaspora, as Eastern electronics giants bought up local start-ups. Faced with the prospect of adapting to stuffy corporate culture or—even worse—relocating to corporate headquarters, the employees of these acquired companies, as one of them put it, "started looking for other pastures."[4]

Yet Don Hoefler was still writing about a very, *very* niche market. If you were taking bets on the place that would become the center of the computerized universe at the start of the 1970s, Northern California remained a long shot. IBM still ruled the business-machine world. Texas pumped out far more microchips. The new technology setting the computer world on fire, minicomputers, was a Boston business. And the vast majority of investment capital—including venture capital operations like the funds headed by Ned Heizer and David Morgenthaler—was still based east of the Mississippi.

Wall Street analysts had no interest in following the semiconductor industry ("The computer industry is IBM," one coolly informed Regis McKenna), and *The Wall Street Journal* refused to write about any company that wasn't listed on the stock exchange. Making Silicon Valley more illegible to the wider world was the fact that its firms sold to other electronics companies, not to consumers. An Intel chip might be inside the computer down the hall or in the calculator on your desk, but you wouldn't know it.

Ten years on, the Valley had vastly increased in size and influence and was very much in the consumer-electronics business. Another decade after that, Don Hoefler's snappy headline had become shorthand to refer to the entire computer hardware and software industry.

Why did "Silicon Valley" not only beat out its regional competitors, but become two words that were synonymous with the entire American high-tech industry? Technology, it turns out, was only part of the story.

THE COMPUTER ON A CHIP

Out of the Cold War cradle, Northern California chipmakers had established a healthy non-defense business by the start of the 1970s, making memory chips for Eastern computer manufacturers as well as for a booming new market: electronic calculators. Only a few months after Intel's founding, it had gotten a commission from a Japanese manufacturer to custom-build a sophisticated chip for a line of desktop calculators. This precipitated a design process that eventually led to a breakthrough rivaling Shockley's transistor and Noyce's integrated circuit: the microprocessor. The device leapt beyond merely placing multiple circuits on a chip: now there were even more of them, and they were *programmable.* With a stored-memory microprocessor inside, any sort of appliance or device—a car, a telephone, a bedside clock—effectively turned into a computer. Fast, powerful, and less expensive than mechanical controls, the microprocessor could "be stuck in every place," as Gordon Moore put it.[5]

Marketed as "a computer on a chip," the Intel 4004 made its public debut with an ad in *Electronic News* in November 1971, less than a year after Hoefler's giddy series gave Silicon Valley its name. Only a few months later, Intel followed on with the release of the twice-as-powerful 8008, followed by the 8080 in 1974. By that time, Intel had marketing as carefully designed as the products it was selling, having brought in Regis McKenna Himself to provide vision and execution. Other shops in town could do ads and brochures, but McKenna understood the semiconductor business like no one else.

"First time technologies required 'education of the market,'" McKenna remembered. It wasn't just about ads and sales brochures; it was about placing articles in trade journals where systems designers could see them, and running educational seminars for corporate managers who didn't know the first thing about semiconductor design and application. But the ads mattered too: crisply modern and speaking a language that regular business people could understand, Intel's were a different breed from the usual kind seen in the industry, which tended to be heavy on technical specs and light on illustrations. "The 8080 Microcomputer is here," blazed one brightly colored spread, "incredibly easy to interface, simple to program and up to 100 times the performance."[6]

Now, a sliver of silicon contained all the computing power of a mainframe or a minicomputer that cost tens of thousands of dollars. Dearly expensive and space-hogging technology was on the verge of becoming accessible to nearly anyone. The microprocessor set the miniaturization of the computer into hyperdrive, turned all sorts of products from analog to digital, and gave Intel and the rest of the chipmakers a conviction that they were truly changing the world. "We are really the revolutionaries in the world today—not the kids with the long hair and beards who were wrecking the schools a few years ago," said Gordon Moore.[7]

By 1975, Intel had 3,200 employees and sales of $140 million. National Semiconductor had sales of $235 million. Northern California's 1950s and 1960s had been about manufacturing bespoke, expensive products for a small number of deep-pocketed customers: the Defense Department, NASA, the mainframe computer makers. Silicon Valley's 1970s were about turning these small electronics into market commodities. Intel operations chief Andy Grove famously referred to the company's products as "high technology jelly beans." But their blueprint for making microchips at scale wasn't the mass-production assembly line of Henry Ford. It was the franchise model of McDonald's hamburgers. Manufacturing grew by building small-to-medium fabrication plants across the country and, increasingly, overseas.[8]

Within headquarters, chip executives grouped their employees into small teams that competed against one another to develop the best product. "Big is bad," Bob Noyce declared in a keynote address to a group of businessmen in December 1976. "The spirit of the small group is better and the work is much harder." Intel avoided hiring people over age thirty. But this wasn't a search for anti-establishment rebels—it was a quest to find people with ambition to create a new industry.[9]

Amid Seventies malaise, semiconductor industry profit soared. Silicon Valley's denizens became more unabashed about the money they were making and their desire to make even more. "The basic thing that drives the technology is the desire to make money," said Robert Lloyd of National. Don Hoefler made big-spending vignettes a feature of his industry coverage. By 1972, he had so much good copy that he began his own weekly newsletter, *Microelectronics News*, chronicling all the happenings at local companies. The biggest personalities got the most attention. "Hardly had the ink dried on Jerry Sanders's order for his $64,000 Rolls-Royce Corniche," Hoefler dished in late 1975, "than the Mercedes-Benz importer phoned him from New York to offer a 7.5 liter bomb for $40,000, which M-B bows next year. Jerry's response: wrap one up; I'll take it. So goes all of Jerry's 1976 salary."[10]

As the cash flowed and the flash increased, more East Coast journalists started trekking out to Silicon Valley. The term "Silicon Valley" very gradually started popping up in the business sections of *The New York Times* and *The Wall Street Journal* (it nearly always appeared in quotation marks). Gene Bylinsky of *Fortune* rolled out a series of euphoric articles about high-tech execs and the venture capitalists who financed them, sketching portraits of risk-taking iconoclasts that sounded a lot like Hoefler's Wagon Wheel chronicles and blind items. This wasn't just another business story: it was an *entrepreneurial* story of people audacious or foolhardy enough to strike out on their own. "If you are a capitalist—and I am—you graduate to the Olympics of capitalism by starting new businesses," one Silicon Valley executive told Bylinsky.[11]

THE SILICON VALLEY STYLE

It certainly seemed to outsiders like this was something different. The shadow of the world's worst boss, the rigid and imperious Bill Shockley, still haunted the industry. The chipmakers didn't want to be my-way-or-the-highway micromanagers; they wanted to give their employees room to test out new ideas. They remained men of the electronics lab, too, choosing their hires on the basis of who was "smart" and priding themselves on their commitment to meritocracy.

Yet Valley "meritocracy" also placed great value on known quantities: people who came from familiar, top-ranked engineering programs, or who had worked at familiar local companies, or whose references came from known and trusted sources. The high degree of job-hopping between companies facilitated this, creating a mobile workforce that often worked in a series of different enterprises, sometimes with the same managers and colleagues.

The hiring habits set in place by the semiconductor companies continued over the Valley's successive technological generations. By the end of the 1990s, dot-com-era firms were filling close to 45 percent of engineering vacancies by referrals from current employees. By the 2010s, software giants were throwing "Bring a Referral" happy hours and offering up free vacations and cash bonuses to employees who helped snag a successful hire. It made sense: topflight engineering distinguished great tech companies from the merely good. From the age of first-generation chipmakers to the era of Google and Facebook, this talent was in chronically short supply. Plus, hires who were known quantities were able to hit the ground running, adapt quickly, and produce results at the speed the market demanded.[12]

And it was a fiercely competitive market. Rising up at a moment when America's postwar boom was giving way to economic precariousness and new global rivalries, the chipmakers of Silicon Valley took the technology-driven, total-immersion ethos of HP and added a topcoat of Darwinian struggle that reflected an insanely competitive business of high risk and high reward. There were no reserved parking spots for top executives—a message of meritocracy, but also a signal of the value of pulling long hours. Everyone knew who came in early and snagged the best spots by the front door. Everyone saw whose car lingered in the parking lot long after dark.[13]

And although most of its leaders exuded genial charm, the chip business was unrepentantly macho. Beyond the secretaries and the quick-fingered women on the microchip assembly lines, the industry was nearly entirely male. The result was a profanity-laced, chain-smoking, hard-drinking hybrid of locker room, Marine barracks, and scientific lab. Meanwhile, as one executive's wife told Noyce biographer Leslie Berlin, the women "stayed home and did your thing so that the warriors could go and build the temple." In firms trying to keep up with a frenetic product cycle, the all-or-nothing nature of hardware and software design—things either worked, or they didn't—translated into business organizations where work overtook family life, unvarnished criticism was the norm, and self-doubt was a fatal weakness.[14]

The high-testosterone vibe reverberated throughout the Valley. When Ann Hardy discovered that she was the only Tymshare manager not invited to an offsite retreat, and confronted the meeting's organizer about the omission, he responded, "If we include you, then we need to include all the spouses." Why is that a problem? she asked. "Well," he said matter-of-factly, "we only go to these offsite meetings so we can spend our evenings with prostitutes." Hardy marched off to CEO Tom O'Rourke to complain. The organizer disappeared. Hardy wasn't sure what happened to the prostitutes.[15]

The culture of 1970s Silicon Valley could be as old-school as its gender relations. At a moment when the Bay Area had become synonymous with the drop-in, drop-out counterculture and the freewheeling Me Decade that followed, the chipmakers' main concession to the changing times was to grow slightly longer sideburns. They leaned toward free-market Republicanism like that practiced by Dave Packard, yet were aware of how government shaped their operations, and paid deference to the system. As Bob Noyce put it in 1970: "This really is a controlled society, controlled out of Washington, and if you're trying to steer around in all the traffic out there, you'd better listen to what the cop is telling people."[16]

The organization charts of these growing companies looked a lot like those of typical "old economy" corporations. They featured all the requisite support functions (sales, marketing, human resources) that had become critical to doing business in the modern era. Yet they differed in important ways. For one, they moved through product cycles far more quickly than other kinds of manufacturing, as the propulsive force of Moore's Law made their products faster, cheaper, and more ubiquitous by the year. For another, from Hewlett and Packard to Noyce, Moore, and Grove, the founders of firms often stayed at the helm as their CEOs or chairmen. They blended the organizational chart of the twentieth-century corporation with the personal sensibilities of the nineteenth-century sole proprietorship.

Of course, when company founders were sober-minded engineers, this highly personalized approach worked well. When they were more freewheeling, it could generate chaos. Take, for example, the video game pioneer Atari.

Founded near San Jose in 1972 by a group led by a charismatic twenty-nine-year-old named Nolan Bushnell, Atari was an early market leader in one of many industries made possible by faster and cheaper microchips. Within months of its founding, Atari was disrupting the slightly seedy world of pinball and Pop-A-Shot with its arcade phenomenon Pong.

Both in its arcade form and in the home-console version that Atari released three years later, Pong was marvelously simple, devilishly difficult, and irresistibly addictive. The tech was straightforward: a black screen, with pixelated white lines on each border representing table tennis paddles, and a white digital dot of a ball that pinged back and forth. You had three choices of game: singles, doubles, and catch—where instead of returning the ball back to a partner, you tried to snag it in a small opening in your paddle.

The semiconductor guys may have surrendered reserved parking spaces and let loose after work at the Wagon Wheel, but Atari took California casual to a whole new level. The very early years were characterized by management squabbles among its young executives, drug use on the manufacturing line, and goofy product ideas that never would have made it past most corporate decision-making structures. There were games that only could have come out of a company whose designers and engineers were all young men, most notoriously 1973's *Gotcha*, whose controls were designed to look and feel like women's breasts. "They didn't have bumps on them or anything," helpfully explained an Atari designer, "but the way they were the size of grapefruits next to each other, you got the picture of what they were supposed to be."

As one Atari employee recalled mistily, the company was "a bunch of free thinking, dope smoking, fun loving people. We sailed boats, flew airplanes, smoked pot and played video games." Atari executives—"known to partake of the ganja as well"—were self-aware enough to realize that their company vibe was a little edgy for its quiet suburban surroundings. The first employee newsletter opened with a plea to "show as much sophistication to the outside community as possible," because "the thought of a company composed of longhairs is frightening to them."[17]

Amid this fast and loose organizational structure, Atari had some stumbles as it tried to turn Pong's success into a lasting business. But in 1975, it hit the consumer-product jackpot with its semiconductor-powered gaming consoles that plugged into living room televisions. Atari wasn't alone, as the established and well-capitalized electronics giant Magnavox entered the home market at the same time. Magnavox's Odyssey and Atari's home-game version of Pong became the must-have Christmas gifts of the year, addicting a generation of children and teenagers to the hypnotic blips and beeps of video games.

Atari's products were exactly the right diversion from inflation-wracked family incomes and oil-embargoed hours spent waiting in line for gas. In big cities and small towns across the country, retail giant Sears placed working consoles at the center of their showrooms for prospective customers. Kids who'd associated Sears with boring family shopping trips for Toughskins and washing machines now lined up three deep for their turn at Pong—the first arcade game that you didn't need a quarter to play. Then they'd run home and plead with their parents to buy them one.

Providing escape from '70s stagflation paid off handsomely. In 1976, after deciding against a Wall Street IPO, Bushnell sold Atari to Warner Communications for $28 million, netting $15 million personally. The video game revolution—one that introduced America's future software engineers to the wonders of manipulating pixels on a transistorized screen—had begun.[18]

APOLITICAL ANIMALS

One critical and often overlooked factor in Silicon Valley's rise was the decidedly gloomy national economic context in which it happened. The silicon capitalists of 1970s Northern California provided a stark contrast to business-page stories about besieged automakers, unemployed machinists, and spiraling inflation. At a time when big business was increasingly unpopular, particularly among young people who'd spent their college years protesting corporations as soul-crushing warmongers, this new generation of companies entered the market seemingly unencumbered by history.

And while all of American society was utterly saturated in politics, the Silicon Valley crowd appeared remarkably (and to many, reassuringly) unconnected to the political. Their politics was an ideology of working hard, building great technology, and making lots of money along the way. Nearly all of them were transplants from somewhere else, their loyalties and social bonds all lay with the industry that brought them there, and they remained remarkably untouched by the local political culture.

The dissonance was on stark display in San Jose. The city was home to Intel and IBM as well as to many people employed by the tech industry. It had grown large—over 200,000 lived there in 1970—but had accomplished that chiefly through annexing unincorporated subdivisions. It retained a small-town soul, a place with its roots in the Valley's agricultural past. When it came to politics, the city might have been on another planet from the tech companies that surrounded it—Democratic-led, with a growing minority population, and a strong labor union presence. The political mobilization of its white middle class was limited mostly to pushing for growth controls and land conservation measures that would limit the pell-mell development in the flatlands from creeping up the coastal hills.[19]

Minority students continued to mobilize on the campus of San Jose State, and up and down the Valley, Latino and Asian American communities were winning new recognition, rights, and political representation. Across the highway on the Bayshore flatlands was predominantly black East Palo Alto, where unemployment was twice the national average and crumbling infrastructure reflected two decades of sharp segregation and unequal allocation of public resources. There, Black Power activists spearheaded the founding of an Afrocentric day school and college and suggested renaming the city "Nairobi."[20]

But these nearby events made not a ripple on the hiring practices and cultural politics of the region's tech world, whose denizens rarely paused to think much about the implications of having engineers who looked and thought so much alike. From where they sat, common backgrounds strengthened common purpose, and success entailed immersive focus on building the best possible product.

The first generation of tech titans who rose up in Fred Terman's penumbra in the middle of the century—Dave Packard, especially—later became deeply involved in regional civic and political affairs. Packard chaired the Stanford Board of Trustees, founded a regional economic development group, and was a donor and mentor to a generation of state and local politicians. He even donated to East Palo Alto's Nairobi College. Leaders of the semiconductor industry, particularly Bob Noyce, ultimately became deeply engaged in politics and philanthropy. But their engagement focused on the national and global, not the local.

To be sure, the men and women of the postwar electronics scene mobilized politically when it came to issues with a direct impact on their homes and neighborhoods. Woodside's early 1960s fight against the power lines occurred about the same time that Palo Altans were mobilizing against a proposed expansion of the Stanford Industrial Park; by the early 1970s, local activism had resulted in a host of local measures up and down the San Francisco Peninsula that controlled growth and protected open spaces. And, as we will see later, semiconductor pioneers politically mobilized when their industry was in peril. But the chipmakers largely remained aloof from broader regional affairs. They were, as Joan Didion later wrote of the restless and rootless Californians surrounding Ronald Reagan, "a group devoid of social responsibilities precisely because their ties to any one place had been so attenuated."[21]

ALL POLITICS IS LOCAL

In Boston, it was a lot harder to remain unencumbered by history. The high-tech companies of Route 128 not only sat amid Revolutionary War battlefields and nineteenth-century mill towns, but also in a regional economy anchored in the past. When the Santa Clara Valley was still nothing but fruit trees, metropolitan Boston had been an industrial powerhouse for over a century. In 1940, two of every five people in the regional labor force worked in manufacturing, clustered in brick-lined mill towns like Lowell and Waltham and Maynard.[22]

Over the next three decades, mills had shuttered and firms moved south and west, lured by lower taxes and cheaper, non-union labor. Surging into the gap: aircraft manufacturing and electronics. By the end of the 1960s, the Lincoln Lab and MIT's other federally funded electronics research units had already spun out more than one hundred companies. Metropolitan Boston boomed. The suburban semicircle of Route 128, which when built in the 1940s had been derided as a "road to nowhere," now filled with office parks housing some of the country's biggest high-tech names, from established names like Raytheon, RCA, and Sylvania to newer, Massachusetts-grown companies like Polaroid and Wang. And then there were the minicomputer companies, led by Digital and its scrappy younger sibling, Data General. Boosters hailed 128 as "the golden horseshoe," "the ideas road," and, of course, "America's Technology Highway."[23]

But the intense dependence on defense spending, combined with a continued reliance on other kinds of manufacturing, sent the Boston economy into a sharp nosedive in the late 1960s. More than 100,000 manufacturing jobs evaporated between 1967 and 1972, and the defense contracts awarded to the New England region shrank by 40 percent. Just as in San Jose, the Vietnam-era cutbacks left scores of scientists and engineers out of work. But Massachusetts had a larger hit to its defense sector than anywhere else in the country. "The resurgence of the '60s was only temporary and disguised an underlying weakness," was the sour assessment of one economic development study in the fall of 1970. "We are now back in phase with the historical pattern."[24]

And that historical pattern, Boston's business establishment argued, was an unconscionably high cost of doing business. Average labor costs had gone down as union jobs declined, but taxes were too high, and the spiking energy costs of the 1970s further added to the burden on businesses. "Look Out, Massachusetts!!!" warned a pamphlet shooting out of the Bank of Boston in 1972: with the state's high taxes and big-ticket welfare spending, the bank argued, businesses and residents were getting crushed.[25]

In the middle of spiking unemployment and dark talk about "Taxachusetts," minicomputers gleamed as the state's great hope. Digital's payroll grew from less than 4,000 in 1970 to more than 10,000 five years later. Data General went from a 200-person start-up to a 3,000-employee public company. Minis became the computer industry's fastest-growing sector, and 70 percent of the nation's mini makers were in Massachusetts.

Other business machine companies made their mark as well. By 1975, Wang saw its sales hit $76 million. The next year, the company relocated its headquarters to the ailing mill town of Lowell, where the company's payroll of 5,500 made it bigger than any of the textile factories that had come before it, spurring a turnaround that made founder An Wang into a beneficent local hero. The unemployment rate in Lowell was 15 percent when Wang moved in. Ten years later it had shrunk to 3 percent. With the surge of companies like DEC, Data General, and Wang, high technology accounted for 250,000 Massachusetts manufacturing jobs by the end of the 1970s, a third of the state total.[26]

But Boston still wasn't "the Olympics of capitalism." Scrappy origin stories aside, the region produced very different things and operated in a context that was vastly different from Northern California. With much of its computer industry made up of firms that paid in salary rather than stock options, it wasn't the land of young millionaires like the Valley. Not only did Boston have only half as many VC dollars flowing through the system as Silicon Valley, there wasn't the young network of tech-focused venture capitalists, so many of whom had come up in the electronics industry themselves.[27]

The regional disadvantage wouldn't be clear for quite some time: the minicomputer's multibillion-dollar glory days were still to come, and Boston later produced PC software companies that ruled the market for a good chunk of the 1980s. Yet the region's ecosystem never had the sustained, multigenerational staying power of the Valley. The volatile high-tech era demanded new agility, and Boston didn't have it. The horizontal networks of Silicon Valley—a webbing of firm and VC, lawyer and marketer, journalist and Wagon Wheel barstool—did.[28]

But there was the bigger history at work here, too, one of politics and economics, of a Boston region that was still a manufacturing hub at heart, and where the wrenching changes of the 1970s were evident on nearly every street corner. The Route 128 scene was never as insulated from national events as the silicon capitalists of Northern California. Its eventual economic comeback—the widely hailed "Massachusetts Miracle" of the 1980s—came partly from a return of defense spending, which by the middle of the Reagan years reached $12 billion, or more than 8 percent of the net state product. "It's convert or die on '128,'" warned *The Lowell Sun* in the early days of the 1970s defense cutbacks. Many Massachusetts electronics firms didn't convert—they just waited things out until the defense budget surged back.[29]

ALTHOUGH DISRUPTIVE, THE SILICON VALLEY SEMICONDUCTOR GUYS WEREN'T revolutionaries. As the chip market grew, the paydays got larger, and start-ups turned into publicly-traded global corporations, tolerance for the Bob Widlars of the world diminished. "The wild-eyed, bushy-haired, boy geniuses that dominate the think tanks and the solely technology-oriented companies will never take that technology to the jelly-bean stage," said Andy Grove. "Our needs dictated that we fill our senior ranks with a group of highly competent, even brilliant, technical specialists who were willing to adapt to a very structured, highly disciplined environment."[30]

The revolutionary cause would have to be taken up by a new generation of technologists, ones in their twenties instead of in their forties, who were coming of age in a very different America, one fractured by wrenching economic change and violent struggles for social justice. The silicon chip had given the Valley its name, and the microprocessor had turned chips into computers.

Now it was time to build a computer from that computer-on-a-chip—one that would be quite different, and far more personal, than ever before.

CHAPTER 8

Power to the People

"To change the rules, change the tools." This was Lee Felsenstein's motto, a revelation from his years at the University of California, Berkeley, where he'd been the nerdy technical guy in the middle of the most transformative campus protests of the 1960s. Surrounded by impassioned liberal-arts types who knew everything about politics but little about technology, Felsenstein realized that he could contribute to the cause of social change by designing better ways for the change-makers to communicate. He developed better printing and distribution systems for the thousands of flyers blizzarding Berkeley with calls for sit-ins and be-ins, walkouts and marches. He built radios so that activists could listen in on police scanners. He made clearer-sounding electronic megaphones for rallying campus crowds. Socially awkward and finding real-life interpersonal connection difficult (later in life, he received a diagnosis of mild autism), Felsenstein decided to devote his life to creating technology that helped people powerfully and efficiently share information—but that was so simple anyone could use it.[1]

Born in 1945 in Philadelphia, Lee Felsenstein was less than a decade younger than captains of the semiconductor industry like Andy Grove and Jerry Sanders. Yet his generation's experience was so different that it might as well have been a century. Like so many boys of his postwar generation, he'd plop down a carefully saved quarter each month to buy the latest edition of *Popular Electronics*, poring over the glorious multipage spreads within that described how to make your own electronic gadgets. At age eleven, he built a crystal radio out of a kit discarded by his older brother. When he was twelve, Sputnik rocketed into space, and he built a small satellite that won third place in the regional science fair. When he was in high school, he'd hop on his bike to pedal down the three miles to the city's great temple of science and engineering, the Franklin Institute, to see the Philadelphia-built

UNIVAC that the museum had proudly enclosed in a glass display. The summer after graduating from high school, he landed a job as the UNIVAC's caretaker, earning $1.54 an hour for untangling its tape and minding its switches. They hired him, he remembered, for his "raw technical competence" and tolerance for an extremely low salary.[2]

When he wasn't busily tinkering, he was marching and picketing. "I was a sort of pathetic child radical," he recalled. "I really didn't know anything about politics, but I knew enough to stand in a picket line." He marched on Washington for civil rights. He picketed Woolworth's to demand the integration of Southern lunch counters. The Franklin Institute came close to firing him after he joined a pacifist protest of visiting museum dignitary Edward Teller, father of the H-bomb.[3]

Felsenstein's family didn't have much money to send him to college, so he chose Berkeley for its relatively cheap tuition, its electrical engineering program, and its reputation for left-wing politics. He arrived in time to witness the explosive birth of the Free Speech Movement in December 1964. It was the peak of the civil rights era. Like college students across the nation, many Berkeley students had traveled south that year to participate in Freedom Summer, returning energized for a fresh round of activism. UC administrators squelched that nearly immediately, banning on-campus demonstrations and other political activity. The students responded with a season of mass protest that became a proxy for a broader struggle emerging between two generations with starkly different worldviews.

Berkeley, jewel in the Californian crown of public higher education, was even more enmeshed in federal defense research programs than its southern Bay Area neighbor, Stanford. A Cold War university *par excellence*, it was the host institution to a major federal research laboratory and home to the chief architects of the weapons of thermonuclear war. Students who came to its campus in the early 1960s found a place humming with top secret research labs and blinking mainframes, leaving many undergraduates feeling like unhappy cogs in a modern technocratic machine. "Clark Kerr has declared," one student newsletter said of the UC chancellor, "that a university must be like any other factory—a place where workers who handle raw material are themselves handled like raw material by the administrators above them."[4]

Yet unlike his compatriots on the Berkeley barricades, Felsenstein didn't think that computers were the problem. The people who *controlled* the computers were the problem. "Building a tool for a Fortune 500 company would tend not to fulfill me," he declared. "Building tools that people use to make Fortune 500 companies irrelevant—that's more my style." Moving languidly through his coursework to avoid the draft, Felsenstein was there as Berkeley's student scene shifted from a focus on civil rights toward protest of the Vietnam War.[5]

Over that time, the campus mood became more radical, more pessimistic, and often violent. A frenetic 1967 filled with increasingly militant demonstrations culminated in a late-October melee at the Oakland Army Induction Center, where more than two thousand "Stop-the-Draft" protesters were met with hundreds of nightstick-swinging and mace-spraying police officers. Felsenstein wasn't among the twenty-seven protesters sent to the hospital that day, nor was he one of the seven antiwar leaders arrested (even though the bullhorns he'd designed had egged on the crowd). He still had a nervous breakdown soon afterward from the stress of it all.

After flunking all his courses, he dropped out. He spent the next few years bouncing back and forth across the Bay between Berkeley and the newly christened Silicon Valley, continuing to pursue his dream of building technical tools that would allow people to escape the establishment's clutches, and possibly overthrow the system altogether.[6]

AT THE SAME TIME LEE FELSENSTEIN WAS PORING THROUGH *POPULAR ELECtronics* and taking apart radio sets, Liz Straus was sitting in a classroom in Dana Hall, an all-girls' prep school in Wellesley, Massachusetts. Math and science were already in her blood: Straus's mother was a science teacher at the school, and her engineer father was deeply involved in computer and radar research at MIT. In the single-sex environment of Dana Hall, a venerable institution with many alumnae who'd gone on to careers in science, engineering, and medicine, Straus developed confidence that few girls of her generation were allowed to possess. "I didn't know girls weren't good in science and math because all my classmates were female and many were technically excellent," she remembered. But there weren't any computers in her high school, and her father's work was top secret. "Although I was pretty much a tomboy, I was still a girl," she admitted. "Girls and electronics didn't mix."[7]

She didn't learn much about computers in college either. Arriving at Cornell as a freshman in the fall of 1963, the closest she got was "key punching for the decks of cards my boyfriend needed for his research project." It was a familiar ritual for plenty of college students in those days: trot over to the computer center, hand the cards across the intake desk, and return the next day to pick up cascading sheets of dot-matrix printouts. Computers remained pretty mystifying until several years later, when Straus decided to go back to school to get a graduate degree in education.

By that time, it was 1971, and Liz Straus was now Liza Loop. She was married with a young son, living in the placid farm town of Cotati, California, amid Sonoma County wine country. She'd missed the fieriest days of countercultural protest, but she'd become a passionate crusader nonetheless: for educational reform. She wanted all children to have a learning experience as stimulating, and empowering, as her own had been at Dana Hall.

Liza Loop had plenty of company. Student demonstrations and teacher strikes had become a global phenomenon, as newly politicized young people pushed against a rote and rule-bound system seemingly trapped in the nineteenth century. Educational experts questioned the relevance of teaching with paper and textbooks in an age when nearly every American family had a television in their living room, and enthusiastically endorsed the idea of introducing programmable "teaching machines" into the classroom. On top of it all, American schools remained civil rights battlegrounds, as continuing integration struggles had given way to fiercely contested court-ordered busing. These measures prompted some white parents to violently resist, and many others to opt out of public school systems altogether. Alternative schooling movements like Montessori and Waldorf spiked in popularity.[8]

A growing number of educational advocates had begun talking about computers as critical features of the new and improved schools of the future. The newly installed Nixon Administration began to investigate how it could build, and fund, a network of computers in schools. Major electronics firms, including IBM, joined in the effort to create specialized hardware and software for education. The Ford Foundation was so committed to the enterprise that it established a separate nonprofit dedicated to the cause. "Learning is the new growth industry," exulted its president, Harold Gores. Berkeley dean Robert Tschirgi proclaimed that the computer was "the greatest thing to hit education since Johann Gutenberg invented movable type."[9]

When the first computers actually landed in the classroom, however, high-flying corporate rhetoric didn't come remotely close to educational reality. The software was clunky. The hardware was as rigid and bureaucratic as the analog systems it replaced. Most of all, teachers didn't know enough about computers to teach students how to use them, and the user interface wasn't designed for students to learn it on their own. Fancy hardware ended up in school basements and storage closets, gathering dust.

Things worked out differently, however, when computer specialists *themselves* brought the machines into schools, and worked with teachers and students to create more-individualized curricula. This happened in the case of Dean Brown, a Stanford psychologist who'd started working with Montessori teachers to teach very young children how to program devices and play rudimentary computer learning games. Brown's vision of how computers might transform education was miles away from—and conceptually far more audacious than—the ideas bandied about by political and corporate leaders. "Education is the realization and the unfolding of that which is latent, already there," he wrote in 1970. "The teacher is a creative artist and the computer can be a chisel in his hands."[10]

Shortly after she enrolled in Sonoma State's master's program, Liza Loop found herself in a class led by Brown. Everything changed. "I spent five minutes in the room with Dean and said, 'That's my career, that's where I'm going,'" Loop remembered. Intensely sociable and expansively friendly, this Sonoma County housewife didn't fit any tech stereotype. "I'm not particularly interested in computers," she jauntily confessed. "It's humans that turn me on." But Loop's lack of formal training in computing ultimately became an asset, enabling her to push past technical jargon and explain to ordinary people—kids, teachers, and especially girls and women like her—how computers could become part of their lives.[11]

THE NEW GUARD

Blending the change-the-world politics of the counterculture with the technophilic optimism of the Space Age, Felsenstein and Loop became two members of a steadily enlarging techno-tribe that emerged in the Bay Area and other college towns and aerospace hubs at the turn of the 1970s.

Many were like Lee Felsenstein: Sputnik-generation boys with science-fair ribbons who'd collided head-on with the cultural liberation of the Vietnam era. They proudly called themselves "hackers," relentlessly future-focused, suspicious of centralized authority, pulling all-nighters to write the perfect string of code. They demonstrated superior technical talent by infiltrating (and sometimes deliberately crashing) institutional computer networks. Overlapping their ranks were the renegade "phone phreaks" who discovered how to use high-pitched signals to break into AT&T's networks and enjoy long-distance calls for free.[12]

But a good number were also like Liza Loop: baby boomers drawn to computers by a passion to change the way society worked, especially in how it educated a new generation. There was Pam Hardt, a Berkeley computer science dropout and co-founder of a San Francisco commune called Resource One; she secured a "long-term loan" of an aging SDS minicomputer, settled it in the commune's living room, and made it the mothership of a time-shared bulletin-board system called Community Memory. There was Bob Albrecht, an engineer who quit his corporate gig at super-computer maker Control Data Corporation to join an educational nonprofit called the Portola Institute, a far-ranging collective operated on a shoestring. Portola spawned the bible of the techno-counterculture, the *Whole Earth Catalog*, created by artist, utopian, and "happening" impresario Stewart Brand. High-tech met hippiedom on the *Catalog*'s pages, which featured fringed buckskin jackets and camp stoves alongside scientific calculators. Its motto: "Access to Tools."[13]

Albrecht's project was the People's Computer Company, started in 1972 as a walk-in storefront for computer training, accompanied by a loose and loopy newsletter "about having fun with computers." Festooned with hand-drawn dragons and off-kilter typesetting, the *PCC* had the rangy look and feel of an underground tabloid like the *Berkeley Barb* (where Felsenstein had become a staff writer). Instead of columns decrying Nixon's bombing of Cambodia, the *PCC* had features on how to learn computer language, with titles like "BASIC! Or, U 2 can control a computer."[14]

If Bob Albrecht was the revolution's Ben Franklin, then Ted Nelson became known as its Tom Paine. A computing-entranced former graduate student in sociology with a prep-school accent and the manners to match, Nelson considered himself a specialist in ideas "too big to get through the door." In the mid-1960s, he came up with a nonlinear system for organizing writing and reading he called "hypertext." In 1974, he applied the concept in a self-published book titled *Computer Lib: You Can and Must Understand Computers NOW!* (Flip the volume over, and there was a second book, *Dream Machines*, which talked about computers as media platforms. Nelson was thinking well ahead of his time.)

"Knowledge is power and so it tends to be hoarded," exhorted Nelson. "Guardianship of the computer can no longer be left to a priesthood" who refused to build computers that could be understood by ordinary people. Released into the world as Richard Nixon was helicoptering away from the White House in disgrace, *Computer Lib* made clear who the enemies were. "Deep and widespread computer systems would be tempting to two dangerous parties, 'organized crime' and the Executive Branch of the Federal Government (assuming there is still a difference between the two)," he wrote. "If we are to have the freedoms of information we deserve as a free people, the safeguards have to be built in *at the bottom, now*."[15]

In that drop-out-and-tune-in place and moment, these men and women began to think and talk about how computers could transform from fearsome weapons of the establishment into tools of personal empowerment and social change. Of course, the fact that Northern California had been such a hub of Cold War science was why many of them were there in the first place. They'd migrated west for college and graduate school, or jobs in government labs and industrial research operations. Their knowledge of computing came from their prior participation in the technocratic system they criticized. And they weren't all kids. Many were professionals in their twenties and thirties, with children, mortgages, and graduate degrees.

Thus the gulf between the scientific Cold Warriors and the techno-utopians was not as great as it seemed. Many of the ideas that animated the personal-computer crusade, like human-computer interaction and networked collaboration, were the same ones that had consumed the Cambridge seminars of Norbert Wiener in the 1940s and the labs of McCarthy and Minsky and Licklider in the 1950s. The new generation believed in the same principle that had animated government science ever since Vannevar Bush celebrated its "endless frontier" in 1945: technological innovation would cure society's problems and build a better American future.[16]

Such technophilia also made this change-the-world movement oddly conservative when it came to disrupting conventional gender roles, reckoning with society's racism, or acknowledging yawning economic and educational inequalities. In this crowd, the Liza Loops and Pam Hardts remained in the distinct minority. Non-white faces almost never appeared. Radical feminism's impact was brief and glancing; Black Power and other civil rights movements rarely were given a nod. For some of these technologists, a singular focus on computing was an escape from identity politics. For others, tech was an answer to social inequities. The overwhelmingly white and middle-class group had faith that "access to tools" would fix it all.

For all their blind spots, the new generation's minds were blown open by the political and cultural earthquakes of the 1960s. And technology *did* set them free, for they were able to make their ambitious and intensely personal vision of a thinking machine far closer to reality because of the existence of the microprocessor. Their core idea—that the computer could be used by anyone for creation and communication and work and play—became the right idea at exactly the right time.

INFORMATION OVERLOAD

The metaphor of the mainframe had coursed throughout 1960s politics as eloquent shorthand for American political and social institutions—governments, armies, corporations, universities—and the stifling conformity they imposed. "There's a time when the operation of the machine becomes so odious, makes you so sick at heart, that you can't take part! You can't even passively take part!" famously declared the Berkeley Free Speech Movement's Mario Savio in late 1964. "And you've got to put your bodies upon the gears and upon the wheels, upon the levers, upon all the apparatus, and you've got to make it stop." Student protesters that autumn pinned signs to their chests bearing a riff on the prim warning that appeared on every IBM punch card: "I am a UC student. Please don't bend, fold, spindle or mutilate me."[17]

Students didn't simply decry computers as tools that stripped them of individuality, but of their privacy as well. Only a few months before Savio and his compatriots mobilized in Berkeley, investigative journalist Vance Packard hit the best-seller lists with *The Naked Society*, which outlined the fearsome extent of electronic snooping in several hundred ulcer-inducing pages. "If Mr. Orwell were writing [*1984*] today rather than in the 1940s," Packard wrote, "his details would surely be more horrifying . . . There are banks of giant memory machines that conceivably could recall in a few seconds every pertinent action—including failures, embarrassments, or possibly incriminating acts—from the lifetime of each citizen."[18]

Right on Packard's heels came the publication of the English-language translation of *The Technological Society*, a grim assessment of the modern condition by French sociologist Jacques Ellul. The transistor and computer had locked modern society in a battle between individual agency and machine conformity. The machine seemed to be winning. When that happens, Ellul concluded darkly, "everything will be ordered, and the strains of human passions will be lost amid the chromium gleam."[19]

Perhaps no author captured the information-age zeitgeist more thoroughly than a journalist and self-appointed futurist named Alvin Toffler. The hyperkinetic New Yorker had started his adult life as a Marxist civil rights activist, which he followed with several years experiencing the workingman's life as a welder in Cleveland. Eventually he moved into journalism, ultimately leaving Marx behind to become an editor at the buoyantly capitalist *Fortune* magazine. With his wife, Heidi, he started up a consulting practice and began writing books (she did not get authorial credit until many volumes later). Prolific, prone to hyperbole, and relentlessly self-promoting, Toffler burst onto the best-seller lists in the spring of 1970 with *Future Shock*, a 500-page treatise on how technology was transforming everything—and blowing everyone's minds in the process.

Forty-one years old and sympathetic to the "post-materialist" priorities and open-minded sexual mores of the younger generation, Toffler released a thundering waterfall of prose designed to hook and excite—and frighten—a general audience. "What is occurring now is not a crisis of capitalism, but of industrial society itself," he wrote. "We are simultaneously experiencing a youth revolution, a sexual revolution, a racial revolution, a colonial revolution, an economic revolution, and the most rapid and deep-going technological revolution in history."[20]

If you already thought life was chaotic, Toffler argued, then you didn't know the half of it. "Change is avalanching down on our heads," he said, "and most people are grotesquely unprepared to cope with it." It was, Toffler argued, a question of too much information: "The entire knowledge system in society is undergoing violent upheaval. The very concepts and codes in terms of which we think are turning over at a furious and accelerating pace." The modern world, he memorably concluded, had an acute case of "information overload."[21]

In a nation already deeply anxious about technology and nearly everything else, Toffler's book was a hit straight out of the gate. Three major book-of-the-month clubs chose *Future Shock*, and it had buoyant sales despite withering reviews. (One called it "a high school term paper gone berserk.") Style aside, the book's dystopic overtones were certainly a lot to swallow; Toffler's own mother remarked: "if that's the way it's going to be, I don't want to be here." Nonetheless, *Future Shock* ultimately sold five million copies, and made Alvin Toffler into an inescapable seer of the information age.[22]

For all its wilder ideas and overstuffed prose, Toffler's book was stunningly prescient. He predicted that technology would break down large bureaucracies into an "ad-hocracy" of many smaller, more agile units that could grow and shrink on demand. He talked about how electronic communications would enable a splintering of mass culture into thousands of different, specialized channels where everyone could get their own, specially tailored news. He talked of how inundation by information would reduce attention spans and increase skepticism toward expert authority. He pointed out how much the U.S. already had shifted toward a service economy, and how information technologies accelerated that shift.

From the Manhattan Project to manned spaceflight to massive Cold War–era interventions in the Third World, Americans had understood technology as a big-organization tool to solve large-scale problems—war, famine, poverty, education, transportation, and communication. Most academic seers of postindustrial society generally operated on the presumption that bigness would still prevail, even if the means of production would change. *Future Shock* reflected a shift in a different direction. Technology might instead become a way to fix the problems of the world, to push against social institutions, and achieve self-actualization. But the path to do so would be by going small. One of the few optimistic notes sounded by *Future Shock* had to do with the destiny of big and unfeeling organizations. "The bureaucracy," wrote Toffler, "the very system that is supposed to crush us all under its weight, is itself groaning with change." Ultimately, he asserted, technology would break down big institutions, restoring individual autonomy in the process.[23]

THE COMPUTER NEVER FORGETS

It wasn't only student radicals and grandiloquent futurists questioning the mainframe status quo—it was establishment politicians too. From the mid-1960s glory days of the Great Society to the scandal-wracked last months of the Nixon era, Capitol Hill lawmakers devoted hundreds of hours to fiery floor speeches and hearings on computers and privacy. The target of their ire wasn't IBM or the corporations that used its products, but government bureaucracies with rapidly growing electronic databases enumerating everything from a person's age and marital status to their medical history and draft number.

Senator Sam Ervin was one of the most prominent and consistent of these critics. A strict Constitutionalist (and the ardent states-rights segregationist who had so deeply disliked 1965's immigration reforms), Ervin gained enduring fame as the folksy chair of the Senate Watergate Committee. Before that, however, he spent the first years of the 1970s helming an investigation into government computers, and his hearings generated juicy headlines. With chairman's gavel in one hand and a densely printed sheet of microfilm in the other, Ervin railed against the encroaching "dossier dictatorship" in Washington, warning darkly, "The computer never forgets."[24]

On the House side, the crusader-in-chief was Neil Gallagher, Democrat of New Jersey, who parlayed a Kennedyesque demeanor and a knack for comparably pungent soundbites to make his name as a privacy advocate. He started holding hearings on "Computers and the Invasion of Privacy" in 1966, bringing in witnesses like Vance Packard. By 1970, Gallagher was taking his pitch directly to the computer professionals themselves, describing "the computer as 'Rosemary's Baby'" in an anguished essay for the journal *Computers and Society*. "The whir of the computer as it digests and disseminates dossiers . . . is frequently the sound of flesh and blood being made soulless," he wrote. "Raw data are now extracted in much the same way teeth are pulled: either under the ether of uninformed consent or ripped out by the roots."[25]

The people building the mainframes might not have put it quite so apocalyptically, but they agreed that old notions of privacy had disappeared nearly as soon as the first UNIVAC hit the market. "In fact, there is at this time very little privacy in the private life of all of us," observed pioneering computer designer Evelyn Berezin in a letter to Gallagher. The thousands of machines abuzz in government agencies and corporate computer banks already possessed an astounding amount of personal information, lightly policed and largely unprotected. Programmers might not yet know what to do with all this data, but accelerating computer power indicated that they soon would. As Paul Baran of RAND reminded one inquiring journalist, "Behind all this creating of records is the implicit assumption that they will someday be of use."[26]

In the grim summer of Nixon's resignation, privacy became the issue that nearly everyone in a fractious and wounded Washington could agree upon. "The data bank society is nearly upon us now," warned conservative conscience Barry Goldwater at hearings convened by Ervin that June. "We must program the programmers while there is still some personal liberty left." After a November election that saw a wave of young, reform-minded candidates (most of them Democrats) elected to Congress as "Watergate babies," Congress hurriedly bundled together a flurry of legislative proposals to pass the Privacy Act of 1974. Shutting the door on a truly dreadful year in politics, Gerald Ford signed it into law on New Year's Eve.[27]

The opening lines of the Privacy Act made it clear that computers had been its catalyst: "Increasing use of computers and sophisticated information technology, while essential to the efficient operations of the Government, has greatly magnified the harm to individual privacy that can occur." And like most of the Congressional hearings on the subject over the years prior to its passage, the Act aimed its powers squarely at the federal government. Even though corporations were among the most enthusiastic practitioners of the dark arts of surveillance, personality testing, and sucking massive amounts of consumer information into giant data banks, the Congressional investigations focused on the villain they could see, and whose budget they controlled: the federal bureaucracy.[28]

In being so relentlessly focused on the government, America's privacy warriors paid little attention to what the technology industry was doing, or ought to be doing. Few regulations fettered the way in which American companies gathered data on the people who used their products. The citizen computer user might be able to put some limits on what the government knew, but she had little recourse when it came to controlling what companies might discover. This regulatory latitude, of course, would ultimately make possible one of modern Silicon Valley's greatest business triumphs: to gather, synthesize, and personalize vast amounts of information, and profit richly from it.

SMALL IS BEAUTIFUL

At the same time as politicians and activists were pointing to runaway tech as a source of society's many ills, however, they embraced new applications of technology as the means to make it right. The computers-in-education philosophy espoused by people like Dean Brown was one manifestation of a broader push to think about computers as tools for social change. The 1970 annual meeting of the Association for Computing Machinery, the ACM—usually a days-long geek's paradise of technical papers—devoted the entire program to "how computers can help men solve problems of the cities, health, environment, government, education, the poor, the handicapped and society at large." Consumer crusader Ralph Nader was a keynote speaker. The Nixon Administration's short-lived Technology Opportunities Program reflected the prevailing mood, asking industry to submit its suggestions for how computer and communications technologies could solve social problems.[29]

This played out in popular culture as well. Yale law professor and privacy advocate Charles Reich's homage to countercultural values, *The Greening of America*, rested atop the best-seller lists for weeks after its September 1970 release. Just as in the calls to action Mario Savio shouted through his megaphone, Reich framed the modern world as a machine in need of a reboot: "Americans have lost control of the machinery of their society, and only new values and a new culture can restore control." British economist E. F. Schumacher had a similarly technological emphasis in his 1973 bestseller *Small Is Beautiful*. The relentless push for economic growth must give way to a new philosophy of "enoughness." Large and inhuman systems needed to be replaced by "a technology with a human face."[30]

Underscoring the turn against bigness, the largest computer company of them all had found itself under siege from competitors and regulators. On the last working day of the Johnson Administration in 1969, the Department of Justice filed an antitrust lawsuit against IBM. It wasn't the first time Big Blue had been in the government's antimonopolistic crosshairs, having been the target of a Truman-era action that had ultimately forced the split between its computer hardware and services businesses. Now, the soaring market share and cash generated by IBM's wildly successful System/360 line attracted the government's attention. The lawsuit formed a backdrop as liberals in Congress began to question business monopoly. In 1972, Michigan Democrat Philip A. Hart, known as the "conscience of the Senate," introduced a sweepingly antimonopolistic "Industrial Reorganization Act" into Congress. Hart described it as "the greatest effort which has been put forth to finding a solution for economic concentration."[31]

The adverse effects of this concentration on consumers seemed obvious when it came to the computer industry. Automation of everything from paychecks to airline reservations may have increased business efficiency, but it created everyday hassles when the system messed up. Every misdirected utility bill or lost hotel booking became a proof point for the computer skeptics. Industry executives found themselves on the defensive. "Perhaps if we look inside a little bit," one protested to Hart's committee, "you'll observe that it is not a computer that ever makes a goof; it's the people who use it and the people who write programs for it."[32]

Baby-boomer activists Lee Felsenstein and Liza Loop never would have imagined they'd have much in common with a suit-and-tie executive of a big computer company. But all were pushing the same message: the power of the computer came from its user.

CHAPTER 9

The Personal Machine

Down at Stanford, similar ideas had been brewing.

As the turbulent politics of the 1960s swirled around it, the once placid, seemingly apolitical Farm had turned into a hotbed of student activism. Like their counterparts at Berkeley, Stanford students mobilized for civil rights, packing Memorial Auditorium to hear Martin Luther King Jr. speak on the eve of 1964's Freedom Summer, mobilizing in support of striking California farmworkers in 1965, and hosting a "Black Power Day" featuring Stokely Carmichael in 1966. By 1967, this activist energy had shifted much of its focus to the war in Vietnam. That February, Stanford undergraduates screamed down a visiting Vice President Hubert Humphrey and held all-night peace vigils in the university's Memorial Church. Not too long after, students burned an effigy of Hoover Institution director Glenn Campbell on the steps of Hoover Tower.[1]

Fred Terman had spent two decades turning Leland and Jane Stanford's sentimental project into one of the nation's preeminent hubs of defense research. Now, his steeples of excellence became the targets of ferocious student anger about the Vietnam War. For nine days in April 1969, several hundred students occupied the Stanford Applied Electronics Lab, demanding that the university put an end to classified research. Soon after, university administrators cut ties with SRI and its controversial portfolio of classified projects. The decision disappointed the students, who had hoped that SRI would be shut down altogether.[2]

Had that happened, Stanford would have squelched an operation that was building an entirely new universe of connected, human-scale computing—the home of Shaky the Robot, of Dean Brown's education lab, and of Doug Engelbart's "research center for augmenting human intellect."

In Engelbart's emphasis on networked collaboration, this low-key member of the

Greatest Generation was completely in sync with the radical political currents swirling around the Stanford campus and the bland suburban storefronts of the South Bay. Just down the road from SRI's Menlo Park facility was Kepler's Books, which owner Roy Kepler had turned into an antiwar and countercultural salon. Beat poets, Joan Baez, and the Grateful Dead all made appearances at Kepler's, and the store's book talks and rap sessions became can't-miss events for many in the local tech community. That included members of the Engelbart lab, who'd drop in on their way to catch the commuter train home. And Engelbart's vision of expanding the mind's capabilities through networked technology had much in common with that of Michael Murphy, a Stanford graduate who co-founded the Esalen Institute on the Big Sur coast in 1962. Esalen's goal, Murphy told a *Life* reporter, was "reaching a terra incognita of consciousness." While Esalen had meditation and encounter sessions, Engelbart used computers to, as his friend Paul Saffo later put it, "create a new home for the mind."[3]

Engelbart's December 1968 demo had been a revelation and an inspiration to the clan of Bay Area programmers and visionaries who were thinking about computers as tools for work, education, and play. Dean Brown's lab used Engelbart's mouse to test how computers augmented student learning. The event also brought new converts to the movement, notably Stewart Brand, who had joined the demo team as a journeyman videocam operator, and left having been turned on to the power of networked computing. Brand and Albrecht's collaboration, the Portola Institute, and the *Whole Earth Catalog* followed. The demo "quite literally branched the course of computing off the course it had been going for the previous ten years," remembered Saffo, "and things have never quite been the same again."[4]

THE IDEA FACTORY

Not too long after, three thousand miles away from the robot-trolled halls of SRI, a group of corporate executives were sitting in a wood-paneled office, trying to figure out where the next generation of their company's products would come from. Xerox was a relatively young company, but its ascent had been rocket-like, and astoundingly lucrative. After developing some of the first office photocopy machines less than two decades earlier, it had cornered the market as thoroughly as IBM dominated mainframe computing.

As the cash rolled in, Xerox decided to follow the example of its more venerable predecessors like AT&T, and set up a stand-alone research facility. And where better to do it than Palo Alto? Top electronics firms had been setting up labs in Fred Terman's orbit for two decades by then, and no other place could match the combination of Class A real estate, Class A engineering talent, and Class A weather. "Xerox Plans Laboratory for Research in California," read the tiny item buried inside *The New York Times* business section in the spring of 1970. Its purpose, said Xerox research chief Jacob Goldman, was "to advance data processing technology."[5]

The innocuous announcement marked the launch of a venture that eventually turned Doug Engelbart's vision into market reality. Ironically, although Xerox bankrolled the enterprise, the company was not the ultimate beneficiary of the breakthroughs it sponsored. It was not a computer company, and the copier business was simply too lucrative to justify creating entirely new product lines and sales channels in order to become one. Instead, its Palo Alto Research Center, or PARC, became the seedbed for new, mostly California-based companies whose profits ultimately would dwarf those of the photocopier and mainframe industries combined. PARC became to personal computing what NASA had been to microchips: a deep-pocketed financier who pushed money toward research and development of blue-sky technology, and then mostly got out of the way.

The people who filled PARC's halls in the first years of the 1970s were tightly networked into the convivial Bay Area ecosystem of computer professionals. At a time and place dominated by the semiconductor industry, "with its famously macho disdain for women," as PARC scientist Lynn Conway put it, there was remarkable gender diversity in its ranks. The new hires took advantage of Xerox's abundant resources and loose oversight to creatively interpret Goldman's definition of "data processing technology," pursuing projects inspired by Doug Engelbart's ideas about augmented intelligence and by hacker culture more generally. Engelbart's SRI operation had drifted after the great demo—investors couldn't figure out the devices' commercial potential—and several members of his team moved over to PARC. Still more came from the academic engineering diaspora of Stanford and Berkeley.

There was Alan Kay, who wanted to develop a computer small enough to fit in a book bag. Down the hall, Bob Metcalfe and David Boggs invented a way to connect multiple computers that they called the Ethernet. Across the way were Adele Goldberg and Dan Ingalls, also computer-education evangelists, who collaborated with Kay on a transformative and classroom-friendly new computer language called Smalltalk. Conway, who'd already made major breakthroughs in high-performance computer architecture in an earlier career at IBM, was recruited from Memorex. And coming in to head PARC's Computer Science Laboratory was none other than Bob Taylor, the man who only a few years before had pulled together that marvelous academic network called the ARPANET.[6]

Xerox had assembled an all-star team. But its Palo Alto facility first entered the broader public consciousness as a looser, more rebellious outfit, courtesy of an article by Stewart Brand that appeared in *Rolling Stone* in December 1972. Titled "Spacewar: Fanatic Life and Symbolic Death Among the Computer Bums," Brand's piece told of techies commandeering the facility's computer networks to play midnight video games, naming their offices after J. R. R. Tolkien characters, growing their hair long, and having little regard for traditional authority. Alan Kay told Brand, "A true hacker is not a group person. He's a person who loves to stay up all night, he and the machine in a love-hate relationship . . . They're kids who tended to be brilliant but not very interested in conventional goals. And computing is just a fabulous place for that. . . . It's a place where you can still be an artisan." The executives back at Xerox headquarters were so horrified by Brand's story that they prohibited the PARC crew from ever again talking to the press. In the meantime, "Spacewar" became a founding document of Silicon Valley legend.[7]

The beards, beanbag chairs, and midnight video games made it hard for some contemporary observers to understand that these men and women were in the process of developing the foundational technologies for desktop computing and networking. Only three months after "Spacewar" appeared in print, the PARC team produced a prototype desktop computer. Called the Alto, the machine featured a keyboard and screen. It had a mouse. It had a graphical interface instead of text. Documents appeared on the screen looking just like they would when printed on paper. The machine even had electronic mail. Unveiled less than five years after the mother of all demos, it took Engelbart's tools of the far-out future and put them in a machine that could fit on an office desk. It was unlike nearly every other computer in existence, in that you did not need to be a software programmer to use it.[8]

BEING TOM SWIFT

Beyond the leafy surroundings of Xerox's dream factory, Lee Felsenstein's quest for people-powered computing continued. Like most everyone, he'd spent quality time in the PARC beanbag chairs. While he was wowed by the technology, the social-justice crusader in him disliked the idea of being beholden to a corporate overlord. The Alto had cost $12,000 to build. The retail price tag promised to be more than three times as much.

Felsenstein wanted to build networks that were grounded in community, and cheap enough to be accessible to nearly anyone. He'd been a co-conspirator with Pam Hardt and the Resource One crowd in building the pioneering electronic bulletin board Community Memory, whose terminals still dotted Bay Area record stores and bean-sprout cafes. But the units were merely screens that relied on data from a central computer, and the finicky network tended to break down easily. What the world needed now, the frustrated hacker concluded, was a "smart" terminal with its own memory system.[9]

The popular science magazines and their "construction projects" for home hobbyists remained a go-to resource for DIY engineers like Felsenstein, and the September 1973 edition of *Radio-Electronics* (subtitle: "For Men with Ideas in Electronics") featured a cover story about a device that might just solve his problems. The "TV typewriter" was the brainchild of Don Lancaster, an aerospace engineer who'd exiled himself to the Arizona desert to become a fire spotter and back-to-the-land outdoorsman. Lancaster's simple device was able to transmit words typed on a keyboard onto a television screen. It took the computer hobbyist community by storm. Here was a connection between keystrokes and on-screen characters that allowed homebrewers to do a featherweight version of PARC's Alto.

Lancaster's idea inspired Felsenstein to post notices on Community Memory and troll for ideas at PCC's weekly potlucks, looking for input on how to build a device that was like the TV typewriter, but that had amped-up intelligence. It might have been possible to do something like this using one of Intel's microprocessors, but those cost thousands of dollars apiece. Felsenstein wanted something cheap, smart, and easy to build.

The result was "The Tom Swift Terminal," named after an old series of adventure books that had been a staple of mid-twentieth-century American boyhoods. Felsenstein published the schematic in the *PCC* newsletter in late 1974. The Tom Swift user could not only scroll up or down on the screen, but could plug in different preprinted circuit boards to add new functions—printing, calculating, playing games—regulated through a "bus" that relayed communications from peripheral to terminal. In essence, the system took apart a computer architecture usually sealed within a mass-manufactured computer or chip, and turned it into standalone components that any electronics nerd could build themselves. "If work is to become play," the spread proclaimed at the top, "then tools must become toys."

Like so much of what appeared in the *PCC*, the Tom Swift article wasn't just a set of engineering specs. It made a pitch for a new political philosophy. Not too long before, Felsenstein's father had passed along a copy of a book by socialist philosopher-priest and countercultural guru Ivan Illich, titled *Tools for Conviviality*. Illich already had made a splash with *Deschooling Society*, a broadside against traditional education that inspired countless students to drop out and tune in. Now he turned his focus to technology. Like Charles Reich and E. F. Schumacher, Illich decried the "radical monopoly" that elites and experts held on the functions of modern life, resulting in a system "where machines enslave men." But now things were different. "We now can design the machinery for eliminating slavery without enslaving man to the machine."[10]

The vision entranced Felsenstein. To one side of the Tom Swift specs, he added his own declaration. "The dollar sign isn't quite where it's at," he wrote, a hat tip to Illich's socialism. "Before there was an industrial system, people were building tools that other people could use without much training. . . . P.C.C. is showing how computer software can be handled in this convivial fashion." He expanded on the idea further in a technical paper published shortly thereafter. The mainframe computer was just like a bureaucratic organization: hierarchical, siloed, the domain of experts. The bus design is "a system of free interchange subject only to simple traffic rules."[11]

The expansive optimism of "convivial cybernetics" that Lee Felsenstein outlined in 1974—just like Ted Nelson's bold declarations of "computer lib" that same year—burned brightly among the community of programmers and social reformers, even as the grandest hopes of the counterculture ebbed. The good vibes of the Summer of Love and Woodstock had been subsumed by the violence of Altamont and the Manson Family and Kent State. Nonviolent campus sit-ins had given way to the bomb attacks of the Weather Underground. Heiress Patricia Hearst, kidnapped from her Berkeley apartment by another group of violent radicals, the Symbionese Liberation Army, had emerged in April 1974 as a gun-toting participant in a fatal bank robbery. Inflation spiked, incomes stagnated, and the president was a crook. The wreckage reinforced the technologists' conviction that pure politics wasn't enough. As Illich had put it, "Changes in management are not revolutions." The inequities of society stemmed from the industrial mode of production, and they wouldn't end unless the platform itself changed.[12]

But unlike Ivan Illich, the computer revolutionaries contained few true Marxists in their ranks. They may not have wholeheartedly approved of the establishment, but their careers had greatly benefited from its bounty. They'd worked in the topflight engineering programs of Stanford and Berkeley. They'd built the future within ARPA and the rest of the Defense Department. They'd enjoyed the ample industrial research budgets of companies like Xerox, which allowed them to build the networked desktop computer of their dreams in less than three years. And on top of all of it, they could look down the road at the Valley's semiconductor companies, and see how new technology might make them very, very rich.

By 1974, the new generation had plenty of evidence showing how the computer had the potential to change everything, but without blowing everything up. Felsenstein summed up the philosophy in a nutshell: "You don't have to leave industrial society, but you don't have to accept it the way it is."[13]

COMPUTERS AS TEACHERS

This message resonated across Northern California suburbia of the mid-1970s, as the men and women of the Age of Aquarius transitioned into ordinary middle-class lives of ranch houses and cul-de-sacs, small children and suit-and-tie office jobs. Despite outward appearances, the generation was remarkably different from that of their parents, continuing on its search for personal fulfillment and freedom from the conformity of their youth. They practiced yoga, retreated to Esalen, and attended EST seminars. They came out of the closet, got out of the kitchen, and marched for women's rights and the Equal Rights Amendment. As gender roles and sexual mores shifted, California's divorce rate more than doubled between 1960 and 1975, eased by the introduction of no-fault divorce laws.[14]

Up in Cotati, Liza Loop wasn't yet getting divorced, but her home life was unfulfilling. Full-time mothering—another baby, the demands of toddlerhood—had consumed her for several years after that fateful encounter with Dean Brown in the Sonoma State classrooms. But she hadn't stopped thinking about computers, and she had stayed in touch with Brown. Sonoma County wasn't all that far away from Silicon Valley, but it lacked networks and organizations where she might connect with other enthusiasts and learn more about computing. She became impatient. No sense in continuing to wait for Sonoma's computer revolution to appear, she reckoned. She should just start one herself.

First, she embarked on her own crash course in building hardware and programming software. Her hands-on computer experience hadn't gotten much further than keypunching those IBM cards at Cornell. She did all she could to learn on her own, subscribing to the *PCC* and going down to Menlo Park to visit the new "People's Computer Center" that had spun off from Albrecht's operation, where both adults and kids could come in to learn how to program and play. She learned BASIC. In order to draw the scattered and reclusive local population of hackers out of their basements and garages, she started her own group: the Sonoma County Computer Club. Like many others popping up around the country, the club's membership blended lifelong technologists and passionate autodidacts, and—reflecting an America where "girls and electronics don't mix"—it was overwhelmingly male. (When the Southern California Computer Society did a survey of its 184 members in the fall of 1975, only five respondents identified themselves as "Ms." instead of "Mr." "Come on, ladies, let's balance out this organization!" the society newsletter exhorted, as if the feeble representation was simply due to women's lack of effort.)[15]

Being nearly the only "Ms." in the room didn't deter Liza Loop. Nor was she intimidated by her relatively rudimentary technical knowledge. Technology was changing so quickly that even the most seasoned computer operator could find their knowledge obsolete. And the inexorable force of Moore's Law was making computers cheaper and faster, truly opening the door to an era when classrooms could have as many computers as they had pencil sharpeners.

She persuaded Brown and computer designer Stuart Cooney to help her bootstrap a walk-in educational center in downtown Cotati, where anyone could come and learn how to use computers. She rented a computer terminal and set up a telnet line, then opened an account at a time-sharing company down in the Valley with access to a hulking HP 2000 mainframe. They called it the LO*OP Center, standing for "Learning Options Open Portal." The play on Loop's last name made it clear who was in charge. Adults and children could learn how to program and use different software. There were computer games for the kids. There even was a publicly available copy machine, which was something of a community gateway drug for those who didn't think computers had anything interesting to offer.

As buzz around computers grew, so did the LO*OP Center. It moved from a second-floor office to a street-level storefront. Loop closed her time-sharing account and bought a Digital PDP-8 secondhand from the People's Computer Center up in Menlo Park. On off hours, she packed up the center's equipment—minicomputer, teletype, peripherals—into the back of her dusty pickup truck and drove around Sonoma County, landing at school after school like a tent-revival preacher. Liza Loop wanted to demystify computers, not glorify them. In her view, human teachers remained central to the success of the computer-aided classroom. "The computer is only a medium of communication between teacher and student," she told the classes she visited. "It can never replace the teacher." Technology finally had gotten its human face.[16]

PROJECT BREAKTHROUGH!

Although life at home had delayed her entrance into the world of computer clubs and educational storefronts, Liza Loop's timing had been perfect. Because just as she was brewing her plans for the LO*OP Center, a kit computer had emerged that was unlike anything hackers and activists had ever seen. It was called the Altair, manufactured by a little company down in Albuquerque called MITS, and it took the hobbyist world by storm after being unveiled in *Popular Electronics* at the start of 1975.

Here was a construction project like no other. There was the usual board, components, and bus, with the addition of a nice blue metal box in which to encase it all. But unlike any other kit, the Altair featured an Intel 8080 microprocessor for its computing power and memory. Ed Roberts, the entrepreneur and designer behind MITS, had struck a deal with Intel to buy some slightly blemished chips at volume for $75 apiece, a fraction of their retail price. In the hands of an experienced hacker with patience and a soldering iron, the Altair kit became a zippy desktop computer that cost little more than $400 to make.

After the prototype made its debut on *Popular Electronics*' January cover—"Project Breakthrough!" enthused the headline—MITS became inundated with orders: 200 the first week, and 2,000 by the end of February. Roberts couldn't keep up with the demand, and hobbyists impatiently waited as the kits trickled off the Albuquerque assembly line. As soon as one lucky person in town got their Altair, they'd gather their fellow enthusiasts and pore over it together. In classic hacker fashion, they'd start talking about making this construction project even better—with peripherals, with software. Soon, the usual informal networking and weekly potlucks weren't enough. More organized meet-ups began. And the tinkerers and hobbyists and countercultural warriors began to think about how they could not only build gadgets to improve the Altair, but start companies to *sell* those gadgets.[17]

Talking and meeting and tinkering happened all over, but the center of the action was Silicon Valley. If someone like Liza Loop wanted to figure out how to get the most out of the Altair and join the personal-computing revolution, then she needed to trek down to Palo Alto. She heard through her PCC connections that a new group had started monthly meetings, drawing in all kinds of interesting people—from PARC hackers to Intel engineers, Stanford researchers to Whole Earth types.

So, in April of 1975, Liza Loop hired a babysitter, hopped in her pickup truck, and headed south, to the second meeting of the Homebrew Computer Club.

CHAPTER 10

Homebrewed

Are you building your own computer? Terminal, TV typewriter? Device? Or some other digital black-magic box? Or are you buying time on a time-sharing service? If so you might like to come to a gathering of people with like-minded interests. Exchange information, swap ideas, talk shop, help work on a project, whatever . . .

So read the mimeographed invitation to the first meeting of the Homebrew Computer Club in March 1975. For Palo Alto in those days, the flyer that Fred Moore hastily drew up and delivered by bicycle wasn't all that unusual. Hackers and computer liberators had been gathering for years over weekly pots of spaghetti at PCC potlucks, at seminars in the PARC beanbags, or all-night *Spacewar!* marathons in Stanford computer labs. Similar things were afoot in Boston, in Southern California, in Seattle—anywhere that had a critical mass of programmers and electronics hobbyists. But the announcement by Moore and his co-conspirator, Gordon French, signaled the start of a new era for high technology.[1]

The two made an odd pair. Moore was a pacifist. As a teenager, he had run away from home in a failed attempt to join the Cuban revolution. As a Berkeley freshman in 1963, he set off an opening salvo for decades of student activism by conducting a hunger strike on the steps of Sproul Hall to protest his forced induction into ROTC. He saw computers as a means to share information about peace activism. French was an engineer of the 1950s generation, a serious-minded guy with a military clearance, whose main concession to changing times had been to grow his hair a little longer. When he built his own kit computer, he named it Chicken Hawk. Moore and French had met through the PCC and bonded over a shared passion for outreach and computer education, becoming driving forces behind the

Menlo Park People's Computer Center. When MITS's Ed Roberts sent the center a review copy of an Altair, the two decided that a show-and-tell was in order, and invited folks over to French's garage.[2]

WELCOME TO THE CLUB

Thirty-two people showed up on that rainy night, sitting cross-legged on the cold concrete floor when they ran out of chairs, sharing technical specs and insider gossip in rap-session style. It was a geek's paradise. The group "argued about everything from which was the best microprocessor chip to the virtues of octal vs. hexadecimal notation for coding computer instructions," remembered participant Len Shustek. "Six people had already built their own computers, and almost everyone else wanted to."[3]

The meeting attracted many of the usual suspects. Lee Felsenstein drove down from Berkeley. But it also drew in some new faces. Coming in from Cupertino was a former phone phreaker who'd spent his college years selling marginally legal "blue boxes" door to door in his dorm with a high school buddy. His name was Steve Wozniak, and his buddy's name was Steve Jobs.[4]

Part swap meet, part intelligence gathering, part networking session, the biweekly Homebrew meetings quickly morphed into a local phenomenon. The second meeting moved from French's garage to John McCarthy's Stanford artificial-intelligence operation, then spilled out to the auditorium at the Stanford Linear Accelerator Center on Sand Hill Road, attracting hundreds of people each month. Conversations that started in meetings continued over beers and burgers down the road at The Oasis (or "The O"), the well-worn college dive on El Camino Real. It took a while to settle on a name for the club. Steam Beer Computer Club, 8-Bit Byte Bangers, and Tiny Brains all got rejected before the group arrived at Homebrew.[5]

French ran the first two meetings, but his droning delivery wasn't the right match for the restless crowd. So Lee Felsenstein took over, drawing on a skill set built up by years of antiwar protests and community organizing. Moore wrote up a newsletter to record the proceedings and share findings with the world beyond. Featuring shakily hand-drawn portraits of club members (beards, long hair, and Coke-bottle glasses predominated), first-name references to meeting participants, and rough-and-ready page layouts, the newsletters echoed the *PCC* in their chatty informality, even as the Homebrew Computer Club grew in size and influence.

Liza Loop stood out in the crowd. She was the only woman on the early Homebrew membership roster, and she was a computer newbie. To encourage swapping and sharing, Moore's newsletters included blurbs from members about their skills and needs. Steve Wozniak's was typical, showcasing dizzying technical virtuosity: "have TVT[ypewriter] of my own design . . . have my own version of Pong," he wrote. "Working on a 17 chip TV chess display (includes 3 stored boards); a 30 chip TV display. Skills: digital design, interfacing, I/O devices, short on time, have schematics." In contrast, Loop wrote: "I am not primarily a computer person. So my greatest contribution is to help professionals communicate with total laymen and kids. Have access to apples, fresh eggs, beautiful countryside. Need: TTY, acoustical coupler."[6]

While high-intensity hackers dominated the Homebrew scene, the stunningly rapid ascent of personal (or "micro") computing—and the enduring legend of the Homebrew Computer Club—had a lot to do with the Liza Loops: people who weren't necessarily lifetime hobbyists, but evangelists passionate about the computer's possibilities and able to translate the insider language of tech to bring the story to a wider world. And they weren't all in California. They were educators like her, working to bring computers into math classrooms and school libraries from New York to Texas to Washington State. Within a few years of the first Homebrew meeting, high school computer labs and after-school computer clubs had proliferated in the Bay Area and beyond, making computers an increasingly common feature of the K–12 years of most middle-class Americans born after 1965.

The evangelists were also journalists and publishers. Wayne Green, a New Hampshire magazine man and ham-radio enthusiast, started *Byte* magazine in September 1975. As polished as *PCC* had been homespun, the first issue had a cover headline out of the dreams of Lee Felsenstein: "COMPUTERS—The World's Greatest Toy!" *Byte* was not alone in spreading the word. Within three years of Homebrew's start, nearly a dozen magazines about microcomputers were rolling off the presses nationwide.[7]

Also spreading the word: the event impresarios who blew the computer-club vibe up to trade-show size. SDS organizer turned MITS marketing director David Bunnell orchestrated the first of these, the World Altair Computer Convention in

Albuquerque in early 1976. New Jersey technical writer Sol Libes launched the Trenton Computer Festival a couple of months later. (Ted Nelson gave a loopy keynote at Bunnell's event, alarming the crowd as he expounded on the marvelous possibilities of microchip-powered sex toys.) By 1977, the New Jersey meeting had become an annual event, joined by Computermania in Boston and the *Byte*-sponsored Personal Computing Expo in New York City.[8]

In the Bay Area, there was Jim Warren, a math professor turned programming enthusiast and Homebrew member, who started the biggest party of them all: an annual West Coast Computer Faire that attracted 13,000 computer die-hards at its inaugural outing in spring 1977. "The impact of the personal computer will be comparable to that of a gun," he remarked. "The gun equalized man's physical differences, and the private computer will do the same for his intellect." Warren, a free spirit with a reputation for throwing all-nude parties at his Redwood City ranch house, knew how to find an attention-getting metaphor.[9]

Warren also became the personality driving another important early platform for the growing population of computer junkies, *Dr. Dobb's Journal of Computer Calisthenics and Orthodontia: Running Light without Overbyte*. The mouthy and fanciful publication grew out of a special three-part edition of *PCC* that reprinted the stripped-down software code that Albrecht and Stanford computer scientist Dennis Allison had developed for the kids to program computers at their Menlo Park storefront. "It was being put together on a sorta spare-time basis by the PCC mob," Warren explained to his readers in the second issue, which came out at the beginning of 1976. "Once we became aware of the information gap that we are now focusing on filling, it took a coupla weeks or more to gather together a staff and organize a full-scale magazine production effort." Warren was no coder, nor was he the next Henry Luce, but as editor of *Dr. Dobb's* he helped fill the gap between increased availability of computers and a lack of off-the-shelf software to go with it.[10]

Then there were retailers, who'd popped up on the scene as distributors of Altair kits, and quickly morphed into far more. Paul Terrell started the Byte Shop in Mountain View at the end of 1975, disregarding the advice of friends who thought he'd never find customers. When sixteen people showed up one day for a seminar at the store on "Introduction to Computers," Terrell realized that computer courses needed to be a regular feature at the Byte Shop. How-to classes translated into sales, which were so brisk that Terrell opened a second location four months later, and had sixty stores nationwide by the end of 1977. Three other chains had grown up, as well as hundreds of independent outfits with names like Computer Shack and Kentucky Fried Computer (whose owner had to change the name after a cease-and-desist order from the fast-food giant). Retail stores became informal marketplaces, just like the computer clubs. "Ten percent of the people that come in," Terrell said, "are there to sell me, not to buy. They represent themselves and what they offer is a better widget, designed in their back bedroom or garage."[11]

CALIFORNIA'S NEW AGE

Homebrewing was a national phenomenon, but only in Northern California was there such a robust combination of things happening to give the new movement momentum and velocity. The Valley had its silicon capitalists of the semiconductor and computer hardware industries as well as its distinctive club of venture investors on the hunt for the next big thing. Certainly, the chipmakers had many other things on their minds in 1975 and 1976. Their big customers were the old-school mainframe and minicomputer makers as well as the car companies, watch manufacturers, and other kinds of companies putting microchips in all sorts of consumer products. The general slowdown in the U.S. economy had forced layoffs and downsizing as well as more overseas deals with Japan. Plus, the semiconductor guys made "computers on a chip." They didn't make actual *computers*.

Yet in the small and cozy world that ran along the spine of El Camino Real, most of the Homebrew hackers had personal and professional connections with the Silicon Valley establishment. A good portion of those at the first Homebrew meeting had day jobs at high-tech companies like HP and Intel; many others worked at Stanford. By the time the club moved over to a big auditorium, more employees from local companies started showing up to hear the buzz firsthand. Some of them were dispirited by rounds of layoffs and stagnant stock options, and were thinking about jumping into their own entrepreneurial ventures. Some were simply curious to see what all the fuss was about.

While many of the top people at the chip companies remained skeptical, it was clear to some further down the chain of command that a new market was poised to emerge. Intel engineer Albert Yu (one of the increasingly common foreign-born faces among the Silicon Valley semiconductor crowd) didn't need to come to a Homebrew meeting to sense that microprocessors were going to end up in more than traffic signals and car alternators. "Where was it going to go?" he thought. "It's going to go into the home. So home computers were going to come and be in the millions." Yu persuaded two of his Intel colleagues to jump ship and join him in starting a home computer company called Video Brain.[12]

Another Intel type who began poking his head into Homebrew meetings was John Doerr. A St. Louis native and Rice engineering graduate who'd followed the breadcrumbs dropped by Burt McMurtry and come out to the Valley, Doerr was marketing the 8080, the chip powering the Altair. As he studied up on the emerging personal-computer ecosystem, Doerr sensed that Intel's microprocessor market might become even larger than his bosses dared to imagine. By 1980, he had made the jump across Sand Hill Road to join the emerging VC powerhouse Kleiner, Perkins, Caulfield, and Byers. There, he'd later shepherd some of the biggest deals in Valley history.[13]

The microcomputing phenomenon gained velocity from broader political and cultural forces. Five months before the first Homebrew meeting, California voters elected thirty-six-year-old Jerry Brown as the state's new governor. The son of Pat Brown, the liberal icon who'd presided over California's expansive postwar boom before he was vanquished by conservative Ronald Reagan in 1966, the younger Brown won mostly on name recognition and the anti-Republican mood in the aftermath of Watergate. He quickly distinguished himself by his national ambitions (his run for the 1976 Democratic nomination commenced almost instantaneously) and by his blending of countercultural ideas with pragmatic fiscal conservatism.

Brown's job description was state-level; his vision was cosmic. He retreated regularly to a Bay Area Zen center. He befriended and sought the advice of Stewart Brand, with whom he shared a taste for the theatrical as well as a belief in the power of technology to change the world for the better. "What we are proposing in California is something different than the cowboy ethic, where people ride into town and wreck it and move on," proclaimed the governor whom Sacramento pundits liked to call "The Free Spirit." At the same time, Brown distanced himself from his liberal father by clamping down on public spending, especially on social services. "Limits impose restraints," he said, "but also create possibilities."[14]

High-tech enterprises fit into Brown's worldview of a society where business and government were efficient and future-minded, but still had a heart and a soul. In the late summer of 1977, as Americans lined up at movie theaters by the millions to see another Northern California–made blockbuster, *Star Wars*, Brown proclaimed a statewide "Space Day" to celebrate technological achievement. While his techno-futurist proposals later earned Brown the derisive nickname of "Governor Moonbeam," they reinforced California's reputation as the place where new ideas—and new movements—began.

New industries too. The countercultural values that seeded feminist bookshops and health food co-ops a decade earlier already were spawning a new generation of baby boomer–led companies that sought to do good and make money at the same time. In the words of SRI psychologists Duane Elgin and Arnold Mitchell, whose research on market "psychographics" later had a big influence on how Valley companies positioned their products, more and more Americans were choosing lives of "voluntary simplicity . . . living in a way that is outwardly simple and inwardly rich." These consumers still wanted to buy things, Elgin and Mitchell observed, but the price tag mattered less than the perception that a product reflected their social values.[15]

Although social idealism and technological passion drove Homebrewers like Moore and French, the explosion of clubs and newsletters and trade shows and stores and microchip-industry interest and California futurism showed the techno-hippie faithful that personal computing wasn't just a new movement. It was a new *market*. In short order, the hobbyists in garages morphed into entrepreneurs.

CHAPTER 11

Unforgettable

MITS had been completely unprepared for its success. The machine Ed Roberts had showcased on the cover of *Popular Electronics* in January 1975 was an empty box, not a functioning computer. When the Altair orders started flooding in, all he had on hand were a few prototypes. The little company could barely keep its head above water in delivering the basic kits—never mind being able to fulfill orders for plug-in boards and peripherals that it had originally promised to Altair customers. But the add-ons weren't bells and whistles; they were essential to making the Altair a functioning machine.

Hobbyists seized the opportunity. Bob Marsh and Gary Ingram, two college friends of Lee Felsenstein's, started up a company shortly after the first Homebrew meeting. Called Processor Technology, or Proc Tech, it built plug-in memory and input-output (I/O) boards. They set up shop in Marsh's Berkeley garage, carving out a corner for Felsenstein to work as an in-house repairman for temperamental Altairs while designing boards and other products. The business took off. As New Jersey enthusiast Sol Libes put it, Processor Technology "made the Altair a real computer rather than just a toy."[1]

In another garage a few miles away in the East Bay town of San Leandro, engineer-hobbyist Bill Millard was determined to go one better. He wasn't just going to build add-ons for Altairs, he'd build and sell an entire 8080-powered microcomputer. Called IMSAI, the company's kits cost $100 more than the ones from MITS, but they proved much more reliable and powerful. Millard's computers soon started outselling the Altair. Meanwhile, other little Homebrew-inspired companies with far-out names (Cromemco, Xitan, PolyMorphic) sprouted across the Bay Area. Many started by selling Altairs and peripherals, but often moved into the business of building whole new microcomputers altogether.[2]

With the blessing of *Popular Electronics* editor Les Solomon, who promised a cover story, Felsenstein joined with the Proc Tech team to commercialize the Tom Swift Terminal. Despite the pains they took not to tread on Altair's toes—the unit was designed as an "intelligent terminal" rather than a computer—the effort ultimately produced an important Altair competitor, the Sol. In another first, the Sol included BASIC software as well as hardware. The user base had grown beyond people who could simply consult *Dr. Dobb's* to program their machine. As an indication of how large *Popular Electronics* still loomed on the early personal-computer scene, they named it in honor of Les Solomon.[3]

Across America, start-ups bubbled up nearly anyplace where there were lots of engineers and active hobbyist chatter. Specialist publications, retailers, and trade shows spread the word to true believers and new converts. California's governor was going on Zen retreats and *Star Wars* was breaking box office records. It was the right time to deliver a message about technology as something enlightening, emancipating, and *fun*.

The entrepreneurial pool expanded to include more-unlikely computer moguls, like Lore Harp and Carole Ely, two stay-at-home wives of electrical engineers in Southern California. They had been toying with starting a travel agency, but when Harp's tinkering husband developed a memory board, they jumped on the microcomputing bandwagon. In the Bicentennial summer of 1976, the two pooled $6,000 and set up production facilities in the Harps' suburban home. The company they named Vector Graphic managed to ship 4,000 units in twelve months, and their initial investment turned into over $400,000 in revenue. By the summer of 1977, they were producing an elegantly designed system that they pitched as "the perfect microcomputer." The design showed what a computer could become when designed by nonspecialists. In contrast to the fussy proliferation of switches and lights on the Altair, the Vector 1 had only two simple buttons on its front.

Five years later, Vector Graphic had annual sales of $25 million. While standing out as female founders in a nearly all-male industry, Harp and Ely were like many other new tech entrepreneurs in that their wild professional success came at a steep personal cost. Both split from their husbands within a few years of Vector's founding. "I was running away from a marriage," said Carole Ely, "into a company."[4]

The original microcomputer generation didn't necessarily go into business expecting to make millions. Lee Felsenstein made about $10 or $12 per unit on the computers he designed for Proc Tech, which he reckoned "added up to a nice chunk of change." Hackers like him initially were just interested in making the Altair *work,* and in swapping parts and sharing know-how in the same way they'd been doing for years at the PCC and elsewhere.

But very soon the computer users weren't only the hackers anymore. By 1977, Jim Warren estimated that there were 50,000 personal computers in use. He might have been exaggerating, but not by a lot. That market translated into money, attracting people who had business smarts as well as engineering talent. The companies that came out of Homebrew and the other clubs usually involved an odd couple like Gordon French and Fred Moore: tech-obsessed hacker paired with entrepreneurial visionary. Of the motley dozens of early start-ups, however, the few that scaled up into million-dollar ventures also involved people like Lore Harp and Carole Ely: people who understood how to run a company, and how to sell the high-tech dream to customers who'd never taken apart a radio set or subscribed to *Popular Electronics.*

RISE OF THE STEVES

This of course became the secret of Apple Computer Co., the most legendary Homebrew product of them all. The company wasn't all that different from the dozens that sprouted from computer-club soil in 1975 and 1976. But it pulled away from the pack because, very early on, it bridged the hacker world of "The O" and storefront computer labs with the Silicon Valley ecosystem of the Wagon Wheel and Sand Hill Road. While baking countercultural credentials into its corporate positioning from the start, Apple was the first personal-computer company to join the silicon capitalists.

At the beginning, Steve Wozniak was just another hacker in Gordon French's damp garage, standing out a bit because he was a few shades more tech-obsessed. The gregarious twenty-five-year-old had a lifetime immersion in the electronics world of the Valley, as his father had been a Lockheed engineer. While other grade-school boys were building crystal radio sets, Woz was messing around with transistors. Designing computers became a preoccupation in high school and through college. When he saw the Altair at Homebrew, he was entranced, but he didn't have the money for a kit. Instead, he set out to build one of his own.

The result was the Apple I, a simple wooden box that looked like it came out of a tenth-grade shop class, encasing a circuit board of supremely elegant design. The machine got its power from a $20 microprocessor from Pennsylvania-based MOS Technology instead of the dearly expensive Intel 8080, making it an intriguing demonstration of hacker ingenuity and price performance. Although everyone else at Homebrew seemed to be going into business, Woz wasn't interested in selling his gadget. He built it because it was cool.

His friend Steve Jobs had to convince him otherwise. Jobs was five years younger than Wozniak, but they already had partnered in a couple of moneymaking ventures—first that blue-box phone phreaking enterprise at Berkeley, then a gig writing a new game for Atari, where Steve Jobs worked at the time. Jobs had helped Woz craft the Apple I ("we liberated some parts from Hewlett-Packard and Atari" to make it, he later boasted). After seeing how the Homebrew crowd responded, he sensed that this circuit board in a wooden box could be the start of something much greater.[5]

The kid of working-class parents in neighboring Cupertino, Steve Jobs exhibited supreme self-confidence and relentless focus from the start. He was a child of the Valley, but of an even newer generation. At age twelve, he'd run out of parts for an electronics project. So he cold-called HP, requested to be put through to Bill Hewlett, and proceeded to ask the tech titan if he had any parts to spare. He got them, and Hewlett offered the cocky middle-school kid a summer job.

As Doug Engelbart was putting on the mother of all demos and prototyping the first computer mouse, Jobs was a floppy-haired regular of the Homestead High School Computer Club. While Xerox was establishing PARC and developing graphical user interfaces, Jobs was phone phreaking and fixing broken stereos for his classmates. In appearance and worldview, he was miles away from the clean-cut engineers who'd peopled the electronics industry for decades. By the time Homebrew started up, he'd dropped out of college, traveled around India, and picked up such unusual hygiene and dietary habits that Atari gladly agreed to give him the night shift so that his funk wouldn't disturb others.

Jobs was a hacker, but not on the order of Steve Wozniak. Instead, his edge came from his tenacity, and his remarkable ability to explain the transformational power of computers to others. As Regis McKenna later observed, "Woz was fortunate to hook up with an evangelist."[6]

On April Fools' Day, 1976, the two Steves and a third partner, Ron Wayne, started Apple Computer Co. The first logo, designed by Wayne, had the retro-hippie design beloved by techie newsletters like the *PCC* and *Dr. Dobb's*. It featured Isaac Newton sitting under a tree, surrounded by words uttered not by Newton, but by William Wordsworth: "A mind forever voyaging through strange seas of thought—alone." The inaugural sales flyer was similarly loopy, with a typo in the first sentence.[7]

Jobs persuaded Paul Terrell at the Byte Shop to buy fifty units of the Apple I, which Terrell agreed to do under one condition: no kits. The machines needed to be fully assembled. In a move that Apple's marketers later made sure to burnish into company legend, Jobs sold his VW microbus and Woz sold two HP calculators to finance the start-up costs. After months of frenetic sixty-hour weeks, Apple ultimately produced and distributed 200 Apple Is. Not all of them were sold. Woz, still the ambivalent capitalist, gave one away to Liza Loop for her LO*OP Center up in Cotati.[8]

By that point, the Steves had a new Apple product line in the works that would rival the still-in-development Sol: a fully assembled computer, with terminal, keyboard, and BASIC software included. They called it the Apple II. In order to reach a market beyond the hobbyists, it was obvious that Jobs needed to go beyond the Homebrew wading pool for advice on how to scale up the business. (Ron Wayne, alarmed by the intensity of Jobs's ambition, had left. Jobs bought out his 10 percent stake for $2,300.)

A great admirer of Intel's Bob Noyce, Jobs wanted to build a campaign for the Apple II that was as jazzy as the one that had propelled the Intel 8080 into the stratosphere. In a replay of his audacious call to Bill Hewlett a decade earlier, Jobs dialed up the Intel switchboard, where someone connected him with the man who'd crafted that marketing campaign, Regis McKenna Himself.

McKenna was unfazed by Apple's garage setting and the co-founders' scraggly looks. He'd worked with "lots of strange people" in the Valley already, and he was familiar with the Homebrew scene and the intriguing little enterprises bubbling up from it. The first meeting, however, was a bust. The Steves wanted help placing a Woz-authored article on the Apple II in *Byte*. It turned out that Steve Wozniak was much better at building elegant motherboards than crafting accessible prose; the piece was a rambling mess better suited for the hobbyist crowd back over at

Dr. Dobb's. McKenna told them it would have to be rewritten, and an offended Woz refused. Then I have nothing to offer you, replied McKenna.[9]

But Steve Jobs was not one to take no for an answer. Ever persistent, he called McKenna "about forty times" to persuade him to take Apple on as a client. Here's the deal, McKenna told Jobs. Good corporate marketing involves lots of different elements, and all of it costs money. Apple needed venture capital to make it work. Jobs already had been making futile attempts to fund-raise: pitches to Wozniak's bosses at HP, then to computer maker Commodore, then to his old boss Nolan Bushnell, who had just hit a $10 million payday by selling Atari to Time Warner. They'd come to nothing.

Both McKenna and Bushnell pointed Jobs toward a financier who might be willing to help: Don Valentine, the hard-driving National Semiconductor executive who'd now become a venture capitalist. Valentine agreed to come by the garage. There, the Mercedes-driving Republican in a rep tie found a skinny, bearded kid who looked "like Ho Chi Minh" and his nerdy, equally shaggy business partner. "Why did you send me these renegades from the human race?" Valentine called to ask McKenna after the meeting was done. The VC wouldn't invest—yet—but he thought enough of Apple's potential to connect the two Steves to Mike Markkula, a Fairchild and Intel veteran who recently had retired as a microchip millionaire in his early thirties. Markkula was intrigued. These two founders were very young and kind of strange, but the Valley was full of strange tech people, and the Apple II was exciting.[10]

With Markkula on board as an advisor, Apple had enough credibility and money for Valley VCs and marketers to start taking it seriously. By December 1976, Regis McKenna had taken on Apple as a client and drafted a comprehensive marketing plan, recorded in tight cursive in a narrow-ruled spiral notebook. (Among the possible distribution channels noted by McKenna in 1976: "Apple Stores." He was thinking big.) The next order of business was some better branding. McKenna's art director Rob Janoff replaced the hippy-dippy Newton etching with the iconic once-bitten apple. Local designer Tom Kamifuji, known for his ability to adapt the psychedelic graphics of a 1960s rock poster into corporate branding campaigns, added the rainbow colors. Clean high-tech lines infused with a countercultural vibe: a perfect new image for a microcomputer company with market-altering ambitions.[11]

McKenna's team orchestrated a similar transformation in the company's print advertising, which morphed from misspelled newsprint into sleek color spreads with clean typography and maximum visual punch. "The people we were trying to reach was very specific," McKenna explained. "The hobbyist looking to the next level, affordable computer, people who had programming skills and built their own computers from kits." Yet the Apple II was also for those who weren't already homebrewing: "professionals such as teachers, engineers, or people who would put in the time and effort to learn how to use this new computer."

In 1977, nearly everyone in these target groups was male. The debut spread for the Apple II, which ran in general-interest magazines like *Scientific American* as well as more specialist publications, pictured a young husband at the kitchen table, checking stocks on his computer as his smiling wife looked over from her work at the kitchen sink. The ad copy actually had been written by a woman, but the picture was worth a thousand words. "Within just a few weeks," the copywriter confessed, "Steve [Jobs] received a letter from a woman in Oregon, complaining that the ad was sexist—which it very clearly was." Whether their ads were homey or slick, all of the personal-computer companies were like Apple: they marketed their products to the men who already were inclined to use them. Girls and electronics *still* didn't mix.[12]

By the time Jobs, Wozniak, and Markkula hit the 1977 trade-show circuit, the company was starting to get a professional polish that belied the fact that this was a tiny little operation of about a dozen employees. The Steves donned collared shirts, ran combs through their hair, and pinned nametags to their chests. Their first stop was Jim Warren's first annual West Coast Computer Faire, a landmark in tech history all on its own. Apple landed a coveted booth near the entrance.[13]

Despite the proliferation of new entrepreneurs, the first Faire had a program and a vibe that was more *Whole Earth Catalog* than *Wall Street Journal*. Panels focused on the change-the-world potential of computing, with titles like "If 'Small is Beautiful,' is Micro Marvelous? A Look at Micro-Computing as if People Mattered" and "Computer Power to the People: The Myth, the Reality, and the Challenge." There were sessions on computers for the physically disabled, and four panels on using personal computers in education (Liza Loop appeared on one of them). Novices were welcomed with their own panel on "An Introduction to Computing to Allow you to Appear Intelligent at the Faire." Business uses were rarely mentioned, aside from considering "Computers and Systems for Very Small Businesses." For $4, attendees could buy an official conference t-shirt that read, "Computer Phreaques Make Exacting Lovers."[14]

But this was no PCC potluck. The times were, at last, a-changing. "Here we are, at the brink of a new world," Ted Nelson proclaimed in his keynote speech at the Faire. "Small computers are about to remake our society, and you know it." He continued:

> The dinky computers . . . will bring about changes in the society as radical as those brought about by the telephone or the automobile. . . . The rush will be on. The American manufacturing industry will go ape. American society will go out of its gourd. And the next two years will be unforgettable.[15]

Nelson's exuberance didn't seem all that irrational as soon as a conference-goer traveled from the breakout rooms to the trade show floor, and saw the crowds clustering around the booths of the Valley's new little computer firms. The booth with the biggest crowd of all was Apple, where the spruced-up Steves proudly showed off the new Apple II. The device finally delivered what the tech evangelists had been promising for so long: a self-contained unit. You plugged it into a wall, hooked up an ordinary TV and tape recorder, and started typing. It had BASIC installed so that you didn't need to write your own software. It had eight—count 'em, eight—expansion slots that allowed the user to add on applications and memory. The Apple II wasn't cheap—at $1,300, it cost twice as much as the Apple I and more than three times as much as the Altair—but it was the friendly little machine that the microprocessor finally made possible. As Apple's first print ad said, "You've just run out of excuses for not owning a personal computer."[16]

The shows unveiled the Apple II to the world, and they also introduced Steven P. Jobs to the national press. As business reporters trolled the trade show halls in Boston and Dallas and New York over the frenetic months of 1977, McKenna's team made sure that the "Vice President—Marketing" was available to talk. A fresh-faced Luke Skywalker who could speak a consumer-friendly language, Jobs reliably provided comments with enough color and interest to make it into the story. He usually was the only personal-computer entrepreneur quoted.

Steve Jobs may not have built the Apple's motherboard, but he knew how to explain it in evocative language, a rare talent in the engineering world. The chip industry brimmed with hard-charging, take-no-prisoners personalities, so Jobs's mercurial temper wasn't at all remarkable. But his skill set was. Here was someone who combined Andy Grove's gimlet-eyed understanding of product with Bob Noyce's charisma, in a youthful package with countercultural appeal. The barefoot vegan with the straggly beard also understood that the sales job at hand involved more than slick logos and zippy sloganeering. "Inventions come from individuals," observed Regis McKenna, "not from companies."[17]

The large-computer industry was legible and familiar to these reporters and their readers. The microcomputer industry was mysterious, and it still was difficult to understand why anyone would drop $1,300 on an Apple II that couldn't do much more than balance a checkbook. Jobs keenly understood this dilemma. "Most people are buying computers not to do something practical but to find out about computers," he told *The New York Times*. "It will be a consumer product, but it isn't now. The programs aren't here yet."[18]

Ah, the programs: Apple and others succeeded in developing personal-computer hardware, but there remained a huge shortage of software programs to run on it. This was a big stumbling block to home and office users who didn't have the appetite to learn BASIC and build apps of their own. For all the hype, the personal-computer business remained tiny. Its total 1977 sales came to about $100 million in a more than $22 *billion* computer industry. To scale up, it needed software makers to catch the same entrepreneurial spirit that had seized the kit-builders.[19]

Software was a very different business proposition, however. Electronics hardware was something that hacker-hobbyists had been used to buying ever since they were kids playing with their first crystal radio kits. To get parts, you went to a hobby store and paid for them. But code wasn't something you could find on a retail-store shelf. It was *know-how*, something you learned by doing. If you didn't write the software yourself, you could borrow code from others. The only entities that sold software were behemoths like IBM, who packaged it with their large, expensive computers. In the moral calculus of hackerdom, stealing software from big computer companies was equivalent to using blue boxes to make free long-distance phone calls. No one got hurt, just corporations that already made far too much money.[20]

By the time Homebrew began and *Dr. Dobb's* published the code for Tiny

BASIC, it was common understanding that personal-computer software was to be shared, "liberated" from corporations, and given away for free. Less than a year into Homebrew's happy swap meets, however, the organizers received a sharply-worded missive from someone who was trying to turn software into a business. "An Open Letter to Hobbyists" wasted no time in laying the blame for the software gap at the feet of the hackers themselves. Paying for hardware, but not for software, kept good programs from being written. "Who can afford to do professional work for nothing?" the author asked. "The thing you do is theft."[21]

The letter came from Albuquerque, penned by another relentless twenty-year-old college dropout. His name was Bill Gates.

JET CITY

William Henry Gates III was a Cold War kid from Seattle, born and bred in a place and time where opportunities abounded for a curious boy to learn about computers. His childhood coincided with the apex of Seattle's Boeing boom, when the aerospace giant employed hundreds of thousands and produced successive generations of innovative commercial airliners at an astonishingly rapid pace. Seattle was also home to the University of Washington and its multiple federally funded science and engineering programs, including in the new academic discipline of computer science. The city of modest bungalows sprawled into a metropolis of millions, as highways snaked through neighborhoods and floating bridges unfurled across the lake to connect the Eastside suburbs.

In 1962, Seattle's space-age boosterism spawned the Century 21 Exposition, a futuristic World's Fair featuring a soaring Space Needle and exhibits of the latest innovations in computing—a UNIVAC that contained a library's worth of information, a "TV telephone" that broadcast pictures along with sound, and the *Freedom 7* capsule that had just carried NASA astronaut Alan Shepard into space. Gates was only six when he visited, and his favorite thing was riding the fair's sleek new monorail. But by the time he hit middle school, he'd become a die-hard math and science guy, so much so that he didn't bother to earn decent grades in other subjects. From an early age, all Bill Gates cared about was technology, and he was quite certain that he understood it better than anyone else.

Luckily, his parents had sent him to the private, all-boys Lakeside School, in the hopes that its reputation for academic rigor would rub off on their son. Exhorted by a resourceful teacher who believed that computers were about to become the future of education, Lakeside parents had raised money to buy the school its own teletype machine connected to a time-sharing system. The computer room became Gates's hangout from eighth grade on. It was where he learned BASIC, and where he met a quiet tenth grader named Paul Allen. Soon the two could be found haunting the University of Washington computer labs, learning the ins and outs of a Digital PDP-10 and becoming notorious for their ability to skillfully hack into places they weren't supposed to be. In Gates's last year of high school, he and Allen embarked on their first business venture, selling a microprocessor-based device they programmed to analyze traffic volume on city streets. They called the company Traf-O-Data.

Gates started his freshman year at Harvard in September 1973, the same month that Don Lancaster's TV typewriter graced the cover of *Radio-Electronics*. His buddy and business partner Allen came to Boston as well, to take a job at Honeywell. Gates didn't stay in college long, but left an impression. "He was a hell of a good programmer," remembered his faculty advisor, but he was "a pain in the ass." Gates was far more interested in extracurricular hacking and video gaming than in his Harvard coursework. (At one point, he became obsessed with the Atari game Breakout—a game that had been built out by the two Steves before they started Apple.)[22]

Then, halfway through his sophomore year, came the Altair. The kit was all hardware, no software. Gates and Allen immediately sensed an opportunity. Even though they had no connections or particular entrée to MITS, they sent a letter to Ed Roberts. We'll write BASIC for the Altair, and you give us office space and royalties in exchange. Roberts, skeptical but desperate, told them he'd take a look. Six weeks of all-nighters later, Allen flew out West, paper tape in hand, to demonstrate their new software. It worked, and the two Seattleites had a deal. Within months, the two had decamped from Boston to Albuquerque, and Traf-O-Data had a new name: Micro-soft.[23]

Bill Gates had been like most hackers, with little compunction about the backdoors and gray areas that would allow him to get things for free. He'd gotten in trouble at Lakeside and Harvard for hogging free computer time and for using school and university computers for his own side projects. But now that he and Allen were devoting every waking moment to building a software business, his outrage grew about those who thought they could get code for nothing. Only a few months into Gates and Allen's MITS adventure, a spool of their Altair BASIC tape got into the hands of a Homebrewer, who, in classic hacker fashion, made fifty copies of it to distribute to other members. Those folks made copies for their friends, and on and on. It became, observed Gates biographers Stephen Manes and Paul Andrews, "the world's first and most pirated software."[24]

The young entrepreneur was furious, and fired off the angry note that would go down in tech history as "The Gates Letter." In it, Gates drew enduring battle lines in the tech world. On the one side, there were the people who believed information—software—should be proprietary data, protected and paid for. On the other side were those who believed in a software universe like Homebrew: where people shared and swapped, iterated and improved, and didn't charge a cent. Two decades later, open-source software evangelist Eric Raymond famously framed the divide as "the cathedral" versus "the bazaar."[25]

As the personal computer revolution gained speed and financial velocity, and as Jobs and Gates morphed from oddball kids into two of the richest and most celebrated business leaders in the world, the space between the cathedral and the bazaar—and the entrepreneurs and the idealists, the mercenaries and the missionaries—grew even wider.

BOYS HAD BEEN TINKERING WITH ELECTRONICS SINCE THE DAWN OF THE RADIO age. The hackers might have stayed in their basements and bedrooms if the enabling technology of the mass-produced microprocessor hadn't come on the scene, and a cadre of ferociously innovative chip companies hadn't created new markets for computerizing nearly everything. The microcomputer might have remained a funny little toy if it hadn't been for the educators and evangelists and marketers who showed how it could work in the classroom, in the office, and as home entertainment.

This did not occur in a vacuum. The same conditions that had made Homebrew possible in the first place—microprocessors and miniaturization; a turn away from the military-industrial complex toward other applications for tech; time-sharing and the spread of programming languages like BASIC; a desire for a more inspiring and interactive relationship with technology—created avenues for Silicon Valley hackers to turn their ideas into viable businesses. A bleak economy upped the desire for escapist, tech-fueled fantasy, from *Star Wars* to video games to Jerry Brown's Space Day.

What's more, the electronics hobbyists came together at a moment when the inhabitants of a prosperous and complacent postwar world were experiencing a period of wrenching, bewildering change. America and Western Europe were awash in data, and increasingly distrustful of the institutions and experts who managed that data. The personal computer held the promise of reclaiming control.

"The truth is that one of the main problems—perhaps *the* main problem—of the time is that our world suffers from information overload, and can no longer handle it unaided," wrote British futurist Christopher Evans in his prescient 1979 bestseller, *The Micro Millennium*. "The world needs computers *now*, and it will need them more in the future; and because it needs them, it will have them."[26]

CHAPTER 12

Risky Business

As hundreds crowded into Homebrew meetings, long lines formed outside Byte Shops and Computer Faire booths, as subscriptions soared to *Byte* and tens of thousands of copies of *101 BASIC Computer Games* flew off bookstore shelves, it certainly felt like a revolution was on its way. But as 1977 rolled into 1978, few of the rangy little companies sprouting out of the computer-club world had managed to do what Apple had done: grab the outside funding and management expertise needed to turn a neophyte's garage start-up into a scaled-up operation aimed at a broader consumer market.

Instead of Silicon Valley's nascent computer industry divvying up the market among itself, experienced and well-resourced operators barged in from elsewhere. At the same moment that Jobs and Woz were revealing the Apple II at the West Coast Computer Faire, a Pennsylvania calculator maker named Commodore was a few booths over, unveiling the Personal Electronic Transactor, or the PET. (The designer thought the gimmicky acronym made a nice play on the 1970s Pet Rock fad.) A few months later, Texas-based recreational electronics giant Radio Shack joined the game with a personal computer called the TRS-80. Both machines had less power and sophistication than the Apple II, but they were half the cost, and threw in a monitor to boot.[1]

An awkward fact remained: whether shoestring start-ups or well-financed electronics firms, the microcomputer makers were changing the world for a relatively small, nerdy group of very early adopters. The micros still didn't have the computing power or the workplace applications to turn them into serious challengers to mainframes and minis in the office market. Personal computers were great for playing games, and playing around, but not a lot more. Even in Silicon Valley, Homebrew remained an after-hours sideshow to the main acts of semiconductors and other electronics. If the new generation wanted to make a market-disrupting commodity product like Andy Grove's microprocessor "jelly beans," they needed more speed, more software—and more investors to help them grow.

The venture capitalists, however, had other worries on their minds.

CASH CRUNCH

The problem began back in 1970, the year Wall Street's ebullient Sixties gave way to tightened markets. "Has the bear market killed venture capital?" blared a headline in *Forbes* magazine that summer. A gush of public offerings slowed to a trickle. More than 500 new companies went public in 1969. Only 4 did in 1975.

By the time that OPEC's oil embargo had American drivers stewing for hours in gas lines, capital-starved tech companies were slowing production and selling off technologies and licensing to foreign firms. In 1969, the national venture capital industry had raised more than $170 million in new investment. In 1975, it raised a paltry $22 million. What's more, only one venture investment in four went to tech companies. The silicon revolution seemed to be stalling just after it had gotten going.[2]

Why had things suddenly gotten so tough? Electronics was a growth market! Microchips were going into everything! Its companies were dynamic and flexible, able to open and close plants as needed. They had already moved manufacturing elsewhere, into the Sunbelt, and availed themselves of cheap skilled labor in Taiwan and Singapore. They weren't weighed down by giant pension plans and outmoded manufacturing processes like those that were waterlogging old-school American manufacturers.

Yet the shiny newness of tech start-ups became a liability when it came to raising money beyond the seed-capital stage. Tech entrepreneurs were beginning to get rich, but the vast majority of the big money still was *old* money: the endowments of Gilded Age fortunes, the pension plans of Big Labor, the in-house investment operations of giant corporations. The new economy depended on the old economy for financing, and the old economy played it safe, especially in a down market. Tech investment was called "risk capital" for a reason. While payoffs could be huge, the odds were steep. From Cupertino to Cleveland, tech VCs couldn't get investors to buy what they were selling.[3]

The Valley's money men were blindsided by the crisis, especially those who were relative newcomers to the investment side. After his first, modest fund with Melchor, McMurtry got together with Reid Dennis and set out to raise money for a new, much bigger fund—right at the height of the OPEC oil embargo. Despite Dennis's connections and reputation, it took them the better part of a year to raise $19 million. Then the Dow Industrial Average plummeted, falling below 600 by the autumn of 1974. Even a preternatural optimist like McMurtry got spooked. "We were so terrified," he confessed, "that we didn't make an investment for fifteen months."[4]

It was bad back East as well. Stewart Greenfield was a former IBM product marketer who had jumped into tech venture investment at the start of the 1970s, leading what was called the "Sprout" fund at New York City investment bank Donaldson, Lufkin & Jenrette. Going into 1974, he, too, was trying to raise a new fund and had already gotten promises for $12 million from his investors. He was optimistic he'd raise more than twice that much. Then, as he put it, "something strange happened." One investor after another called him over that summer and fall, "withdrawing their interest." Pension funds had been some of his biggest investors, but Greenfield wouldn't see another dime of pension money for three years.[5]

The tech world still pulsed with young companies, but new entrants couldn't get off the ground. A bad economy and high interest rates dampened investors' enthusiasm, to be sure. And after the explosive growth of the chip business at the start of the decade, it wasn't all that surprising that the industry hit a plateau. The next big thing, microcomputers, were still a niche market. But as the VC community saw it, the biggest thing scaring off investors wasn't any of these things. It was the U.S. tax code.[6]

DEATH AND TAXES

The roots of their belief ran deep. In 1921, less than a decade after the passage of the Sixteenth Amendment creating the federal income tax, Congress responded to business-sector pressures to bestow special, lower tax rates on investment income, also known as capital gains. Through the Great Depression and New Deal, through wartime austerity and postwar prosperity, capital gains taxes remained low despite escalating levels of government spending and taxation. Effective tax rates hovered around 15 percent; the highest the rate ever got was 25 percent—even as taxes for the top bracket exceeded 90 percent.[7]

Even though rates were relatively modest, investors and their political allies hated the capital gains tax with a white-hot passion. In 1928, the American Bankers' Association pronounced it "unfair and economically unsound" and blamed it for excessive stock market speculation. In 1930, President Herbert Hoover declared that the tax "directly encourages inflation by strangling the free movement of land and securities." Throughout the 1930s, New York Stock Market president Dick Whitney—who, like Hoover, became a blistering opponent of Roosevelt's New Deal—blasted the capital gains tax at every opportunity, asserting that economic recovery would come from "the initiative of private enterprise, the intelligence of private management, and the courage of private capital."[8]

There was another reason investors fiercely resisted raising the capital gains tax, of course: it was a tax on a big chunk of their income. This was particularly true for the limited partnership model that became so prevalent in the tech VC world. Regular management fees were only a piece of their take; most of the upside came from their hefty 20 percent slice of fund returns—i.e., money subject to capital gains tax. Yet this was rarely mentioned in the battle cries that rang out every time rumblings of tax reform stirred at either end of Pennsylvania Avenue.

The bankers and investors made supply-side arguments decades ahead of the Reagan Revolution: cutting taxes would unleash investment in new industries, new companies, new waves of entrepreneurship. Any hit to the Treasury would be more than repaid in new tax revenues created by this growth. By the 1960s, maintaining the capital gains differential had become an unquestioned tent-pole of tax policy for politicians of both parties, despite scant evidence that capital gains tax rates had the outsized economic effects that Wall Streeters contended they did. A few brave investors hazarded the opinion that the low rate encouraged speculative investment simply for tax benefit, and a raise in the rates would move stock-pickers "back to fundamentals." Yet the capital gains differential stuck.[9]

This changed in 1969. A Democratic-led Congress inched the rate up, arguing that the rich needed to pay their fair share. By 1972, the capital gains rate had risen to more than 36 percent, and the rhetoric on that year's presidential campaign trail indicated it might go even higher. "Money made by money must be taxed at the same rate as money made by men," declared Democratic nominee George McGovern. After his landslide victory, Richard Nixon remained focused on the "silent majority" of the middle class who might swing into the Republican column. Slashing capital gains taxes wouldn't earn many of those votes.[10]

Swamped by the populist mood, businessmen once again rolled out arguments that capital gains and other corporate tax cuts were actually things that helped the little guy. The king of the venture capitalists, Ned Heizer, was outraged about Congressional inaction. As the economy soured, even the deep-pocketed Chicagoan had taken heavy losses. "This country was founded by entrepreneurs," he declared, "but Congress is killing this process." The future of the economy wasn't a matter of propping up U.S. Steel or General Motors. "If we could assure a flow of capital to young companies, then if General Motors in time didn't do a good job, new companies could compete against it."[11]

In the middle of all of this came the Employee Retirement Income Security Act (ERISA) of 1974, which placed strict regulations and performance standards on private pension plans, including a "prudent man rule" that increased fund managers' personal liability for how they invested retirees' money. Here was the reason Stew Greenfield had suddenly lost all of his investors. A couple of years after ERISA's passage, an industry survey found that most trustees had become "unwilling to invest in anything but blue-chip stocks and bonds." VCs had lost one of the biggest sources of money out there.[12]

THE RED CARPET ROOM

As those clouds were gathering in early 1973, San Francisco–based VC Pete Bancroft found himself commiserating with some colleagues over lunchtime martinis. "Sending representatives of the Rockefellers, Phipps, and Jock Whitney to Washington to lobby Congress about the need for tax reduction would surely get us laughed out of town," he lamented. They had to organize themselves differently, and they had to change the conversation.

That two-martini lunch led to an impromptu summit of the biggest players in the industry at a place that was an easily accessible middle ground between East and West, and that happened to be in Ned Heizer's backyard: the United Airlines Red Carpet Room in Chicago's O'Hare Airport. Out of that strategy session came a new, grandly titled trade group, the National Venture Capital Association, or NVCA. Not everyone agreed that politics should be the NVCA's business; one member cheerily suggested that the group didn't need to organize anything more formal than the occasional golf outing. But David Morgenthaler didn't think that way. When Heizer called for volunteers to go to Washington to lobby for a tax cut, he stepped right up.[13]

Morgenthaler was only a few years into his second career as a venture capitalist, and he had been enjoying it immensely. He was in a different spot than the McMurtrys and the Greenfields of the world, having enough personal wealth in the bank to not worry quite as much about raising outside money. With a career that ran from the New Deal through the heyday of the military-industry complex, Morgenthaler was old enough to have an appreciation for what government intervention could do. Yet he'd been a corporate executive long enough to have a strong distaste for excessive regulation and high taxes. By the early months of the Ford Administration, Morgenthaler was regularly hopping on the short flight from Cleveland to pound Washington's marble hallways. Pete Bancroft flew in from San Francisco to join him.

It didn't go very well at first. The two didn't understand the Byzantine hierarchies of the committee system, the legislative dance, and the fact that getting a meeting with a Senator or Congressman was no guarantee of getting legislation introduced. They didn't have high-priced lobbyists; they did most of the talking themselves. They looked on helplessly as lawmakers pushed the rate up even further in 1976. The wealthiest now had to pay capital gains taxes of nearly 50 percent. "We were so green back then," Morgenthaler recalled later. "We sort of mentally associated ourselves with the big boys from New York, until we got there and we found the congressmen were seeing us as these nice little guys from the country."[14]

As it turned out, however, the country boys had landed in Washington at exactly the right time. The post-Watergate moment was all about reforming the system and embracing outsiders. The president had resigned. Manufacturers sheared off jobs. Big, liberal cities like New York were on the verge of going bankrupt. The

Keynesianism that drove generous spending and high deficits from Roosevelt's New Deal through Eisenhower's military-industrial complex had abruptly fallen out of favor. Austerity politics—cutting taxes, cutting spending, balancing budgets—gained traction.[15]

With the new mood, the feds became more interested in how high-tech enterprises could give the economy the boost it needed. In early 1976, Ford's Commerce Department released a study championing "technical enterprises" as powerful job creators, finding that young high-tech firms had job growth rates that were nearly forty times that of "mature" companies. Setting aside that the sample was tiny—only sixteen companies in all—and that it perhaps wasn't surprising that tech jobs were growing like wildfire during the early-'70s microchip boom, Ford's team saw a juicy election-year talking point.[16]

Morgenthaler and Bancroft had found their opening. If tech companies were job creators, then Washington must act to fix the horrible climate for venture investment. It was time for the NVCA to issue a white paper of its own. *Emerging Innovative Companies—An Endangered Species* landed on the Hill in November 1976, just a few weeks after another self-styled entrepreneur and Washington outsider, Jimmy Carter, won the presidency. Pete Bancroft had written most of it. Morgenthaler made sure every congressional office had a copy. The Republican electoral loss didn't deter them. They had high hopes for the Georgia peanut farmer, a centrist reformer who'd sung the praises of small business on the campaign trail, and they took pains to point out that tech companies were small businesses too.

"It is the successful new small company of today which becomes the important innovative company of tomorrow," the report declared. "They keep America at the cutting edge technologically and help our foreign trade balance." Six months later, they published a follow-up featuring a comprehensive tax plan to combat the "erosion of capital investment" that was killing innovative companies. Just as bankers had been doing since the 1930s, the papers made a crisp argument for the supply side. Any investments made through tax cuts would be earned back in new jobs and "increasing revenues in the form of income taxes"—billions more, in fact, than from the "mature companies" of the manufacturing economy. America's economic future depended on its high-tech companies, and tax cuts were the way to make sure these companies would thrive.[17]

DINNER IN WOODSIDE

Silicon Valley's little Galapagos hadn't been much involved in these first missions to the marble halls of Washington. The lobbying business wasn't an easy sell for engineers-turned-chipmakers who considered themselves a breed apart from stodgy old-economy types. WEMA, the trade association co-founded by Dave Packard back in the 1940s, remained focused mostly on military procurement policies—something with little relevance to the firms now ruling the Valley. Even though business lobbying was surging in the 1970s, WEMA had only just hired its first Washington representative. Headquarters remained in Palo Alto, three thousand miles away from the corridors of power.[18]

That suited the Valley just fine. The chipmakers were highly skeptical of people who looked to government for help. Beleaguered Detroit automakers might be going hat in hand to D.C., but the Valley wouldn't ever stoop to that. Sure, they gave money to Republican candidates when Packard asked them to, but National Semiconductor's Charlie Sporck summed up the sentiments of a fair-sized minority when he declared: "I was anti-government and viewed all politicians as a bunch of bastards."[19]

The one exception to that rule might have been their Republican Member of Congress, Pete McCloskey. And that was largely because McCloskey didn't give a rip about what the Washington establishment thought of him. The medal-draped Marine had been an early, vocal critic of the war in Vietnam, going so far as to run against President Nixon in the 1972 primaries as a protest against the bombing of Cambodia. He was an ardent environmentalist, hewing to the center-left at a time when the party had begun a sharp tack to the right. Governor Ronald Reagan and state party leaders tried and failed to find someone to challenge McCloskey in 1972 and 1974. In the aftermath of Nixon's resignation, the congressman seemed a little less traitorous, but he still couldn't get a decent committee assignment.[20]

Despite his strong history in the Valley, however, the earthy Irishman wasn't a tech guy. He was a lawyer. Wilson Sonsini had become a tech powerhouse after he left, and he hadn't spent much time on economic issues during his stint on the Hill. This needed to change, thought Reid Dennis.

Unsurprisingly for someone who'd been organizing the Valley VC pack since the start, Dennis had been among the first to join the NVCA. Before too long, he'd serve as the organization's president. And, being in the home of silicon chips, he had a clear view of another looming high-tech threat: Japan. While America was stagflating, the Japanese government was pumping up its economy, including making major investments in semiconductor R&D. Under their swagger, Dennis could tell that his politics-averse friends in the chip industry were getting nervous.

Everyone knew everyone in the intimate world of the early Valley, but Reid Dennis had known Pete McCloskey for an especially long time. They had been Stanford classmates two decades earlier; they were longtime Woodside neighbors. So he decided to invite the congressman—and a few other old friends—over to dinner to talk politics.[21]

The shambolic swap meets of Homebrew were happening only a few miles down the road as the dinner party gathered around the candlelit table that warm California evening, but the contrast could not have been starker. Instead of mop-haired engineers with motherboards, there were millionaires in sport coats, all staring down the evening's guest of honor. Intel's Bob Noyce was there. So was Tom Ford, the real estate developer who was in the process of turning Sand Hill Road into the premier address for venture capitalists. The one card-carrying Democrat among the guests was Mel Lane, a lifelong Californian and publisher of *Sunset* magazine, whose glossy pages had brought the sun-drenched postwar California lifestyle to a national audience. The only woman in the room was the host's wife, who quietly glided around replenishing wineglasses and bringing in new dishes from the kitchen.

Four decades later, McCloskey still hadn't forgotten how quickly the dinner-table conversation turned confrontational. He needed to stop hugging trees and protesting wars from the backbench, and start behaving like "a businessmen's congressman." The tech industry needed tax cuts. They needed less regulation. They needed help competing with Japan. What they really needed was a reality check, McCloskey thought. "I'm a Republican in a Democratic House," he snapped. "I'm not on any important committees. Everyone in Nixon's world wanted to put me in jail." He paused, looking at the unsympathetic faces around the table. There really was only one thing he could do to help them. He could introduce them to his friends who had more-powerful committee assignments, and he could help them learn the ways of Washington.[22]

That combative first dinner party was the beginning of a new push—this time by the company CEOs themselves—to get Washington to do something about tech's bottom-line problems. Many of the executives around the table swallowed their laissez-faire distaste and began shuttling east to Washington, just like David Morgenthaler had, to lobby personally for policies that could ease the capital crunch and help tech grow.

WEMA established a "Task Force on Capital Formation" and tapped Ed Zschau to run it. After its 1969 founding, Zschau's System Industries had grown into a $17-million-a-year business. It helped that the former business-school professor could tap his old students for capital, and this gave the company the cushion it needed at the start of the lean years. But it became harder and harder to attract new investors to give him the capital he needed to keep it going. He licensed some of the technology he developed to Japanese firms, but even that wasn't enough. In 1974, it took him six months to raise $750,000. "We were thirty-six hours away from bankruptcy," he recalled.[23]

Zschau hadn't been particularly engaged in politics before, but his company's near-death experience had given him political religion. He also had an academic's appetite and aptitude for research, and developed a punchy four-page white paper making a case for the job-creating power of high-tech companies. Zschau soon had company; another candlelit dinner produced a lobbying group specifically for the Valley's chipmakers: the Semiconductor Industry Association, or SIA.

Even though he and Reid Dennis now worked side by side, Burt McMurtry didn't attend any of these dinners. He was as willing to pitch in for the industry as the next guy, but he thought his business partner was wasting his time. The market was bleak, but there were signs that it would rebound. One of the brightest signs was ROLM, a company that his old partner Jack Melchor had seeded and nurtured since its founding in 1969.

McMurtry was closer to ROLM than to any other company in his portfolio. The connection was personal—the firm's four young founders were members of the Rice Mafia, three of whom had worked under McMurtry in his final months before becoming a VC. The fourth was someone he'd tried to recruit. ROLM stood out among its peers for the diversity of its product line: its early marquee products included a "rugged" field-ready minicomputer for the military and the first fully computer-controlled telephone exchange. (Venturing into telecom-hardware territory was particularly gutsy at a time when AT&T controlled all the long-distance phone lines and forced customers to rent clunky home telephone units by the month.)

Despite these odds, the company stayed in business, growing slowly but steadily, even if it wasn't providing any gains for its investors. Needing to raise more cash, it went public in 1976—and proceeded to sell under its offering price for fifteen months. But right as Reid Dennis was gathering folks around his dining room table, ROLM's stock price started to bob up from under the water. "The market was beginning to arise," McMurtry remembered. "The early birds were coming out." In the case of Burt McMurtry, patience paid off. Ultimately, he and his partners reaped a fortyfold return on their ROLM investment, making him a fortune, and a reputation as one of the sagest VCs in town.[24]

Even though these promising signs meant that some Valley insiders weren't convinced that politicking was necessary, the true believers dove into the task with single-minded focus. Another phase in Silicon Valley's relationship with Washington had begun.

THE BATTLE OF 1978

It was Jimmy Carter who turned the capital gains tax from a boutique issue into a mainstream political debate—and it wasn't because he sided with the venture capitalists. For all his campaign-trail talk of cutting red tape and promoting small business, Carter did not prove to be the reformer of the VCs' dreams.

For one, his Administration wouldn't budge on the hated "prudent man" rule. Stew Greenfield had become the NVCA's point person on that particular issue, and he was gobsmacked by Department of Labor (DOL) diffidence about the rule's unintended consequences. Tiring of Greenfield's repeated visits, his government contacts were blunt. "This isn't what Congress intended," they acknowledged. "But our union friends are happy because the ventures you guys start are always non-union." A furious Greenfield left convinced that they "wanted to stick it to the NVCA."

Dennis and Morgenthaler met with similar resistance. Leaving DOL after another fruitless visit, the two found a sardonic *New Yorker* cartoon taped to the back of an Assistant Secretary's door, featuring two men cackling, "Venture capital! Remember venture capital?"[25]

Then, in early 1978, the president unveiled an unabashedly populist tax reform plan that went straight after the fat cats. "The privileged few are being subsidized by the rest of the taxpaying public," declared Carter, "when they routinely deduct the cost of country club dues, hunting lodges, elegant meals, theater and sports tickets, and night club shows." Nestled within the red-meat proclamations was a proposal for capital gains: not to cut the rate, but to eliminate the differential altogether, taxing the earnings at the same rate as ordinary income.[26]

The tech community sent its lobbying efforts into overdrive. The Californians beefed up WEMA into a national group, rebranding it as the American Electronics Association, or AEA. Boston tech mobilized as well, with leading mini makers helping found the Massachusetts High Technology Council and both Digital and Wang joining the Massachusetts Business Roundtable. Zschau expanded his task force to bring in more of the national contingent, and started working more closely with Morgenthaler, Dennis, and the venture capitalists.[27]

The energetic platoon bombarded the Hill with white papers. They testified at hearing after hearing that high taxes on tech were "killing the goose that lays those golden eggs." Zschau in particular displayed uninhibited zeal in his pursuit of a tax cut, going so far as to compose and sing an original ditty titled "Those Old Risk Capital Blues" and pressing cassette tapes of the recording into the hands of every Congressman and Senator. Those that listened reported back that Zschau sang in "a plaintive moan that sounds somewhere between Leadbelly and Frank Sinatra with a head cold."[28]

New and savvy Washington operators soon joined the cause. One of them was a House Ways and Means staffer named Mark Bloomfield, who would soon join a new lobbying group devoted to supply-side goals, the grandly titled American Council for Capital Formation. A long-distance runner who spent 1976 working on Ronald Reagan's bid for the Republican presidential nomination, Bloomfield had tenacity and sympathy for underdog causes, and he joined Zschau in the hunt for sympathetic legislators who could write tax bills. Before the spring of 1978 was out, they'd found one: an up-and-coming young Republican legislator named Bill Steiger, who had a seat on the all-powerful House Ways and Means Committee. And as luck would have it, Steiger also happened to be friends with Pete McCloskey.

William Albert Steiger was a pink-cheeked, spit-and-polish Republican from Oshkosh, Wisconsin, a man who appeared so youthful when first elected in 1967 that he often was mistaken for a Capitol Hill page. His 4-H Club looks masked a serious mind and prodigious work ethic, and he soon secured a plum post on tax-writing Ways and Means. Still, for his first several terms, he remained an obscure member of the minority party. He represented a rural district of dairies and clapboard farmhouses, and he seemed most interested in education and agricultural issues.

Yet he was among a cluster of rural legislators—both Democratic and Republican—who cared deeply about keeping capital gains rates low. Property-owning farmers and dairymen were entrepreneurs, too, and a good portion of their income could come from property appreciation or company stock. Steiger's own family owned a development company; he understood how a capital gains tax rate could affect the bottom line. With Bloomfield and McCloskey making introductions, Steiger soon was sold on the tech leaders' pitch. He agreed to introduce legislation to return the capital gains tax to 1969 levels.[29]

The venture capitalists and the Valley could not have found a better spokesman. With heartland credentials and friends across the Hill, Steiger was earnest, methodical, and knew how to land a political punch. He also grasped that the tax issue was as much an issue of psychology as it was of economics. "Capital gains taxes are even more important as a determinant of investment decisions than they are as a producer of revenue," he explained. "This point is fundamental, and it is one that, to my sorrow, the President's advisers fail to recognize."[30]

The bill thrust Steiger from obscurity to inside-the-Beltway fame, and it also raised the political visibility of the tech executives who rallied to Washington in increasing numbers. Since his first, stumbling visits to the Hill three years before, David Morgenthaler had transformed from a political novice into an eloquent evangelist for the entrepreneurial class. "Before 1969, our society's clear message to the entrepreneur was that it valued highly the creation of new companies with their new jobs and new technologies," he testified at one hearing that year. The tax hike had changed all that. "The high-grade, successful people who make the best entrepreneurs are not slow to understand such a message." While there was scant evidence that the general public actually cared about capital gains tax cuts—polling in fact found that Carter's tax plan had good support among voters—official Washington became consumed by them.[31]

Importantly, Democrats, who'd long considered these tax breaks as something that unfairly benefitted the rich, began to buy into the narrative. One *Washington Post* op-ed in support of the bill did a clever retake on New Deal rhetoric by calling the American investor "today's 'forgotten man.'" Underscoring the mood was the passage of the property-tax-cutting Proposition 13 in California, which ended the Golden State's expansive postwar era of public spending and infrastructural growth.[32]

Worried legislators looked at the brewing tax revolt and didn't want to be on the wrong side of history. "It's the old Horatio Alger vision," said one analyst. "People want to keep taxes on investments low in case they strike it rich themselves someday." "It's obvious," observed the *Post*'s editorial board that June, "that there has been a profound change in basic opinions about the tax structure—particularly among Democrats."[33]

The Carter Administration struck back with force. Treasury Secretary Michael Blumenthal scoffed that the bill should be called the "Millionaire's Relief Act." But other Democrats were rallying to the tax cut cause. Using legislative language helpfully supplied by Zschau's task force, Senator Clifford Hansen, conservative Democrat of Wyoming, pulled together sixty co-sponsors to pass a companion bill to Steiger's House measure. As the summer session neared its end, it was clear that the tax-cut advocates had a veto-proof majority. The White House had to back down. In the autumn of 1978, Carter signed a sweeping tax reform law that looked very different from the populist package he'd introduced only months earlier. And there within it was the work of Bill Steiger: a capital gains tax cut to 28 percent. The venture capitalists and the tech industry had won, and their emissaries had become seasoned political operators in the process.[34]

Official Washington learned a potent political lesson in 1978. High-tech business was now a virtuous, even populist, political actor. They would score no points by appearing to stand in the way of its growth. Then, the following year, another tech-industry ally, Texas Democrat Lloyd Bentsen—"a gem," declared

Stew Greenfield—wrangled the DOL to loosen its interpretation of the "prudent man" rule. The firehose cranked open. Between 1978 and 1980, a staggering $1.5 billion in additional venture investment flowed into the tech world.[35]

The tech industry looked back on 1978 with the utter conviction that the tax cut was the catalyst for the high-tech boom. The same players continued to lobby an even lower capital gains rate throughout the 1980s. But was the capital gains tax rate the decisive factor? Tax cuts certainly made a difference, but other factors undoubtedly fueled the fire. The economy began to sweeten. The Federal Reserve under Paul Volcker gave the economy a high-interest-rate shock treatment that began to put the brakes on inflation. And the personal computer finally reached the technological and price point that made a large consumer market possible.

Subsequent analyses have indicated that the liberalization of the "prudent man" rule actually had a greater impact on enlarging the venture capital investment pool, pointing out that investments subject to capital gains make up only a minority fraction of overall venture investment. The biggest boon that the 1978 tax cut provided may have been a psychic one, giving both VCs and entrepreneurs more confidence to enter the start-up game once more.[36]

BILL STEIGER NEVER SAW THE EXTRAORDINARY TRANSFORMATION OF THE TECH industry that was about to come. After Jimmy Carter signed the tax bill into law, the AEA threw a huge party in Palo Alto to toast Steiger and Hansen, the legislators who had carried their cause to victory. Over 400 Valley executives crammed the ballroom at Rickey's Hyatt House on El Camino; Ed Zschau was the emcee.[37]

Tragically, the California celebration turned out to be Steiger's last. Back in Washington on the following Monday, the congressman was struck dead by a heart attack. He was thirty-nine, had been a diabetic since adolescence, and was in more fragile health than his workhorse habits let on. Congress grieved en masse, lining up to pay bipartisan tribute on the House floor. "I think the sky was the limit for Bill Steiger," said a sorrowful George H. W. Bush, who had been one of the Wisconsonite's closest friends. The venture capitalists showed their gratitude in other ways, establishing a college fund to support Steiger's young son.[38]

The GOP had lost a rising star, but Steiger's wife, Janet, carried on the family legacy with a two-decades-long career as a political appointee in both Republican and Democratic administrations. She became the first woman to chair the Federal Trade Commission. There, in 1990, she instigated the first antitrust investigation in the burgeoning software industry. Her target: Microsoft.

ACT THREE

GO PUBLIC

This is the West, sir. When the legend becomes fact, print the legend.

The Man Who Shot Liberty Valance (1962)[1]

Arrivals

TUCSON, 1979

Trish Millines's mother cleaned houses for a living, day in and day out, in a New Jersey beach town whose part-time residents were white and wealthy and whose townies, like her, were mostly black and working class. Born in 1957, Millines was her single mother's adopted only child, tall, athletic, and impatient. She was an honors student in nearly everything, but especially in math. It had taken teachers a while to discover her talent, as a gangly black girl wasn't expected to have those sorts of skills. By the time the grown-ups around her figured this out, Millines was far behind the rest of the honors students and had to spend long hours after school and in the summer playing catch-up. Exhausted by eighth grade, she declared to her mother that she was not going to college. A furious Mrs. Millines responded by bringing her daughter to work, showing the scrubbing of floors and cleaning of toilets that awaited her if she didn't get a college degree. Lesson learned.

A basketball scholarship got Millines to her local school, Monmouth College, whose campus sprawled along the boardwalks of the Jersey Shore. She intended to major in electrical engineering, but the required lab classes happened at the same time as basketball practice. So she decided to go with computer science instead.

Trish Millines's entry into the tech world was far different from that of the bright young things who breathed the rarefied air of Stanford and MIT, but it was a path followed by thousands of Americans in the 1970s and early 1980s—white and black, male and female, immigrant and native-born. Computer science programs had sprouted up at plenty of mid-range public institutions like Monmouth, responding to an ever-constant demand for skilled programmers. There were more courses than there were university mainframes; to get her punch cards analyzed, Millines had to shuttle over to the more well-resourced computer labs of Rutgers. Her tech scene wasn't that of the homebrewers and computer faires. It was one of working-class kids at ordinary schools, learning a skill that could get them out of Asbury Park or Wilkes-Barre or Utica. It remained a world of mainframes and teletypes and management information systems and programming in B, of IT jobs in big companies or month-to-month contracts as coders for hire.

The personal-computer revolution was catching fire by the time she graduated in 1979. But Trish Millines lived in the world of the big machines, a young person without Valley connections. She was female, black, and gay, never having had an internship or a summer job that might open doors and lead to a real paycheck. She scanned the trade magazines, put her résumé in where she could, and ultimately found her first jobs. First came a defense contractor in the Philadelphia suburbs. Then, a few months later, came an offer from Hughes Aircraft in Tucson, Arizona.

Packing her suitcases and boxing up all her worldly possessions, Trish Millines flew west to the desert. She'd never live in New Jersey again.[1]

WALL STREET, 1980

Although he had never worked in Silicon Valley, Ben Rosen's career had intersected with the place since the start. Born in New Orleans at the bottom of the Great Depression, Rosen revealed technical smarts that propelled him westward to Cal Tech's prestigious electrical engineering program by the early 1950s. First-year graduate student Gordon Moore was his freshman chemistry instructor. Rosen later joked about his feeble grade in the class from the co-founder of Fairchild Semiconductor and Intel, "who obviously had never heard of grade inflation." The next stop was Stanford, for a master's in engineering, then off to Columbia for an MBA.

Unlike most of his straight-arrow compatriots, Ben Rosen had a hard time figuring out what he wanted to be when he grew up. He did tours of duty at a couple of big electronics companies. Then he dropped out for a bit to work on his tan and sell Frisbees on the beaches of the French Riviera. He at last returned to the world of suits and ties in the early 1970s, ending up at Wall Street broker Morgan Stanley. There, he forged a reputation as the Street's foremost electronics analyst, his star rising along with the semiconductor industry.[2]

Those companies needed someone like Rosen, desperately. The California chipmakers might be making incredible technical products, but Wall Street wasn't all that interested in hearing about it. “In our opinion,” a Merrill Lynch analyst report noted breezily in 1978, “future developers of promising technologies, new products and new services are likely to be well-financed divisions of major corporations.” That view was typical. Analysts not only didn’t promote the Valley in the 1970s, Regis McKenna remembered, “they bet against it.” In contrast, Rosen was consistently bullish about microchips, even when the industry’s growth had its mid-decade hiccups. By the end of the decade, he was turning his analysis into a business empire, publishing a must-read industry newsletter and throwing an annual semiconductor conference that had become an essential event for anyone involved in the microchip business.[3]

So when entrepreneurs out in California started putting microprocessors on motherboards and building computers around them, Rosen was about the first person on Wall Street to notice. There was a new industry here, Rosen realized, and he organized a “Personal Computer Forum” in early 1978 to showcase personal computers to his investor clients. His high hopes quickly came back down to earth. Only about twenty people showed up. Speakers nearly outnumbered attendees. “Ben,” one of Rosen’s banking clients told him, “when you have a conference on an industry with real investments, let me know.”[4]

Then, rather suddenly, things started to get real.

CHAPTER 13

Storytellers

"When we invented the personal computer, we invented a new kind of bicycle." The boldfaced headline blazed out at readers opening their copies of *The Wall Street Journal* the morning of August 13, 1980, sitting atop a full-page advertisement for Apple Computer. Below came a page crowded with print, accompanied by a portrait of its credited author, a professorially bearded Steven P. Jobs, "talk[ing] about the computer, and its effect on society."

The copy repeatedly referred to Jobs as "the inventor" of the personal computer, an artful fabrication that glossed over the fact that the elegant innards of the Apple came from the inventive mind of Jobs's media-shy co-founder, Steve Wozniak. No insider technical specs here. The ad used simple, evocative language. "Think of the large computers (the mainframes and the minis) as the passenger trains and the Apple personal computer as a Volkswagen," Jobs wrote. A Beetle might not be as powerful as a passenger train, but it could take you anywhere you wanted to go, on your own schedule. It brought liberation; it unleashed creativity. It was the future.[1]

Here was the same story that the countercultural computer folk had been hyping for years. Now, Steve Jobs was delivering it to a much broader audience of Wall Street dealmakers who thought of computers as IBM behemoths, considered word processing the work of secretaries, and who had never wielded a soldering iron nor darkened the door of a hobby shop. Even though the financial world didn't understand the technology, however, its inhabitants salivated at Apple's sales figures. The company now had upwards of 150 employees. Its 1980 revenue was about to hit $200 million—double what the *entire* microcomputer industry had been only three years before. Apple was readying for its IPO, and brokers were keen to get in on the action. "It's like bears attracted to very sweet honey," remarked one.[2]

WALL STREET AWAKENS

Much to Jimmy Carter's election-year distress, recession still gripped the American economy, but the flush times were returning to Wall Street, and tech was a big driver of the boom. At the start of 1980, analysts had greeted the new decade with some uncertainty—the year could be "either the beginning of the long-awaited recession or the beginning of the long-touted Electronic Eighties," speculated Ben Rosen.

It turned out to be the latter. Median stock gains that year ultimately topped out at 40 percent, and electronics-industry performance proved even more impressive, with high-flying stock prices that averaged 65 percent gains. Enthusiasm for tech stocks started spiking so high that some analysts were getting worried about a repeat of the irrational exuberance of the late 1960s. For the semiconductor industry, the stock surge defied economic logic. Chips were getting cheaper and the surge of demand from established industries had eased up, but, as Rosen observed, "the prices of technology stocks have done nothing but soar."[3]

A number of factors contributed to this boom. One was the logic-defying nature of the chip business itself, as the Moore's Law flywheel delivered products that became vastly more powerful as they became less pricey. Another was the river of capital flowing into the tech industry, accelerated by the confidence boost of capital gains tax cuts and newly loosened rules on pension-fund investing. It suddenly seemed as if everyone was starting a new venture fund: brokerages, established corporations, old and new VCs, and entrepreneurs themselves. Tech now made up the majority of all VC investments. "We've gone from one extreme to another," Gordon Moore remarked, a little anxiously. The rush of newcomers spawned a sardonic, and hard-to-shake, nickname for the industry: "vulture capitalists."[4]

All of this drove the buzz around personal computing, and Apple in particular. Frenetic traders ran up the stock prices of Apple's chief rivals even though no one had new product releases on the horizon. Further whetting their appetite was the emergence of a fresh set of offerings from the growing field of biotechnology, notably the Northern California–based Genentech, a firm co-founded by Stanford and University of California scientists and Silicon Valley VCs.

Although Genentech had made a profit in only one year of its four-year history, dealmakers salivated at the prospect of getting in on the ground floor of a new and hugely lucrative field. *New Issues*, a stock-industry newsletter, called the company "the Cadillac, Mercedes and Rolls-Royce of the industry rolled into one." Swirling into the soaring valuation was the U.S. Supreme Court's June 1980 decision in *Diamond v. Chakrabarty*, which ruled that new life forms created in laboratories were patentable inventions. Another factor was legislation on the cusp of being signed by Jimmy Carter, the Bayh-Dole Act, that allowed universities and their researchers to commercialize inventions that sprang from government-funded research—a move of particular benefit to the health sciences. The prospects for biotechnological commercialization made Genentech seem like only the tip of a very large iceberg.[5]

Biotechnology was profoundly different from computer hardware and software—it was far more anchored in basic research, was more tightly regulated, and had much slower product development cycles—but investors rightly recognized that the two sectors shared the same venture-capital DNA. Genentech in fact owed its existence to Eugene Kleiner and his venture partner Tom Perkins, who had adapted an "incubation" model of recruiting young associates, then giving them a mandate to hunt down promising tech and build new companies around it. When Genentech went public on October 14, trading opened at $35 a share and shot upward to a peak of $88 only an hour later. It was the biggest run-up in Wall Street history. Yet the spike was brief, and the stock ended up at only a few more dollars than its initial valuation. The IPO had made Genentech's founders rich, but it hadn't been as good to other investors.[6]

Observing disapprovingly that most Wall Street brokerages still lacked analysts with enough knowledge to properly understand either tech or biotech, "or to put proper valuations on these issues," *BusinessWeek* quickly pronounced Genentech "the perfect example of how investors can overreact to a stock." Yet company co-founder Bob Swanson recognized something else at work as well. "The market was ready for a risky, small company," he observed. Biotech "was the kind of technology that can capture people's imaginations. You had people say, 'We'll put this stock in the drawer for our grandkids. This is something for the future.'"[7]

The fever for Apple could have easily been interpreted as just another market overreaction. The personal-computer industry was younger than a kindergartener. While sales were roaring, it still wasn't clear whether these devices would be merely a passing fad. None of the newcomers had yet managed to crack the big-money office market dominated by IBM mainframes and Digital and Data General minis. They weren't pumping out commodity products like Texas Instruments or Intel.

Plus, Apple was still a one-hit wonder at the end of 1980. The II had sold by the tens of thousands, but its most recent product release, the Apple III, had fallen short of expectations. More-seasoned investors doubted it could keep up its astounding rate of growth. If Wall Street was ready for a risky little company, there were plenty of other candidates out there with comparable, if not better, prospects.

But Apple—and Silicon Valley—had something the established computer companies, and many other new-era competitors, did not. It had a great story. Straight out of Northern California came a tale that fit right in to American legend: a story of invention, of scrappy do-it-yourself entrepreneurship, of thinking different. At the same time, the story was delivered through a polished, multi-pronged campaign that positioned Apple as a serious business enterprise. It was a countercultural message that capitalists could love.

SELLING APPLE

Despite the splash of the Apple II's 1977 debut, the tech world took a little while to warm up to the company, its founders, and the business potential of micros. Digital's Ken Olsen dismissed the notion that there would be "any need or any use for a computer in someone's home." Venture backers were skeptical in 1977 too. "It may have been a while since he had a bath," thought Arthur Rock upon meeting Steve Jobs. Nonetheless, Rock was impressed by the hordes crowding around Apple's computer-fair booths as well as the fact that Mike Markkula was providing adult supervision. He agreed to invest $60,000.[8]

David Morgenthaler thought microcomputers were an iffy business proposition, and when it came to the Steves, "I was turned off by the fact that the two were pretty much kids." But Rock's involvement sparked his curiosity, and he dispatched one of his junior associates to hear Apple's pitch. It didn't go well. "They kept me waiting for a half hour," the colleague reported back. "And they were really arrogant." Morgenthaler had little patience for that nonsense. He walked away. A year later, after researching twenty-five personal-computer companies, the Clevelander realized that "those two guys were going to win" and joined a second round—at a much higher price per share—in 1978. Although the deal made a fortune and sealed his reputation as one of the industry's premier dealmakers, Morgenthaler never stopped regretting that he missed his first chance. "That's cost me a lot of money."[9]

Apple wasn't the market leader, either. It was Tandy/Radio Shack and Commodore—two companies from far outside the Silicon Valley and computer-club circuit, with no venture backing—that became the mass market's gateways into personal computerdom, initially generating far larger sales numbers than the II. Price and distribution were big reasons why. Tandy's TRS-80 was much cheaper and was widely available through the deep-pocketed Texas company's national network of Radio Shack stores. True computer nerds scoffed at the stripped-down machine, calling it the "Trash-80," but the depth of the market "knocked us off the wall," one Radio Shack executive remarked. "We're still catching up." So was Commodore, which had a backlog of thousands of orders for its humble PET, a computer that purists also dismissed as little more than a spiffed-up calculator with Chiclet-like keys.[10]

Regis McKenna knew that if Apple was going to enlarge the market for its more expensive product, the company needed to appeal to the heart, not the head. Those curious buyers wandering into Computerworld and the Byte Shop needed to be convinced that Apples were not cheap playthings, but indispensable home appliances, well worth the sticker price. The start-up founded by two college dropouts needed to look like "a large, stable computer manufacturer." So out rolled the slick, four-color magazine ads, promising potential buyers a whole new world of household efficiency and discovery. "Other personal-computer companies emphasized the technical specifications of their products," noted McKenna. "Apple stressed the fun and potential of the new technology."[11]

The two-page spread heralding Apple II's debut—that one with computer-pecking husband and admiring, dish-washing wife—was just the beginning of a stream of print ads and corporate brochures using friendly headlines to drill right into the psyches of consumers who were pragmatic, price-conscious, and more than a little apprehensive about how to use computer technology. "PR was an educational process," said McKenna firmly, "not a promotional process." Apple was here to educate.[12]

"Sophisticated design makes it simple," reassured one headline. "How to buy a personal computer," invited another. "We're looking for the most original use of an Apple since Adam," ran a tagline above a winsome photo of a handsome surfer type, naked except for a strategically positioned Apple II. It wasn't just about a better method of home bookkeeping. It was about freedom, about creation, about *revolution*. The ads also indulged in a little creative license, calling Apple "the best-selling personal computer," which only worked if you considered the TRS-80 and the PET glorified calculators. Simple, non-technical copy was essential. "Don't sell hardware," Apple reminded its partner retailers. "*Sell solutions.*"[13]

Ad placement further trumpeted the company's bold ambitions. In addition to the usual computer-geek outlets, Apple bought spreads in *Playboy*, *The New Yorker*, and airline in-flight magazines. Other companies marketed to the small demographic sliver of Americans who'd spent boyhoods in basement workshops and science fair booths. Apple went bigger: "Men 25–54, $35,000+ Household income, College Graduate+." (Even though there soon were senior women both in Apple's marketing ranks and in the offices of Regis McKenna Inc., it still did not occur to anyone to pitch the friendly little product specifically to female consumers.)[14]

Over time, the company became even bolder in its product claims as it chased the hearts and minds of upwardly mobile American males. By the time the decade was out, its magazine ads featured actors dressed as great men of history, with both imagery and ad copy hammering home the message that Apple was changing the world. "What kind of man owns his own personal computer?" asked one, accompanied by a photo of an actor dressed as Ben Franklin, delightedly gasping at the wonders of an Apple II. "If your time means money," the ad continued, "Apple can help you make more of it." Others in the series featured Thomas Jefferson, Thomas Edison, the Wright Brothers, and Henry Ford. Apples were tools for legendary men of ingenuity and innovation, and they were tools for men like you. "Don't let history pass you by," read the copy.

The twenty-four-year-old Jobs now looked nearly as natty as his PR man, having trimmed his hair, put on a sport coat and tie, and assumed the breezy confidence of a new-generation California capitalist. McKenna steered his client for one-on-ones with key reporters he'd gotten to know over his years in the business, not only the usual suspects from the science and technology beat, but also the general-issue business reporters who covered Wall Street. McKenna was particularly attentive to seeding good coverage in *The Wall Street Journal*, which the team considered a particularly important "positioning/educational tool" because it would readily print the quotes of company executives and supportive analysts, especially in industries—such as personal computing—where the reporters knew little. Over lunch or dinner, Jobs not only gave the reporters his pitch about Apple, but also provided a crash course on the history and philosophy of computing.[15]

"Man is a toolmaker [and] has the ability to make a tool to amplify the inherent ability he has," Jobs liked to say. "We are building tools that amplify human ability." First there had been ENIAC, he explained. Then came commercial mainframes and time-sharing. Now the march of progress would deliver a computer to every desk.[16]

In his accessible retelling of how computers evolved, and where they were going, Steve Jobs never mentioned government's role. The war work that had inspired ENIAC, and the supercomputers that cascaded in its wake, never figured in the story. Although Jobs often cited Bill Hewlett and Dave Packard as Apple's entrepreneurial inspiration, he never talked about the military-industrial complex that had helped HP and the rest of the Valley grow. He didn't mention the moon shots that had propelled the microchip business or the research grants that built Stanford into an engineering powerhouse. He and his micro brethren didn't mention that Lockheed still hummed away as Silicon Valley's largest employer.

Steve Jobs, the adopted son of a man with a high school education, was indeed a product of the American Dream—and a testament to what was possible amid the postwar abundance that, particularly in Northern California, came out of considerable public investment in people, places, and companies. But if the government was anywhere in his story, it lurked in the background as a dull menace, responsible for nuclear anxiety, misguided foreign wars, enforcement of an outdated status quo.

Although too young for the counterculture, Jobs confessed that the tumult of the 1960s had been personally transformative. "A lot of ideas that came out of that time focused on really thinking for yourself, seeing the world through your own eyes and not being trapped by the ways you were taught to see things by other people." And what Jobs found, as he told his rapt listeners, was that the way to change the world was through business, not politics. "I think business is probably the best-kept secret in the world," he offered. "It really is a wonderful thing. It's like a razor's edge."[17]

The ability to tell a story—and to frame America's extraordinary high-tech history—made Steve Jobs a new-model business celebrity, the premier evangelist for the personal-computer business. "He was one of us in a bigger, alien world, explaining our immature little industry and products to a much broader public than we could reach on our own," explained Esther Dyson, a junior Wall Street analyst who was one of the few tracking the personal-computer business at the time. "Our small industry had lots of its own stars, but only Steve had the charm and eloquence to be a star to the outside world."[18]

COMPUTER COWBOYS

Steve Jobs's tale of the high-tech world wasn't new, of course. He was just reflecting back what he had heard his entire life, from the halls of Homestead High School to the night shift at Atari and the show-and-tells of Homebrew. It was the latest iteration of the free-market narrative that had propelled the Valley from the start.

The mythos had only intensified as the semiconductor generation matured, and the personal-computer industry gained market velocity. Now the independent zeitgeist was topped with a dollop of high-net-worth self-satisfaction. Reporters mingling among a trade-show crowd in 1979 remarked on the Valley's "singular sense of frontierism, its self-awareness that says, 'Hey, we're the people making it all happen.'" Tech was the domain of rebels, of cowboys, of revolutionaries. It was business with an authentic soul.

The trade press for the microcomputer industry became a critical amplifier of this message in the last years of the 1970s. As the consumer base grew, there now was serious money to be made in this kind of publishing. How-to textbooks with titles like *BASIC Computer Games* sold in the hundreds of thousands, providing pragmatic and friendly navigation through the often-bewildering world of personal computing for a new set of users and neophyte programmers.

One textbook impresario was Adam Osborne, a British chemical engineer who had caught the hobbyist bug after moving to California in the early 1970s. Standing out among the shaggy Californians with his trim mustache and posh accent, Osborne became legend for his how-to manual, *An Introduction to Microcomputers*. Unable to find a trade publisher interested in so esoteric a subject, Osborne published the text himself, which proved to be a lucrative decision after eager hobbyists snatched up more than 300,000 copies. With that, he launched an entire publishing imprint devoted to the field, which he sold to textbook giant McGraw-Hill for a healthy sum.

The humble tabloids of the microcomputer tribe had morphed into much glossier trade magazines, enlarging their subscription bases along the way. West Coast Computer Faire impresario Jim Warren's *Intelligent Machines Journal* became *InfoWorld*. Adam Osborne wrote a column for it, where his acerbic product reviews and blunt takes on company prospects earned him the moniker of "the Howard Cosell of the industry." Computer-club stalwarts like Sol Libes and Lee Felsenstein became regular contributors. New titles like *Personal Computing* and *Compute!* joined *Byte* on the newsstands, all bulging with full-page ads for the newest micro models. The shelf devoted to computer specialist publications got longer. Nearly every major personal-computer company had at least one magazine or newsletter dedicated to its fans and users. These special-interest magazines also showed readers how to charge up their computing power through peripherals and software, further driving sales.[19]

It was clear from reading these magazines that personal computing was growing up. Gone were the strings of DIY software code. Instead there were stock forecasts and product announcements. But *InfoWorld* wasn't quite *BusinessWeek*, yet: the micro-generation evangelism was still on display in how reporters and editors talked about the computer and its potential, further binding machine to myth. "The personal computer operator is the Electronic Man on Horseback riding into the (sinking) Western sun," declared *InfoWorld* columnist William Schenker in early 1980. "He is the last of the rugged individualists, and the personal computer is his only effective weapon."[20]

Such big ideas also gained exposure and velocity courtesy of the popular subgenre of business books that alternately despaired about the state of the "old" American economy and made optimistic forecasts about its tech-driven future. Looming largest was Alvin Toffler, whose *Future Shock* had become a touchstone for information-overloaded Americans during the volatile 1970s. Toffler's ardent fans included Regis McKenna, who readily admitted to reading Toffler "over and over."[21]

The ideas in the dog-eared pages of *Future Shock* simmered just below the surface of Apple's cheerful ad copy and Steve Jobs's techno-evangelism. "Important new machines," Toffler had written back in 1970, "suggest novel solutions to social, philosophical, even personal problems." A decade later, Jobs shared similar ruminations: "I think personal computers will promote much more of a sense of individualism, which is not the same as isolation. It will help someone who is torn between loving his or her work and loving the family."[22]

By that point, the admiration had become mutual. Toffler, seeing the personal computer surge as evidence that his predictions were right on the mark, issued a follow-up volume, *The Third Wave*, in early 1980. In 500-plus pages bursting with gloriously operatic Toffler-isms like "techno-sphere" and "info-sphere," Toffler declared the coming of an entirely new era in political economy, fueled by computer technology. Giant institutions of the Industrial Age—including American-style big government—would become "de-massified" and diversified. The market would become one of personal autonomy, of near-infinite consumer choice. It was a future, he wrote, "more sane, sensible and sustainable, more decent and democratic than any we have ever known."[23]

THE ROSEN LETTER

McKenna could get reporters to reprint the gospel of Steve Jobs, and Toffler could persuade curious readers to buy books about a marvelous techno-future. But convincing corporate managers and investors to be bullish on personal computing required credible voices from *inside* the financial industry as well. Because it was all fine and good for a computer to be a tool of personal empowerment, but the real money in the industry came when a computer was an essential tool for *business*. Could Apple be another Digital? Another Wang? Another IBM? Wall Street wasn't sure.

Enter Ben Rosen. Despite investors' skepticism about microcomputing, the geeked-out analyst had kept up his careful tracking as the industry improved on early models with more-powerful machines. He noted with interest when Texas Instruments jumped into personal computing by November 1979, releasing the TI-99/4, a natty desktop designed to go head-to-head with the Apple II. When you added up all the peripherals, it would cost about $2,000 to turn the TI computer into something that could do a decent job of calculating your personal tax returns. Yet it showed that micros might be getting one step closer to taking a bite out of the minicomputers' small-business base.

Rosen watched closely as Tandy/Radio Shack upped its game as well, releasing a higher-end TRS-80 II, and following up within the year with three more product releases in quick succession, each with more computational oomph and wider market appeal. "Companies began to view a favorable report from Ben Rosen as the key to success in the personal computer business," observed Regis McKenna. "A bad review from Rosen was the kiss of death."[24]

Happily for McKenna, Ben Rosen reserved his five-star reviews for Apple. One of the semiconductor guys that Rosen had gotten to know through his conferences was Mike Markkula, and one of the first things that Markkula did once he joined Apple was to introduce Rosen to Steve Jobs and to the Apple II. Rosen was immediately hooked, lugging the machine Markkula had given him between home and work, because the Morgan Stanley IT department refused to buy him one. He soon became, in his words, "the self-anointed evangelist of personal computers in general and Apple in particular."[25]

Rosen brought his Apple II along when he visited his investor clients. He provided demonstrations to visiting financial journalists, earning Apple—and Rosen—valuable publicity among business readers. "Apple for Ben Rosen," read a headline in an August 1979 edition of *Forbes* magazine, atop an article that showed how the analyst used an Apple II to do his job. He talked up the company and its products inside Morgan Stanley—an establishment broker if there ever was one, representing some of the largest, best-known brands in the country. The Apple II wasn't a toy, he'd tell his colleagues, and Apple was a serious business. In payment for his loyalty, Rosen got world-class customer support. "When he didn't understand some feature of the Apple," *Time* reporter Michael Moritz noted, "he called Jobs or Markkula at home."[26]

Then, two months after Rosen's appearance in *Forbes*, a small Boston-based software company released VisiCalc, a full-blown business application for a personal computer—and they made it for the Apple II. The duo behind this electronic spreadsheet program were another example of the magic that could happen when you married engineering and finance: Dan Bricklin was a Digital veteran and Harvard MBA, and Bob Frankston was a computer scientist with multiple degrees from MIT. In a personal-computer software world dominated by space-invader games and rudimentary educational programs, Bricklin and Frankston delivered a piece of software that proved the micro could be a serious business machine. Adam Osborne proclaimed grandly that VisiCalc had finally produced something "that allows these silly little boxes to do something useful."[27]

By the time Apple's "new kind of bicycle" ad splashed on the pages of the *Journal* late that summer, Ben Rosen was sounding positively triumphant about where both semiconductors and personal computers would take Wall Street. "The market is looking ahead, well ahead, over the recession's valley," he declared. "The salutary effect of the Golden Age of Electronics will be dramatic and long term—certainly extending through the rest of this century and probably well into the next." Further boosting Rosen's optimism, fusty old Morgan Stanley eagerly signed on to underwrite the coming Apple IPO. However, the revolution had not won over all converts—yet. Regulators in Massachusetts still deemed Apple stock "too risky" and barred residents from buying it. Tellingly, the denizens of Route 128 would not get a piece of Silicon Valley's most glorious deal to date.[28]

GONE PUBLIC

On December 12, Apple made its stock market debut, offering 4.6 million shares at $22 apiece. Hambrecht & Quist joined Morgan Stanley as the deal's underwriters. Within weeks, the company had a valuation of nearly $2 billion: larger than those of Ford Motor Company, Colgate Palmolive, and Bethlehem Steel. Burt McMurtry hadn't seen anything like it. "We are living in a goofy time," he mused to a reporter.[29]

Apple's IPO enriched the Silicon Valley ecosystem up and down the line. Early venture investors like Arthur Rock and David Morgenthaler made multiple millions, as did the two Steves and their founding team. (A few years later, a fanciful drawing of the perennially media-shy Rock, wearing a suit made of money, appeared on the cover of *Time* magazine.) Morgan Stanley's delight at the results prompted the firm to dive wholeheartedly into the technology business. The broker that wouldn't buy Ben Rosen an Apple II became the favored dealmaker for Silicon Valley companies, brokering some of the biggest IPOs of the next two decades.[30]

Being bullish on personal computers—and Apple—paid off handsomely for Rosen as well. The next year, he handed off his newsletter and conference business to his young deputy Esther Dyson and teamed up with chip entrepreneur L. J. Sevin to start a new venture capital operation based out of Dallas. Sevin Rosen went on to have giant paydays with hits like Compaq Computer and the pioneering word processor Lotus, turning a $25 million fund into $120 million in three years. Amid this stunning wealth creation, other evangelists of tech couldn't bear to stay on the sidelines either. *Time* reporter Michael Moritz parlayed his early acquaintance with Jobs into unparalleled insider access to Apple, and then left journalism altogether to become a venture capitalist, joining Don Valentine's Sequoia Capital in 1986.[31]

Other journalists turned into entrepreneurs. In 1981, Adam Osborne took the earnings from his McGraw-Hill sale and started his own computer company. The machines weren't the prettiest or the most powerful, but they had one asset few possessed—they were small enough for a business traveler to fit under an airplane seat. It was the first portable personal computer. "I saw a truck-sized hole in the industry," preened Osborne, "and I plugged it."

Capitalizing on his personal brand, Osborne put himself at the center of his company's story, and didn't hesitate to draw grandiose, Jobsian comparisons. "Henry Ford revolutionized personal transportation," proclaimed one ad. "Adam Osborne has done the same for personal business computing." Clocking in at over twenty-three pounds, the "luggable" Osborne computer didn't quite live up to the revolutionary hype. Yet tech insiders appreciated its patrimony. Knowing very little about technology—he was trained as a chemist, after all—Osborne had sought out one of the Valley's best technical minds to design the machine: Lee Felsenstein. The countercultural computer had come full circle.[32]

The earlier generation of tech entrepreneurs were engineers by training, men possessing multiple degrees and many thousands of hours logged in research labs and machine shops. So too were many of the original Valley VCs. A company founder like Adam Osborne—textbook author, magazine scribe, showman—would have been unimaginable in the earlier era. But now tech was no longer only about engineering. It had become a business of storytelling and salesmanship.

It also had become something more than just computers and electronics. It was shorthand for the new economy itself. Steve Jobs landed on countless magazine covers during his lifetime. The first major one was *Inc.* magazine, October 1981. Next to Jobs's beatifically bearded face, the headline read: "This Man Has Changed Business Forever."[33]

CHAPTER 14

California Dreaming

Three things happening in quick succession in the second half of 1980—the euphoria over Apple, the stunning biotech debut of Genentech, and the election of Ronald Reagan to the U.S. presidency—marked the start of a new, and even more intense, phase of America's long fascination with California, a place of new starts, new ideas, and dreams coming true.

Reagan was another maverick turned mainstream, a candidate written off as far too conservative for the White House, but who won the day with an energetic, telegenic presence that shook off the Seventies malaise and looked eagerly forward. "Someone once said that the difference between an American and any other kind of person," Reagan declared in his first speech of the campaign, "is that an American lives in anticipation of the future because he knows it will be a great place."[1]

Silicon Valley came to embody that future. It still was one of several major tech regions in the U.S., but the explosive growth and media buzz at the turn of the 1980s had made its business culture so influential that even Bill Gates in soggy Seattle and Ken Olsen in snowy Boston could get swept up into the all-purpose journalistic descriptor of "Silicon Valley entrepreneur." Its variation on the California dream attained velocity and altitude not only because the technology had reached a critical inflection point—a computer on every desk! A video game in every living room!—but also because so many Americans wanted and needed to dream anew. For a society whose idols had been felled by assassins' bullets and sullied by corruption and scandal, the new heroes had arrived.

As both of the major American political parties embarked on projects of reinvention—the Republicans triumphantly waving the flag of entrepreneurship and free enterprise, the Democrats scrambling to redefine their party for a postindustrial age—they found the future-tense allure of the Valley irresistible.

America's political class wanted to write themselves into Silicon Valley's story, and make its California dream a model of where the nation might go next.

Yet the Valley was still small, still young, still feeling out its identity. It was unparalleled in its ability to produce tech companies, but not much else. Its denizens were outliers: engineers with many letters trailing their names, who'd spent lifetimes obsessively working on things that very few people cared about or understood. Success had come from the sharp-elbowed, inwardly-focused pursuit of building better products faster than the guy down the hall or down the street. The Valley wasn't ready to be a model for the rest of the world. "If this area has that much influence on our ideologies and our philosophies and our way of life," one resident remarked as Election Day 1980 neared, "God help us."[2]

ATARI DEMOCRATS

The same August that Apple fever seized Wall Street, pinstriped stockbrokers flipping through *The Wall Street Journal* on the morning commuter train would have encountered a front-page article about Colorado Democrat Gary Hart, who was running for reelection to a second term in the U.S. Senate. Hart was an odd cover boy for the conservative paper, as he had burst onto the scene as campaign manager for the most liberal presidential nominee in modern memory, George McGovern. Yet as the *Journal* explained, the Colorado senator was no bleeding-heart liberal. Nor was he a populist/centrist like Carter. He was a new kind of Democrat: one of the huge wave of "Watergate babies" elected in the wake of Nixon's resignation, who tended to be younger, more suburban and Western, and single-minded about changing a broken political system.[3]

Many, like Hart, represented places with thriving electronics industries. Senator Paul Tsongas was the son of a Greek-American grocer from Lowell, Massachusetts, who had seen how much high-tech enterprises had revived his hometown economy after Wang moved its headquarters there a few years before. Once rated a "perfect liberal," Tsongas believed his party was increasingly out of step with an economy that ran on specialized skills instead of union power. The future lay neither with hard-nosed conservatism nor traditional liberalism, but a middle ground of "free market forces softened by compassion." Congressman Tim Wirth was another of the new breed. A telegenic and outdoorsy Coloradan with the vibe of someone who'd embark on a miles-long mountain hike at the drop of a hat, Wirth had made a name for himself by crusading for alternative energy and the breakup of AT&T. He was sharply observant, unafraid to talk out of school, and a reliably quotable source for Washington reporters.[4]

Then there was the young legislator of chiseled jaw and resolute demeanor, Albert Gore Jr. of Tennessee. While not from a high-tech district, Gore was both a passionate environmentalist and an unabashed Toffler-style futurist. A long-range thinker stuck in a relentlessly right-now political world, Gore organized a "Congressional Clearinghouse on the Future" that met for monthly brown-bag seminars to learn about mind-stretching topics like cloning, climate change, and computer networking. The group was bipartisan: joining Gore in the room was another Toffler enthusiast and fan of very big ideas, Georgia Republican Newt Gingrich.[5]

The new wave took over state capitals as well. A bushy-haired former McGovern campaign worker named Bill Clinton became Arkansas governor at age thirty-two, in 1978. Another youthful lawyer, Bruce Babbitt, became governor of Arizona. Some started out as centrists; others moved right amid the new mood. The fierce anti-tax sentiment of 1978 swept out of office Massachusetts' liberal young governor, Michael Dukakis, replacing him with a more business-friendly Democrat, Edward King. (King was so conservative that Ronald Reagan praised him as "my favorite Democratic governor." He later became a Republican.) A chastened and distinctly more centrist Dukakis beat King in 1982 to return to office. A similar thing happened to the Arkansan: knocked out by a Republican after one term in the Governor's Mansion, he won his old job back after a move to the political center. It also involved acquiescence to the state's traditionalism: Hillary, his wife and partner in politics, changed her last name from Rodham to Clinton.[6]

High technology hadn't been on the minds of these men when they entered politics. These were the kids who ran the student council instead of building radio sets in their basements. But they, too, had been transformed by the 1960s, giving them a similar change-the-world sensibility. Lee Felsenstein and Liza Loop had bucked the establishment; these men had decided to change it from within. The group's collective enthusiasm for the high-tech sector eventually prompted the Washington press corps to bestow a sardonic moniker: "Atari Democrats." Those so named didn't like it much. "We prefer *Apple Democrats*," Wirth remarked wryly. "It sounds more American." But once the pundits got going, it was a hard label to shake.[7]

Although the rumor mill buzzed with speculation about which one of these ambitious young leaders might run for president, only California Governor Jerry Brown had been feckless enough to jump his place in line and bid for the 1980 Democratic nomination. Although Brown's celebrity ties were stronger in Hollywood than in Silicon Valley—his high-glamour posse included Warren Beatty and Francis Ford Coppola—he eagerly hitched his campaign to the industry's rising star, arguing that the nation's "reindustrialization" hinged on what places like the Valley were selling.[8]

Brown's White House bid proved flailing and short. His exit was cemented by a speech televised live from the steps of the Wisconsin State Capitol, directed by Coppola himself. Only two weeks away from winning the Best Picture Oscar for *Apocalypse Now*, the director got inventive. He placed Brown before a blue screen, projecting a background of bold images intended to echo the governor's futurist agenda. But the chroma-key technology didn't work, and on television it looked as if Brown's disembodied head was floating before the photos in the background. The subzero temperatures, a pot-toasted crowd, and Brown's awkward delivery compounded the damage. The weird display was one of the most disastrously staged speeches in American political history—a high-tech candidate sunk by fickle electronics. Political reporters gleefully labeled it "Apocalypse Brown."[9]

DEMOCRATS MAY HAVE TALKED A GOOD HIGH-TECH GAME, BUT IT WAS THE REpublicans who had the electoral edge in Silicon Valley in 1980. The Valley felt the same as voters elsewhere: Carter had been a disappointment, and it was time for a change. Regis McKenna was a lifelong Democrat, but he worried deeply about the future. "I don't believe that the present administration understands economics," he admitted ruefully.[10]

Reagan's courting of the Religious Right turned off many in the laid-back Bay Area, and the way he'd slashed public higher education spending when California governor hadn't gained him many fans in the brainy confines around Stanford and Berkeley. Yet here was a candidate who sounded like he'd deliver all the tax cuts and regulatory reforms they'd been wanting for years. Dave Packard, who pointedly had endorsed Gerald Ford instead of Reagan in the 1976 race, joined a high-profile group of corporate leaders who were raising millions for Reagan in 1980. As their political *capo* stepped up, normally reticent Valley types went on the record to commend Reagan's "clearer grasp of how to effectively stimulate the productive forces in our economy." By the eve of the election, Jerry Sanders was getting straight to the point, telling his chip-making colleagues: "I would like to make an unabashed appeal for all of you to vote Republican!"[11]

It really shouldn't have been much of a surprise. For all the hype around the countercultural personal-computer crowd, Silicon Valley remained in the hands of patriotic midcentury men who'd grown rich in the Cold War economy. The Midwest-born movie star who rode horses and cleared brush on his ranch became the perfect candidate for a group of California transplants and self-made millionaires who liked to think of themselves as cowboy capitalists. The tech industry's late-1970s adventures in Washington lobbying had only reinforced its conviction that government bureaucrats were bad for business, and deepened its faith in supply-side economics.

Plus, many people simply sat politics out. By and large, the next-generation newcomers of the personal-computer industry were too busy building things in 1980 to pay much attention to the political world, even as it paid increasing attention to them. "I've never voted for a presidential candidate," Steve Jobs admitted a few years later, without a speck of embarrassment. "I've never voted in my whole life." *InfoWorld*'s election-season commentary was limited to a single editorial cartoon. "I was going to keep track of all the candidates' significant statements," one man remarked to another as they stood in front of a computer terminal, "but there's no way to process an empty disk."[12]

FREE-MARKET REPUBLICANS

November 4 was a glorious Election Day for the Republican Party. Not only did Reagan have his landslide victory, but the U.S. Senate swung to Republican control for the first time in three decades. The next day, Wall Street had its busiest trading day in history, with more than 84 million shares bought and sold. Defense contractors could barely contain their enthusiasm. "I fixed myself a vodka martini and woke up the next morning with a fat head," confessed an executive at Southern California–based Rockwell International, whose recently cancelled contract for a multimillion-dollar bomber jet would be restored by Reagan. The win, proclaimed an ebullient U.S. Chamber of Commerce head Richard Lesher, was "the end of a forty-year period of economic liberalism where you would turn to the government bureaucracy to get results."[13]

The new Democratic generation agreed. "The New Deal died Tuesday," declared Paul Tsongas. "The old Democratic slogans and solutions aren't selling any more," said Gary Hart. "Traditional liberalism, the old pragmatic approach, is not marketable, at least in this period of time."[14]

The election of Ronald Reagan turned Washington, D.C.'s slow-brewing affection for entrepreneurs into a passionate love affair. These were Reagan's people: some of his earliest fundraisers and advocates as he entered politics in the 1960s had been self-made Southern California moguls like conglomerate king Justin Dart and nursing-home millionaire Charles Wick. Small-business people squeezed by inflation and government regulations had supported the tax-cutting, government-shrinking Reagan with zeal. In contrast, buttoned-down corporate America had thrown its lot in with nearly every other Republican candidate until it was clear that Reagan would become the nominee.

While Reagan's White House aggressively courted business support in general, his staff and supporters talked nearly nonstop about the entrepreneur as a special case, an example of a higher and better form of American capitalism, and a model for a better kind of government. "The goal of the entrepreneur is to be successful," declared Wick. "He's a risk taker." The country is "just a giant business. Other people who have run the country—social scientists—have never had to meet a payroll." It made sense for the White House to do all it could to support, and learn from, entrepreneurs.[15]

High-tech leaders weren't the same breed as the Darts or the Wicks, nor did they have much in common with Main Street shopkeepers, but Reagan paid little heed to those distinctions. "Entrepreneur" was a loose and expansive category for the Great Communicator, and a potent rhetorical strategy to advance his agenda. The entrepreneur was at the center of his idealistic framing of what America once was, and what it would again become. Americans were starting their own businesses in record numbers, and the new president was here to support it. "Two centuries ago in this country," Reagan proclaimed, "small business owners . . . rebelled against excessive taxation and government interference and helped found this nation." Now it was time for "another revolution" against red tape and high taxes, the president continued, to support the imaginative risk-takers of the new generation. A few years later, Reagan would declare that America had entered "the age of the entrepreneur."[16]

The message hit a chord with voters. Starting your own thing was risky, but it could put a person back in control after a decade of economic unpredictability. "It is you," enthused one Boston business owner, "who has your finger on the pulse of everything." Business no longer meant just pinstripes and briefcases; it no longer meant selling out to The Man. "Entrepreneurs are the rock stars of business," quipped Ben Rosen from his new venture-capitalist perch.[17]

High-tech lobbying groups seized the moment. The American Electronics Association had morphed from a dusty Palo Alto field office into a K Street force, its membership stretching from coast to coast, 1,600 companies strong. Chief lobbyist Ken Hagerty became a master of what he liked to call "grassroots lobbying," sending a regular stream of CEOs to Washington to testify personally before Congress. "Their experience," Hagerty boasted, "is something no smooth-talking Washington lobbyist can duplicate."[18]

Then there were the venture capitalists. Reagan promised more tax cuts; the VCs wanted to see him go one step further, and make the tax disappear. "Capital is seed corn," proclaimed the NVCA's then president, swashbuckling Valley venture capitalist Tom Perkins. "Capital should not be taxed."[19]

Yet the new "entrepreneur lobby" consisted of men from the earlier generation: the bankers and electronics manufacturers and established venture capitalists. They were the Burt McMurtrys and the David Morgenthalers and the Ed Zschaus; they'd been around for a while and believed in working within the system rather than blowing it up.

Meanwhile, the part of the industry that was a chief catalyst for all this political hyperventilation—personal-computer hardware and software makers—continued to mostly ignore Washington. Read *InfoWorld*; go to one of Dyson's or Rosen's PC conferences; walk into the Wagon Wheel or The Oasis: you'd hear plenty of talk of new chip designs and user interfaces and operating systems and video games. But you wouldn't hear much about politics. And that seemed to suit the new generation just fine.

NERDS' PARADISE

The swiftness with which the personal-computer market grew was truly remarkable. In 1980, Americans bought 724,000 computers from a couple of dozen computer makers. Two years later, in 1982, the market had ballooned to 2.8 million units, manufactured by more than 100 companies. In the 1981–82 school year, 16 percent of American schools had a microcomputer. That figure more than doubled, to 37 percent, the following year. Mainframes and minicomputers still moved more product and had bigger sales revenues, but the distinctly different thing about personal computers was that they were *personal*—the majority of Americans encountered these machines nearly on a daily basis, whether at home, at work, or at school. And all those sales meant lots of profit, much of it flowing into one ten-by-twenty-mile patch of Northern California.[20]

As the personal-computer boom created millionaires by the minute, journalists descended on the Valley to document its nerdy ruling class with anthropological exactitude. "Low, tasteful, glass-and-concrete buildings are sprinkled among the brown hills, like some enormous landscaped junior college," reported *Esquire*. "No smokestacks, no railroad sidings, no noise, only 'the world's most beautiful freeway' and high-tech industries with sales in the billions." Some looked upon the same panorama with barely disguised horror. Gazing at the beige tilt-ups and bland shopping centers, British architectural critic Reyner Banham saw rapacious Reaganite capitalism meeting hippie-Californian earnestness, a strange kingdom ruled by "keen, thrusting, socially responsible, ecologically aware Porsche-driving PhDs."[21]

Even middlebrow American publications could not resist getting in some digs amid the general praise. "Driving through Silicon Valley, I am flanked by a monotone sprawl of low rectangular buildings, on which corporate nameplates display fusions of high-technology words that give few clues to what goes on inside," wrote the correspondent from *National Geographic*. Hovering over it all in this intensely car-dependent place was an "opaque veil of pink-brown smog."

Outsiders responded to the Valley with the same mix of fascination and fear Americans had displayed about computers since the days of the electronic brain. No matter how alien, how relentlessly geeky, how drearily flat and full of freeways, Silicon Valley was still a marvel—a place with unmatched capacity to pump out one innovation after another, and fistfuls of dollars along with them.[22]

The personal-computer years sent what one engineer called the Valley's "predatory Calvinistic ethos" into overdrive. Eighty-hour weeks were the norm; utter immersion in work was a badge of honor. When your industry's poster boys were Bill Gates and Steve Jobs, it seemed normal that success should entail surrender of a personal life, regular bathing, and even purchasing proper living-room furniture. Workaholism had become endemic to late-twentieth-century American capitalism. In the personal-computer business, however, it gained additional punch because it rested on a total-immersion hacker culture that was about putting all else to one side to build a motherboard or write the perfect string of code. "Having friends," as one Apple engineer put it, "is orthogonal to designing computers."[23]

But the race to work the hardest wasn't mere vanity. Just as in the semiconductor industry fifteen years earlier, the pace of technological change was breathtaking. Companies had to sustain a punishing pace in order to keep their product lines from sinking into obsolescence. As always, a veneer of California casual encased staggeringly high expectations for every worker, who labored under the knowledge that they could be exiled from tech paradise at a moment's notice. "The goal is not utopia," observed one columnist in *Esquire*, "it is profits."

The big hits still hadn't improved the long odds of striking it rich in high tech, but a newly conspicuous consumption had made the gulf between the winners and the losers far more visible. Now everyone seemed like a miniature Jerry Sanders, cashing in stock options to buy outrageous sports cars. Overworked engineers stepped out to pick up their morning newspaper, saw the Ferraris and DeLoreans filling their neighbors' driveways, and doubled down on their resolve to work even harder. Technologists mumbled about being "burned out by thirty," but simultaneously held the not-so-secret hope that they'd have earned enough to retire by then.[24]

The workforce changed in other ways. The rapid growth of the tech industry and its frenetic search for skilled engineers, occurring at the same moment that corporate offices in all kinds of industries were at last opening up to women, meant that there were more female faces in the Valley than there had been in the decade before.

Technical women who had been part of the industry for decades, like Ann Hardy, worked their way up to the executive suite. Newer arrivals found opportunity in a newly consumer-facing industry that no longer demanded a technical background in order to gain entry. Enlarged marketing and PR operations became the most likely places to find a woman in an executive role. One of those early stars was Jean Richardson, a very early Apple hire who soon rose to lead marketing operations—an immensely powerful position in a company whose product was its brand, its story. Regis McKenna might have become a father figure to Steve Jobs, but he and his team reported to Richardson. There had been nearly no women at Apple when she arrived; within a few years her team was mostly female.[25]

Regis McKenna Inc., or RMI, became another important first stop for women who sought to make their careers in the Valley. Ellen Lapham was one of them, having been lured out from the East Coast to Stanford's business school in 1975 by the early buzz around microcomputers. "I saw the microworld as young, open-minded, entrepreneurial. It had a missionary spirit," she said. "The micro revolution was also a cultural revolution." With MBA in hand, Lapham landed at McKenna's shop, then jumped to Apple. By 1981, Lapham had cashed in her Apple stock and become CEO of a start-up selling Apple-powered music synthesizers. Jennifer Jones joined RMI about the same time Lapham left it. She, too, spent nearly all her time on Apple. "The '80s were the best," said Jones. "It put the firm on the map." Another member of the team, Andrea "Andy" Cunningham, started her own high-powered PR shop by the middle of the 1980s. Taking a page from Regis McKenna Himself, Cunningham's business card simply said "Andy."[26]

This was not to say things were easy. RMI executives regularly pulled 80- and 90-hour weeks and felt that they didn't always get the credit they deserved. Regis was a one-name celebrity in the Valley; they weren't. At one company event, the firm had T-shirts made up for the entire team that read "I Am Regis McKenna." It was typical Valley team-building schtick. For a group of professionals who were trying to be taken seriously on their own terms, however, it hit a little too close to home. On top of it all was the dismissive regard with which many technical types in the Valley treated marketing and the people who performed it, something reflected in the disparaging nickname that trailed the powerhouse women of RMI for years: the "Regettes."

For both technical and non-technical women, life in the Valley required a thick skin, tenacity, and a willingness to work absurdly hard. Few had glittering MIT and Stanford engineering degrees, or Harvard MBAs. They had to learn the business on the job. "You plot your own course, you work like heck, you improve yourself," said Jean Richardson, who after seven years running marketing at Apple left to do the same thing at Microsoft. "Anything's possible, if you're willing to put out a hundred and fifty percent. And you must love your job."[27]

A hotly competitive market and endless working hours meant that companies piled on the amenities to keep employees happy during their labors. Employee discontent could mean losing top talent to competitors or, heaven forbid, workers succumbing to the siren call of labor organizers and forming a union. Freshly paved jogging paths encircled Intel's San Jose campus. HP offered free coffee and donuts every morning. ROLM proudly branded itself "A Great Place to Work." Across the Valley, volleyball courts filled in the afternoon with young employees taking breaks from long days staring at computer screens.[28]

Lines between professional and personal lives blurred as each twelve-hour workday passed, and tech firms became legendary—and celebrated—for social events that were extravagant expressions of how much cash they now had on hand. Gone were the wholesome cookouts of HP's early days, when Bill Hewlett and Dave Packard donned aprons to do the burger-flipping themselves. Now, company social events were closer to the "Business is fun!" ethos of Atari, with an emphasis on beer kegs and the occasional recreational spliff. The youthful energy and casual extravagance reminded employees that they were special people, working in a special place.

JIMMY T'S POOL PARTY

Few places epitomized the Valley's new zeitgeist at the start of the 1980s more than Tandem Computers, one of the era's enormous success stories. Like Apple, Tandem had close ties to earlier Valley generations.

The man behind Tandem came from Burt McMurtry's prolifically entrepreneurial Rice Mafia. James "Jimmy T" Treybig had a career that reflected the connections still binding Texas and the Valley: after Rice, there was a short stint as a salesman at Texas Instruments, a Stanford MBA, then over to Hewlett Packard to market its minicomputers. Treybig blazed a spectacularly successful path through HP, eagerly soaking in the wandering-around managerial wisdom of Bill Hewlett. He then decamped to Kleiner Perkins to start sniffing around for ways to start his own company. By 1974, he'd found it: a new and better kind of mini, a "fail-safe computer" with a built-in backup system so it never stopped running.

Recruiting three of his former HP colleagues and persuading Eugene Kleiner and Tom Perkins to invest, Treybig started Tandem. For corporate clients like banks, who couldn't afford to have their electronic databases go on the fritz, the machine was the answer to their prayers, and Treybig's machine became a giant hit. Going public in 1977, Tandem ultimately became a massive win for Kleiner Perkins, and one of the two "early birds" giving Valley VCs hope that the Seventies slump would one day be over. (The second was the Burt McMurtry–backed ROLM.)

By the early 1980s, Tandem had become one of the fastest-growing companies in America. The way it grew drew nearly as much attention as the products it sold. Flashy and Texas-sized, Jimmy T believed in both working and playing hard. He insisted on taking a month's vacation every year and maintained a not-so-secret life as a ham radio operator. His public persona was as far as you could get from the low-key Hewlett and Packard, but he was determined to build as loyal and open a company culture as the one he'd seen at HP—and the perks he offered his workers became nearly as famous as the computers he made.

Everyone at Tandem got stock options, of course. But they also got six-week sabbaticals every four years, and company-paid vacations with their spouses. Headquarters had a swimming pool that was open all day and into the evening. The volleyball court had a locker room and showers. No one wore name tags. No one punched a time clock. Most legendary of all were the beer busts Tandem held every Friday night after work, which not only drew in thousands of employees but made Tandem's Cupertino offices a destination for others in the Valley who wanted to network and get a couple of free drinks.

"This 'people-oriented' management style emphasizes complete informality, peer pressure, and open communications," reported *BusinessWeek* in a glowing

July 1980 profile that was illustrated with a photo of a smiling Jimmy T beside the company pool. The keys to Tandem's success, remarked an executive, were "our attitude that people are responsible adults and our willingness to spend money to keep people happy." There of course was hard work expected in exchange for all these goodies. "Because Tandem's growth demands high productivity, there simply is no room here for people who cannot be depended upon," said another. "Tandem is a *society* in which everybody is important."[29]

Even if some gimlet-eyed competitiveness peeked out from beneath the happy-hour vibe, the overall picture presented was one of a capitalist utopia far removed from the grim economic realities of nearly everywhere else. Flip a few pages ahead in the same issue of *BusinessWeek*, beyond the peppy "Information Processing" section that told of the high-tech miracle workers of the computer industry, and you'd find a relentless sludge of dire news about plant closings, striking workers, and fearsomely efficient Japanese cars. The contrast couldn't have been clearer.

There was also a stark contrast when it came to the press treatment of Route 128. To be sure, there were plenty of upbeat stories about the men and machines of Boston to be found in the pages of national magazines and newspapers. And there was plenty of good economic news: after a recessionary dip at the decade's very start, the region's tech industry roared ahead thanks to the magic combination of minicomputer revenue and prime defense contracts. By 1982, well over 200,000 people were employed in tech in Massachusetts, more than in the Valley. Led by the troika of Digital, Data General, and Wang, total sales revenues approached $20 billion.[30]

Yet the way reporters wrote about Boston's tech ecosystem rarely reached the heights of bubbly effervescence that characterized so much of the coverage of the early-'80s Valley. Even referring to Boston's tech boom as "the Massachusetts Miracle" made it a triumph-over-the-odds story of a Rust Belt state that clawed its way out of crisis. It was a take that glossed over the region's longer, uniquely entrepreneurial history: of MIT and Harvard as the 800-pound gorillas of the federal science complex, of Boston as the original high-tech start-up capital, of Doriot as the first venture capitalist, of Digital and many other companies with roots in that extraordinary moment of electronics innovation buzzing outward from Cambridge in the decade after World War II. Reporters missed the point that the "Miracle" wasn't a comeback story. It was a regional tech industry that started out far ahead of the pack, and never stopped running.

Missing, too, were the personality-driven cover stories and lifestyle pieces. Route 128 lacked both Jimmy T and the pool parties. Instead, it had buttoned-down Ken Olsen and bow-tied An Wang, sober-minded engineers of an older generation. It had old-school defense contractors like Raytheon. Whether in *The New York Times* or *BusinessWeek* or *Time*, the stories written about Boston's tech titans rarely deviated from the bland industry standard for CEO profile in any sort of industry—admiring, deferential, not a lot of color. Boston entrepreneurs had built colossal, hugely influential tech companies. They employed tens of thousands. They made comparably innovative products. After all, Tandem was just another minicomputer maker, following in their wake.

They just didn't make for as good a story. And they weren't in California: the land of sunshine, land of celebrities and mavericks, land of America's future.

WORLD-CHANGERS

The seeds of Silicon Valley legend already had been sown prior to Steve Jobs's press tours and Jimmy T's beer busts. The legend had percolated through media coverage since the 1950s, of voyages to sun-drenched patios and late-night hackathons, of tales of Jerry Sanders's cars and Bob Noyce's airplanes, of odes to venture capital's "Olympics of capitalism" and the chipmakers' "high technology jelly beans." Those stories, too, were products of masterful PR, from the unflagging boosterism of Fred Terman to the artful networking of Regis McKenna. Like all enduring legends, they had power because they were rooted in truth.

Silicon Valley *was* marvelously entrepreneurial, a unique ecosystem that encouraged risk-taking and self-reinvention. It *had* produced stunning technological innovations in remarkably short periods of time, including the stunningly fast evolution of the personal computer from DIY prototype to mass consumer product in less than five years. It *was* a different way of doing business, one that proclaimed that you could make money and change the world at the same time. "To many," observed *The New York Times*, the new tech companies "embody the magic combination of progress without penalty, economic growth without upheaval—the very seeds of the industries of the future."[31]

At the same time, the Valley was less of a mold-breaking domain of mavericks than the press coverage made it seem. The business culture of the Valley had been forged in a Cold War world of crew cuts and missile tests, where audacity was rewarded in engineering, but really nowhere else. It bloomed as a meritocracy of closely-bonded men with advanced degrees, where nearly all women were wives and secretaries. The Valley had added on a layer of countercultural flair thanks to the microcomputer generation, yet the conservatism remained.

The people providing the money and the managerial "adult supervision" were products of that earlier generation as well, their sensibilities shaped by an era when electronics meant hardware (not software), when markets were other companies (not a diverse range of consumers), when corporate executives were white and male. And the Valley could talk about changing the world all it wanted, but it now had become a domain dedicated to the pursuit of money. The dreamers and do-gooders of the micro generation—the Lee Felsensteins, the Liza Loops—had largely fallen out of the picture. The new leaders of personal computing were much like the chipmakers that came before: passionately technical, ferociously competitive.

The Valley was uniquely suited to push the boundaries of the technologically possible. It was not able, or willing, to change the world.

TIME MAGAZINE NAMED THE PERSONAL COMPUTER "MACHINE OF THE YEAR" for 1982, only after first considering Steve Jobs for the honor. The switch irked the publicity-hungry young mogul, and the whole package made him utterly furious at Mike Moritz, whose editor (who usually wrote about rock stars) had heavily revised Moritz's piece on Jobs into a gossipy celebrity profile. The cover package obliquely acknowledged Jobs's grievance, but pushed against the Great Man narrative that he and McKenna had been selling since 1977. "It would have been possible to single out as Man of the Year one of the engineers or entrepreneurs who masterminded this technological revolution," noted *Time*, "but no one person has clearly dominated those turbulent events."[32]

Because a few other things had been happening on the way.

CHAPTER 15

Made in Japan

The boxy rectangle of blue aluminum weighed fourteen ounces, a little more once you put the tape inside. Add foam-eared headphones, sleek controls to play and rewind and stop, and a stubby orange button to press pause if you bumped into someone and wanted to have a conversation. The Sony Walkman was made for being on the go, for taking your music with you in a way more portable and personal than ever before. Unveiled in 1979, just as fitness-crazed members of the Me Generation surged onto jogging paths and laced up their roller skates, the Walkman became a consumer electronics phenomenon. It sold in the hundreds of millions despite its $200 price, becoming the iconic accessory for the upwardly mobile 1980s. NASA sent a specially outfitted Walkman into space. Britain's Princess Diana owned a gold-plated model. Sony's print ads announced: "There's a revolution in the streets."[1]

As the nation's sidewalks bristled with plugged-in music lovers, American electronics executives started losing sleep. The Japanese economic miracle had already upended Detroit's auto industry, pushed aside Pittsburgh's steelmakers, and hacked away at market share for RCA televisions and Whirlpool refrigerators. Using modern manufacturing methods and a great deal of automation, Japan made things at lower cost without sacrificing quality. Its workforce was famously loyal and hard-working; their productivity outpaced the Americans two to one. Hondas trawled suburban highways, sushi restaurants sprouted on big-city corners, and bookstore shelves heaved with titles describing how Japan was doing everything better. U.S. high-tech companies appeared to be the only ones who'd escaped the battering, but there were worrying signs of a coming storm. The Walkman was one of them.

The reasons for Japan's success were many, but one was its manufacturers' ability to smartly adapt and improve on existing innovations and business practices. Sony's portable stereo was no exception. Silicon Valley wasn't in the music business, but the Walkman was the grandchild of technologies that had grown there: the transistorized miniaturization of Intel, the magnetic-tape technology of Ampex, the synthesizers that created the sound of '80s pop. Yet the actual microchips inside the Sony Walkman weren't made in the U.S.A. They were made in Japan. So were the microchips that powered other Japanese products, from televisions to home stereos to newfangled things called VCRs. And they were powering an increasing number of American-made products as well.

Japanese chips were like Toyotas: cheaper, functional, abundant. And Silicon Valley's chipmakers were scared to death.

JAPAN AS NUMBER ONE

Jerry Sanders was one of those who worried. Outwardly, everything looked fine. The stock price of AMD, his eleven-year-old company, was soaring. The company had sold $225 million worth of product in 1980, about $75 million more than Apple, that darling of Wall Street. The semiconductor industry had grown its customer base by more than ten times since the start of the 1970s, and the market's future seemed limitless. "Semiconductor processing technology is today's crude oil," Sanders liked to say, "and the people who control the crude oil will control the electronics industry."[2]

And that's what gave Jerry Sanders heartburn. America still made the vast majority of the world's microchips, but Japanese firms were gaining—fast. Even worse, Japan had started hacking into the most technologically advanced end of the business. Silicon Valley had invented and commoditized 4K RAM memory chips, but now the state of the art was 16K RAM. Japanese firms now accounted for more than forty percent of that market.[3]

More ominously, Japan had started a government-funded research consortium a few years earlier to develop the next wave of super-powerful chips, cramming many thousands more integrated circuits on one sliver of silicon to create computers thousands of times more powerful than any seen before. The method they employed to do this was Very Large Scale Integration, or VLSI. Developed in the second half of the 1970s by a group of researchers led by PARC's Lynn Conway and Cal Tech's Carver Mead, the methodology simplified and standardized the process Intel had used to build its "computer on a chip," and—by making design separate from manufacturing—made it scalable. The how-to manual Mead and Conway released on VLSI became simply known as "the book" by a worldwide tribe of computer scientists, and their method had massive ripple effects on the industry. Chip design no longer required the resources and manpower of a large corporation; it now could be executed by small teams. Standardized, complex chip designs increased automation of assembly plants. Making design a stand-alone process gave rise to a whole set of companies that simply manufactured chips of others' design. Most were in Asia.[4]

Sanders had a big personal stake in the new super-chips, having positioned AMD as a merchant supplier to telecommunications companies whose equipment demanded powerful new designs. It frustrated him to no end that technology developed in California—and freely licensed to Japanese companies, he'd admit—now had propelled Japan to a commanding position in the market.[5]

Even more frustrating: thanks to the Japanese government, these competitors could seriously undercut the Valley firms on price. It was the latest manifestation of the vaunted "Japanese miracle" that had been giving America and Europe economic heartburn for two decades, and that already had brought the U.S. steel and auto industries to their knees. Long the home of cheap transistorized electronics, Japan had set out to move up the microchip value chain several years earlier using what the American press called "Japan, Inc.": the immensely powerful Ministry of International Trade and Industry, or MITI. The new VLSI consortium was one example of MITI's power to jump-start markets by huge investments of R&D dollars.

Like autos, steel, and textiles before it, semiconductor production became a Japanese "target industry," pumped up by special programs and trade subsidies designed to increase global market share. Along with creating the semiconductor research consortium, MITI heavily subsidized production and early sales of next-generation chips, enabling "forward pricing": companies would reduce prices early on, in order to capture a large share of the market. On the U.S. side of the Pacific, the strategy looked like dumping: illegally flooding the market with cheaper chips, then raising prices once a customer base had been established.[6]

Ironically, the problem originated in part from the venture-capital crunch of the 1970s, which had prompted many U.S. firms to license their inventions to Japanese companies in exchange for desperately needed cash. That had been fine, of course, when the U.S. commanded the market for sophisticated chips. Now, some of the industry's most valuable intellectual property was in Japanese hands.

Adding to the pile-on was a growing problem of theft. As microchips, and the technology behind them, became more valuable, there had been a wave of high-profile chip heists across the Valley, the stolen goods funneling into a growing "gray market" where buyers rarely asked questions about where things came from. The details were as wild as any Hollywood crime caper. Over the Thanksgiving holiday in 1981, thieves managed to get past closed-circuit cameras, motion detectors, and several layers of wire mesh to swipe half a million programmable circuits from the Sunnyvale warehouse of advanced chipmaker Monolithic Memories. Gray-marketers and foreign buyers had covert meetings in the San Francisco airport, where they traded stashes of chips for suitcases stuffed with cash. The leader of one of the Valley's biggest electronics crime rings was nicknamed "One-Eyed Jack." Firms in countries under trade embargo restrictions—East Germany, China, South Africa—became particularly eager customers, adding a national security dimension to the problem. Rumors swirled that one of the biggest buyers of stolen Valley chips was the KGB.[7]

Ever since creating the Semiconductor Industry Association (SIA) in 1977, the chipmakers had been trying to get Washington to pay attention to their problems. As had Regis McKenna, whose Pittsburgh roots had given him a keen understanding of how a place could be hugely successful in one generation and in sharp decline in the next. The Californians testified before Congress, supplied industry data to federal regulators, and even bent the ear of *Washington Post* publisher Katherine Graham when she came out to one of McKenna's carefully curated industry dinners with key journalists. Noyce, who had decided to step back from the day-to-day running of Intel, turned over the bulk of his time to the cause, commencing a perpetual shuttle to D.C. to plead the case to lawmakers. It was the hyperkinetic inventor's version of "retirement."[8]

At first, even the dash and charisma of Bob Noyce had a hard time breaking through. Politicians didn't know RAM from ROM, and it was hard to convince them to care about jobs that *might* be lost in the future when so many American jobs were disappearing *right now.* Detroit had cut 30 percent of its production. More than 200,000 autoworkers were unemployed. In October 1979, Chrysler CEO Lee Iaccoca had come, hat in hand, to Congress to plead for over a billion dollars in federal loan guarantees to keep his beleaguered company in business. Despite some reluctance—this is a "con job on Congress!" railed Florida Republican Richard Kelly—lawmakers gave Chrysler what it wanted by Christmas.[9]

In contrast, Silicon Valley was a boomtown, from the bustling assembly lines at the chip fabrication plants to the overflowing cubicles at growing personal-computer companies. It was a landscape of factories without smokestacks, without unemployment, with sunshine and opportunity. Life was so good that people had the luxury of not paying much attention to the national news. Many didn't even bother to subscribe to a newspaper. "Everyone was on a high-tech surfboard," quipped Larry Stone, a councilman in prosperous Sunnyvale, a city that had such a huge budget surplus that its leaders contemplated giving $1 million in tax revenues back to its residents.[10]

High stock valuations reflected Wall Street's assessment that Valley chipmakers presented a case of how to do American manufacturing right. Detroit's Big Three had been slow to wake up to overseas competition, and they had balance sheets weighed down by pensions and benefits for a large and aging workforce. Not in the Valley. Firms like National and Intel had offshored production early, setting up factories in East Asia more than a decade earlier. Nearly half of the industry's employee base already was overseas. High labor costs at home weren't an issue, either, as the chipmakers had unwaveringly opposed worker unionization. A bad deal for blue-collar workers made for a great deal for the chip companies, enabling them to upsize and downsize their manufacturing workforce as demand shifted.

Plus, not every American chip company felt the same way about the looming threat. Japan's surge opened up a rift between Silicon Valley and the rest—reflecting fundamentally different ecosystems and ways of doing business that separated the Californians from their competitors in other parts of the country.

The leading Valley firms had never strayed far from their stripped-down start-up roots. The idea that technology would stay within one company went against the grain. When talented engineers job-hopped, aided by those California non-compete laws, technology often came with them. It was "a structure that allows a hundred flowers to bloom," as Intel's Andy Grove put it: entrepreneurs leaving bigger firms to start their own companies. But start-ups didn't have the bandwidth to do next-generation research. The region's de facto industrial research lab had federally funded operations at Lockheed and NASA Ames and Stanford and SRI—and the government wasn't making the kind of research investments it used to.[11]

In contrast, deep-pocketed, diversified companies like IBM, RCA, and Texas Instruments had the cash to pour into research and into new plants. TI also had established a formidable market advantage after catching Japanese firms violating one of its patents in the late 1960s: in exchange for letting Japan license the technology, TI had wrangled permission to make chips there—the only American company allowed to do so. By 1980, it had three wholly owned factories in Japan.

TI President Fred Bucy, a sharp-tongued Texan, had little patience for the Californians' whining about Japanese competition. "Those fellows on the West Coast sort of have schizophrenia," he chided. Everyone had the same opportunity to get ahead of the Japanese competition. The Californians were just too slow on the uptake. "To say we shot our way into Japan is hokum." If Silicon Valley wanted to complain to Washington, it was on its own.[12]

For the semiconductor CEOs, the first months of the Reagan Administration were no better than they'd been under Carter. In fact, things seemed to get worse. "Washington was really a disaster," remembered Regis McKenna. Instead of asking questions about how and why the Japan problem might be fixed, tetchy congressmen only wanted to know "why we don't get any money out of people" in Silicon Valley when it came time to fundraise. The chipmakers couldn't believe that their corporate star power hadn't been enough to change lawmakers' minds.[13]

The more time the Valley executives spent in Washington, the more they realized what a heavy lift it was going to be to change lawmakers' minds on the issue. "The Japanese challenge the traditional politics of this country," observed one. "The Republicans are confounded by this new breed of international competitor and don't realize we need a new kind of private enterprise system where government points the way to the future. The Democrats don't know how to deal with an industry like ours that is non-union." Tax cuts weren't going to be enough. Nor were blanket subsidies for all industries, regardless of whether they were growing or shrinking—an approach gaining currency in some Rust Belt Democratic circles.

"We all believe in motherhood and the open market," an exasperated Charlie Sporck told a reporter. "I dig the laissez faire, free market approach myself but the world isn't going along with it." It was time for America to pick "target industries" just like Japan had done with MITI, providing research support and some carefully applied trade protection. It was time to have an industrial policy.[14]

AS GOES CALIFORNIA

Jerry Brown had problems too. The bids for space satellites and dalliances with Eastern philosophy had left him with a kooky reputation, amplified when he adopted a vegetarian diet and started dating pop singer Linda Ronstadt. His parsimony when it came to public spending had put him too far to the right of many Democrats on policy, and his personal image had put him too far to the left for most anyone else. The man one reporter called "a 41-year-old oddity" returned to dusty Sacramento after his defeat in the 1980 presidential primaries, in search of a new issue to define him, increase his approval ratings, and—please oh please—help him shake that "Governor Moonbeam" label.[15]

Enter, once again, Regis McKenna. Coming off another inconclusive lobbying jaunt back East, he decided to make a call on the governor to explain the dire situation of the Silicon Valley chipmakers—a problem that put all of California's high-tech miracle in a precarious state. You don't need to hear this just from me, McKenna insisted as he sat amid the plush carpet and wood paneling of the governor's office. You need to come down for dinner with the industry people.

Soon enough, all were gathered around the comfortable dinner table at McKenna's Sunnyvale farmhouse, telling the governor what he should do. It was a blending of high-tech generations, linked by the marketing guru who had made all of them into business-world celebrities: Bob Noyce next to Steve Jobs, Charlie Sporck across from Jerry Sanders, and, down the table, a rising-star software-company CEO named Sandra Kurtzig, one of the few female founders in the Valley. No surprise that Noyce and Sporck would clear space on their calendars, but only Regis McKenna could persuade the Valley's apolitical personal-computer crowd to take an evening out of their lives for dinner with the governor. Things went well. "Jerry was very smart," remembered McKenna, "and he really bought into it."[16]

It didn't take long for signs of the tech titans' influence to appear. "We can't be complacent," Brown said in his January 1981 State of the State address. "Other states are trying to persuade many of our high technology companies to expand outside California and the industries themselves face aggressive competition." Amid cutbacks and flatlined spending, he proposed a $10 million microelectronics research center at UC Berkeley, to be paid for jointly by state and corporate funds. As if to underscore Brown's warnings about interstate rivalry, Massachusetts and North Carolina soon followed with plans for similar research projects.[17]

In the autumn of 1981, right on the heels of Reagan's signing of his game-changing economic package, Jerry Brown formed a "California Commission on Industrial Innovation," appointing a Regis-curated list of bigwigs to come up with how California could get ahead of D.C. on economic policy. The commission was a curious mix: joining luminaries like Dave Packard, Sporck, and Jobs were university chancellors and labor union representatives. No Southern California aerospace moguls or Inland Empire fruit kings. Silicon Valley, it seemed, would be the place from whence all of California's future industrial innovation would spring.

It seemed like a pretty logical strategy. After all, the places that weren't Silicon Valley already were trying to figure out how to become Silicon Valley. State and local politicians' quest to replicate the region's sun-dappled magic had continued nearly unabated ever since Charles de Gaulle's tricolor-festooned limousine cruised through Palo Alto twenty years earlier. The booming fortunes of companies like Apple, combined with the increasingly desperate state of U.S. manufacturing, spurred a raft of fresh efforts to turn crumbling Rust Belt cities and foreclosure-wracked farmland into gleaming Silicon Somethings.

Across the nation, state funds were tight, but leaders eked out budgetary room to fund research parks and high-tech campuses, which mushroomed up in nearly every state. Everyone assumed that if you built it, tech would come. And that it would save the day—even if, as HP CEO John Young observed, tech was merely the nation's ninth-largest industry. "If the eight above are sick, it obviously isn't going to offset the aggregate of all of them. It isn't magic." Never mind Young's dour observations. State and local officials were ready for some magical thinking. "We're the Brooke Shields of the economy," chuckled Howard Foley, chief lobbyist for the Massachusetts High Technology Council.[18]

The comparison to America's favorite teenage supermodel was an apt one. Politicians might not know a microchip from a motherboard, but they began shifting spending priorities toward an industry that was glamorous, young, and seemed to be brimming with economic potential. And as state initiatives surged, officials started realizing that microchips were the fuel that kept the whole thing going. Now that Japan had 40 percent of the 16K RAM chip market and a staggering 70 percent of the 64K market, its threat to the American electronics industry was no longer a hypothetical.

So, when the members of the California Commission on Industrial Innovation issued their final report in September 1982, politicians had started listening.

The commission members strongly endorsed a "picking winners" approach—one based on the presumption that America's winners were, unquestionably, high tech companies. In keeping with the complicated relationship Silicon Valley had with the government that had helped create it, their language mixed celebrations of free enterprise with pleas for more-aggressive state planning and subsidy. "California shows that the spirit of risk-taking is alive and well in America," the report proclaimed, and government "must do whatever is necessary to guarantee that our cutting-edge industries—like semiconductors, computers, telecommunications, robotics, and biotechnology—retain their competitive lead." The need for action didn't stop at the state border: "This California experience must become America's experience in the 1980s. The United States must set out on a conscious path of fostering technological innovation and creativity if we are to foster increased economic growth."

Having pegged the stakes this high, the commission delivered an eye-popping list of fifty policy recommendations necessary to make this entrepreneurial juju happen, ranging from doing away with capital gains taxes to renegotiation of international trade agreements. There wasn't unanimity in the ranks—rumbles of laissez-faire dissatisfaction from conservatives like Packard tempered the more expansive calls for new spending and protectionism. But the overarching tone and ambition were clear: Jerry Brown and the Californians had decided to write a blueprint for a national industrial policy.[19]

Brown's blueprint was perfectly timed. The recession lingered; Reaganomics' big tax breaks and spending cuts hadn't delivered the boost the White House promised. The new-breed Democrats were enjoying increased airtime and attention from party elders. For pols like Paul Tsongas, Gary Hart, and Tim Wirth, an agenda that blended free-market principles with education and training strategies was an unquestioned political winner: they represented places filled with college-educated constituents, who had Apples in their family rooms and copies of *Japan as Number One* on their bedside tables. Industries such as semiconductors and personal computing presented themselves as just the kind of business enterprise this small-is-beautiful crowd wanted to see: entrepreneurial, meritocratic, and without a smokestack in sight.

The White House blanched at the sight of a Democratic governor and a bunch of barely reformed Congressional liberals shilling for corporate tax cuts and lowered regulations, and decided to get competitive about competitiveness. By the summer of 1983, the president announced the creation of his own commission on industrial competitiveness, appointing Silicon Valley bigwigs like Noyce and HP's John Young to the roster. The group was charged with addressing "the technological factors affecting the ability of United States firms to meet international competition at home or abroad."[20]

The semiconductor crew was elated by all the political attention. "We are gaining champions in Washington," crowed Jerry Sanders. "We're no longer considered Cowboy Capitalists, Boutique Manufacturers or California Crybabies." Before Reagan's time in office was out, the SIA had morphed from an odd little band of political outsiders into a lobbying powerhouse. Its member companies joined with computing heavy hitters like Control Data and Honeywell to create the Microelectronics and Computer Technology Consortium (MCC) in 1983, a Japan-like private research collaborative to push forward next-generation computer technology. Cities across the nation competed furiously to become MCC headquarters, throwing bouquets of tax breaks and public subsidies; the winner, Austin, enjoyed a hefty assist from computer mogul Ross Perot, who offered MCC's executives unlimited flights on his corporate jet as an enticement to move to Texas. Old Valley hands, familiar with Perot's tactics, looked on with annoyance and amusement. It was so predictable. Perot was not going to be denied.

Consortia bloomed in MCC's wake, thanks in good part to the lobbying muscle and political connections of the SIA. A bill loosening antitrust enforcement to allow joint research between companies sailed through a Republican-controlled Senate in 1984. A few years later, the chipmakers won approval for a landmark public-private research consortium, called Sematech, to develop the next generation of semiconductor technology.[21]

But the most striking change came from the party of the New Deal and the New Left, the party trying to scrabble its way back into the White House, the party whose brightest new stars believed the party's future depended on moving to the business-friendly political center. As the 1982 midterm elections neared, the House Democratic Caucus had decided to sponsor a report on long-term economic policy, and asked Tim Wirth to write it. The result, titled *Rebuilding the Road to Opportunity*, demonstrated how much things already had shifted. "It is up to our party to rekindle the entrepreneurial spirit in America," the report declared, "to encourage the investment and the risk taking—in private industry and in the public sector—that is essential if we are to maintain leadership in the world economy."

While including a healthier dollop of job retraining programs and other measures targeted to the Democrats' beleaguered Rust Belt constituencies, the report was remarkably in tune with what Jerry Brown's Californians had suggested. "The clearest feature of the emerging world economy," concluded the Democrats, "is that the future will be won with brainpower."[22]

AN APPLE FOR THE TEACHER

Ah, brainpower. That was another issue where politicians fretted about Japan. For the Japanese challenge wasn't just about Walkmans and microchips: it was about the pipeline of smart math and science students to one day build the next generation of high-tech products. The Japanese and Singaporeans had year-round schooling, highly rigorous curricula, soul-crushingly high standards; the Americans had struggling students, aging buildings, cash-starved school districts. But the U.S. had Silicon Valley and its high-tech wizardry. Which is how Steve Jobs ended up bending Pete Stark's ear about computers in classrooms.

Stark wasn't an Atari Democrat in the least. The California congressman from the East Bay was a staunch liberal, quick to criticize what he saw as the pretensions of the new breed. "Tim Wirth talking about economics is preposterous," he scoffed, "and Gary Hart is saying the same thing Ronald Reagan is." Yet Stark was quite literally closer than most in Congress to the center of the high-tech world: his East Bay district was next door to Silicon Valley. Plus, he was a subcommittee chairman on the tax-writing House Ways and Means Committee, giving him power that few Atari Democrats could match. Those two things brought Steve Jobs into Stark's orbit in early 1982, making a pitch for a new tax break.

By this point, Jobs had reached new heights of celebrity. Yet, as was always the case in the ever-volatile tech industry, the smooth facade covered a bumpy reality. IBM was diving into the personal-computer business. Other Silicon Valley companies were scrambling to retain their slivers of market share. Now, Apple's education-sector business became more important than ever before to the company's bottom line—and, for Jobs, a critical proof point for why Apple was a more creative, altruistic, and forward-thinking kind of company than anyone else out there. When students use computers, Jobs explained earnestly to ABC's Ted Koppel during one appearance on the wildly popular news magazine *Nightline*, they see "a reflection of the creative part of themselves being expressed." The result is "something quite democratic."[23]

There was a limit to how much the education market could grow, however. Places that had been early and enthusiastic adopters of personal computers approached market saturation (Minnesota had a staggering 97 percent of its high school seniors taking computer classes). The property tax–capping Proposition 13 passed in 1978 in California had sent that state's school budgets into free fall. Similar measures in other states choked off schools' main sources of revenue. Reagan had come into office promising to make deep cuts to the Department of Education's budget, and he had been largely true to his word. The money simply wasn't there for schools to buy computers, especially fully tricked-out Apples. The way to get more computers in schools was for the companies to give them away. But that came with a steep price tag. Didn't it make sense, Jobs reasoned, to ask Congress to help pay for something that would bring so much to so many?

Fortunately, there were plenty of people on the Hill who believed in "computer literacy" as strongly as Steve Jobs. (Al Gore cared about it enough that he had put on computer classes for his fellow congressmen.) The U.S. tax code already featured a big tax write-off for companies that gave computers to higher education institutions. It didn't seem that unreasonable to extend the break to K–12 schools.

After all, argued Jobs, "the kids can't wait." Stark was hardly one to cuddle up to big corporations, but he was readily won over by Jobs's powers of persuasion. In short order, the congressman introduced the legislation written in large part by the Apple team as "The Technology Education Act of 1982." Insiders called it by a less official title: "the Apple Bill."

Steve Jobs, the political neophyte who'd never bothered to vote, focused all his star power on getting Congress to pass the Apple Bill. Following in the footsteps of the chip-making entrepreneurs who had been his mentors, "I refused to hire any lobbyists and I went back to Washington myself," he boasted. "I actually walked the halls of Congress for about two weeks, which was the most incredible thing. I met probably two-thirds of the House and over half of the Senate myself and sat down and talked with them." It was a whirlwind introduction, but Jobs wasn't terribly impressed. "I found that the House Members are routinely less intelligent than the Senate and they were much more kneejerk to their constituencies," reported the 28-year-old mogul. "Maybe that's what the framers wanted. They weren't supposed to think too much; they were supposed to represent."[24]

The Apple Bill had enthusiastic support from the expected quarters—Al Gore testified in support of it, as did Jim Shannon, who had inherited Paul Tsongas's old House district along Route 128—but, much to Jobs's annoyance, many members expressed skepticism. Wasn't this just a giant giveaway to the personal-computer makers? Why did a rich company like Apple need a tax break? In front of Stark's committee, Jobs testily took on the critiques point by point. There were more efficient ways to increase Apple's education sales, he asserted. Plus, the bill would benefit all of Apple's competitors as well as the company itself. He conveniently didn't mention that Apple held a very large share of the education market. "Congress would be crazy not to take us up on this," he concluded.[25]

The House ultimately agreed, and the Apple Bill passed there by an overwhelming margin. Everyone was worrying about the state of American education and how to give their kids the skills they needed for the modern workplace. In a midterm election year, putting computers in schools was a good vote to have on your record.

Then—failure. On the Senate side, the bill didn't come up until a lame-duck session after the 1982 election. There it stalled, dead in the water. Stark promised to introduce it again the next year. A fuming Jobs, unschooled in the glacial legislative dance, threw up his hands. He'd given up *two whole weeks* to personally lobby for this bill, and he had nothing to show for it. Washington was as useless as everyone always said it was.

Fortunately, Apple had California. At the same time that he had been lobbying Congress, Jobs was successfully ensuring that the Apple Bill became a key recommendation of the California Commission on Industrial Innovation—and that California support a similar kind of tax measure as a backstop if the federal legislation failed. With bipartisan support in the legislature and Jerry Brown in the governor's office, the measure had become law by September. Steve Jobs may not have conquered the nation, but he had conquered its largest state.

The computers-in-education bid marked the visible entry of Silicon Valley's latest generation into politics, and many observers liked what they saw. "Apple has exercised responsible entrepreneurial citizenship—a regrettably rare quality," wrote *Inc.* columnist Milton Stewart approvingly. By the start of 1983, Apple's new "Kids Can't Wait" program—a four-person operation personally set up by Jobs—was shipping close to ten thousand Apples to California schools. Thanks to the California tax credit, the ultimate cost to Apple was about $1 million, a rounding error in a year that Apple was raking in hundreds of millions of dollars in revenue. For a fraction of what it spent each year on advertising, the company had made the Apple II the first computer thousands of California schoolchildren ever touched.[26]

STEVE JOBS MAY NOT HAVE GOTTEN HIS FEDERAL TAX BREAK, BUT ENTHUSIASM about computers in education continued to surge. For in that same spring of "Kids Can't Wait" came *A Nation at Risk*, the scathing assessment of the state of American schooling produced by President Reagan's commission on educational excellence, which memorably pronounced that "the educational foundations of our society are presently being eroded by a rising tide of mediocrity." All the educational gains made after Sputnik had been squandered, trumpeted the report. Students were lazy, undisciplined, unprepared for the global challenge before them. Things had to change, and it had to happen fast. "History is not kind to idlers."[27]

As educational competitiveness turned into a five-alarm fire, the cause of "computer literacy" became a touchstone for politicians of all stripes. Many of the problems enumerated by the commission would entail massive, structural fixes—from lengthening the school year to changing how schools were funded to (thorniest of all) rectifying the deep racial and economic inequities that persisted in American education. All of these were hard, expensive, and politically fraught. In contrast, increasing access to computers was easy, fast, and politically beloved.

Give American kids computers, and they could learn math and science as well as their Japanese peers. Teach them BASIC and have them play computer games, and they'd be ready to compete in the new economy. Put computers in struggling inner-city classrooms, and the vast gulf between the haves and the have-nots would disappear. What's more, you'd improve American education the *American* way. Computers were tools for creativity, for intellectual exploration, for access to information and independence from rote learning. The ideas that had percolated around computers and education since the first days of the PCC and the LO*OP Center now had gone mainstream—and become publicly subsidized.

Over the course of the 1980s, fired up by the prospect of computer literacy and not wanting to lag behind, one state legislature after another passed mandatory computer education programs, which meant that schools had to cough up the money to buy computers for the classroom. Realizing the opportunity presented by a large and captive market of future customers, computer companies began offering schools huge discounts in the hopes that their products would become the platform of choice. But Apple remained king. The presence of so many Apples in the schoolhouses of California reinforced its dominance of the education market. By the 1985–86 school year the company provided 55 percent of the close to 800,000 microcomputers in American classrooms.[28]

Yet computers weren't the magic elixir for American education. In fact, they reflected and magnified some of its biggest problems. In a highly unequal and socioeconomically segregated system, some schoolchildren got the gift of computer literacy sooner than others. Districts that were poorer and those with large minority enrollments received computers later than majority white and affluent districts, and training and software lagged behind as well. What's more, in districts rich and poor, the rivers of public money putting computers into schools rarely came with adequate training programs for teachers.

In California, training came via a rather cursory series of seminars offered at independent computer stores. "The donations are coming whether the schools are ready or not," reported *The Los Angeles Times*. "Some principals report that their donated Apple computers remained stored in their boxes for months because they didn't have trained staff, anti-theft devices or the money to buy educational programs." Students who already were computer-savvy made the school computer lab a favorite after-school hangout. For those who didn't already have access to computers and video games at home, computer time wasn't unleashing creativity—it was an exercise in frustration. "To err is human, but to really foul things up requires a computer," quipped one.[29]

The greatest beneficiaries of the whole business seemed to be the computer companies and software makers who had tapped into a large and newly computer-literate group of educated citizens. The shortcomings of the rush to computer literacy were abundantly clear to many on the front lines of American education. World-weary teachers hoped that the tech craze would abate so "we can go back to real teaching." But it never did.[30]

Three decades later, the seeds planted by Steve Jobs's brief adventure on Capitol Hill had blossomed into a giant, multibillion-dollar business. As one reform wave after another washed over America's educational policy landscape, the one constant was bipartisan agreement that computer literacy and student access to tech were essential to twenty-first-century schooling. Tech giants made their presence known in nearly every American classroom, as cash-strapped states and school districts eagerly accepted discounted software from Microsoft, free cloud apps and Chromebooks from Google, e-readers from Amazon, and—yes—iPads and iMacs and iEverythings from the PC era's original ed-tech champion, Apple.[31]

CAPITOL GAINS

As the chipmakers gained traction and Steve Jobs evangelized for computers in education, the venture capitalists who'd long been championing high-tech entrepreneurship on Capitol Hill were busier than ever. Leading the pack was one of the most familiar figures from the Carter-era tax fight, now an elected official ready to out-Atari the Ataris: Republican Congressman Ed Zschau, who arrived on Capitol Hill in early 1983.

The blues-crooning CEO had caught the political bug in the Steiger Amendment fight, and he'd never looked back. When Pete McCloskey left his seat to embark on a failed bid for the Senate, Zschau waltzed in, the favorite-son Republican in a district engineered to stay a GOP seat. Zschau was the kind of candidate that district voters liked. He wasn't a conservative like Reagan and all those defense hawks down in Orange County. He was a millionaire on a motorcycle, commanding Sand Hill Road boardrooms and Stanford classrooms, equally at ease at the Wagon Wheel or The Oasis. He was one of them.

Silicon Valley's congressman approached high-tech advocacy differently than the Atari Democrats. People like Tsongas and Wirth might represent high-tech districts, but they'd never run high-tech companies. Their vision of industrial policy was big-government liberalism dressed up with a silicon exterior. They were frantically trying to copy Japan without appreciating the Silicon Valley miracle already in their midst. For Zschau, the answer wasn't an elaborate scheme of subsidies and training programs; it was in keeping the cost of capital low and opening up export restrictions so American companies could compete freely. "One of my great concerns is that in the mad dash to show concern for high technology, somebody will screw it up," he told one reporter earnestly. "I come from an industry that has grown like Topsy and has never asked for government help."[32]

True to form, Zschau dove into his new job at full force, working furiously to get his message heard despite being a freshman member of the minority party. He went to the hearings that no one else attended, he sat in on committee meetings and took copious notes. He got the band back together, joining David Morgenthaler and lobbyist Mark Bloomfield in a "Capital Gains Coalition" to argue for another tax cut. Now they had a much easier case to make, and they'd won over converts on both sides of the aisle. Paul Tsongas had voted no on the Steiger Amendment, but he had been so impressed with its aftermath that by 1984 he was declaring that the "bill had more to do with the economic renaissance of my congressional district than I have."[33]

Zschau might have been a rookie backbencher, but his tireless advocacy made an impression. "If he has done nothing else," one GOP colleague observed, "he now has Members of Congress pronouncing it 'Silicon' instead of 'Silicone' Valley, and I think many of you will recognize that as quite an improvement."[34]

A new face joined the crowd too: Burt McMurtry. By the time Ronald Reagan entered the White House, McMurtry commanded one of the most valuable portfolios in the Valley, surging to more than $200 million by 1983. The Steiger Amendment victory had changed his mind about politics, so much so that he spent a term as the NVCA's president and became one of the most plugged-in Valley financiers in 1980s Washington.

Shortly after Reagan took office, McMurtry started shuttling groups of legislators and executive-branch officials out to the Valley, where he'd lead them on tours of his favorite portfolio companies. He was particularly fond of taking them to a start-up called Sun Microsystems, founded by three Stanford graduates to build a powerful mini-meets-micro hybrid called a workstation. Co-founder Scott McNealy—"such a showman," remembered McMurtry—never failed to wow his audience. "Who told you that you could do this?" one Senator asked the entrepreneur in wonderment after he learned about Sun's products. "How did you get permission?" McMurtry shook his head ruefully. "It was a perfect example of the disconnect between D.C. and Silicon Valley."[35]

Despite this, the Texan remained impressed by the political world. He hadn't been a big fan of Reagan at first, but when he attended his first small gathering of business leaders at the White House, the president was charming, engaged, and "ran that meeting as effectively as any CEO I had ever seen." With access to the White House and energetic advocates in Congress, venture guys like McMurtry and Morgenthaler and right-leaning CEOs like Young and Noyce won another capital gains tax cut, further tax credits for R&D, and more. Steve Jobs and Regis McKenna might be playing footsie with liberal Democrats, but the money men and the chipmakers were mostly sticking with the GOP, and it was reaping dividends.

AMERICA'S MITI

While outside observers (then and now) tend to think of Silicon Valley as an undifferentiated unit, and "the tech industry" as a singular noun, the Japanese challenge and the politics that spilled out from it made it clear that there were multiple Valleys, and multiple tech industries, existing side by side. And in the early 1980s, the business interests of the different parts of Silicon Valley were not aligned. The semiconductor crowd wanted trade restrictions so that their more expensive chips could regain market share. The personal-computer makers like Apple relied on cheap chips to keep their prices competitive.

All the Silicon Valleys were united in their desire to pay as little in taxes as possible, but their tax-cutting priorities were different. Capital gains remained the chief preoccupation of the older generation, but you'd rarely find anyone from the younger crowd adding their voices. The Bob Noyces and Burt McMurtrys lobbied lawmakers in the road-tested manner practiced by nearly every other industry: visits to Congressional offices, blizzards of issue briefs, hours of testimony, sitting on presidential commissions. The Steves and Bills, only a few years removed from phone phreaking and stealing computer time, had neither interest in nor need for that kind of politicking.

Regis McKenna—*consigliere* of the new generation, a lonely Democrat in a strongly Republican place—tried to get the personal-computer crowd to engage in good old retail politics. He threw fundraisers for Gary Hart's 1984 presidential bid; he played matchmaker between his industry friends and the new generation of tech-friendly politicians. But it was hard to get traction in the middle of the Reagan Revolution, no matter how many Atari Democrats popped up on Capitol Hill. Jobs's misadventures with the Apple Bill hadn't helped. And having "Silicon Valley's Congressman" hard at work in D.C. made little difference to McKenna. He and Ed Zschau had very little patience for one another, both being strong partisans with very different ideas about what Washington needed to do to support the tech industry.

SO WHATEVER HAPPENED TO INDUSTRIAL POLICY? JERRY BROWN PUSHED IT; Atari Democrats couldn't stop yapping about it. Republicans like Ed Zschau could barely stand the idea, even while agreeing that something had to change. Japan was eating the United States' lunch because of MITI. Yet the idea of another MITI made members of both parties uneasy. MITI was winning the war for chips and refrigerators and Walkmans because it closely controlled the national economy, subsidizing markets, choosing favored producers, and most definitely picking winners. Japanese "capitalist developmentalism," as economist Chalmers Johnson dubbed it, just wasn't America's style, and certainly not in the Reagan era.

What was in style: defense spending. The Reagan Revolution swung the military-industrial complex back into gear, waking up the beast of defense contracting that had lain dormant since the drawdown of the late Vietnam era. Austerity might have ruled the day across most of the executive branch, but not within the Pentagon, where budgets swelled with every year Reagan was in office.

And just like in those months after Sputnik, the defense surge involved a great deal of money for university research. But this time it wasn't for missiles and rockets: it was for supercomputers and AI and scenario modeling and cybersecurity. ARPA had added on a "D" for defense and become DARPA, yet it remained the Lady Bountiful of computer science. In 1983, as Democrats rebuilt the road to economic opportunity and Ed Zschau and David Morgenthaler urged another capital gains tax cut, DARPA began a push for a new program: one to meet the national security challenges that the Japanese economic threat represented. And it was all about computers.

September 1982—the same month that Jerry Brown's innovation commission issued its fifty-point plan for the economic future—the people of DARPA released "A Defense Program in Supercomputation from Microelectronics to Artificial Intelligence for the 1990s." The title was clunky, the content dry, the public awareness nonexistent. Half a year later, the unveiling of the program the report inspired—the Strategic Computing Initiative, or SCI—barely made a ripple in the press. But the proposed budget was hefty: $650 million, more than the entirety of the DARPA budget only a couple of years before. And its pedigree was unmatched. DARPA recruited PARC's dynamic and detail-oriented Lynn Conway, co-creator of VLSI, to help steer the program. Contracts went out on a sole-source, noncompetitive basis to the most elite departments in computer science: MIT, Caltech, Stanford, Carnegie Mellon.

Japan was working on its "Fifth Generation" computing program, an ambitious push for supercomputing, AI, and machine learning. DARPA's new push aimed to go one better, and to harness the power of cutting-edge computing to more ordinary defense challenges along the way. SCI's opening agenda included projects to build autonomous vehicles, a computerized copilot for fighter jets, and AI software to aid in battlefield decision-making.

The program's first months were rocky, to say the least—Conway and other Class A technical talent hightailed it out of there quickly, and scholars had decidedly mixed feelings about the program's military aims—but the computing push ultimately had a tremendous impact. Between the early 1980s and the mid-1990s, federal funding for computer science research more than tripled, to close to $1 billion per year flowing to academic laboratories. The Pentagon was the largest funding agency by far, supporting advanced research in supercomputing, chip design, and AI. All told, federal grants accounted for 70 percent of the money spent on academic research in computer science and electrical engineering between the mid-1970s and the new millennium. The majority of that federal total came from DARPA.[36]

Government money turned the academic halls of computer science into the seedbeds of Internet-era innovations in search and social networking and cloud software, not to mention funding a generation of computer-science graduate students who would go on to design the future of software in Silicon Valley and beyond. Politicians continued to argue about what path was the best one to take when it came to countering the 1980s challenge of Japan. But as they squabbled, the military brass and the bureaucrats decided the matter for them.[37]

America's MITI would be DARPA, and the work it performed over the next decade not only put the U.S. back in the lead in the global high-tech race. It also made Silicon Valley richer and more influential than ever.

CHAPTER 16

Big Brother

While the leaders of DARPA were thinking about the computers of the future, however, the people of Silicon Valley were preoccupied with the computers of the present. For the ominous Borg that most Valley tech folk worried about in the first years of the 1980s wasn't the Pentagon. It was IBM.

Conventional wisdom long mocked IBM for being asleep at the wheel when it came to the personal-computer revolution. In reality, Big Blue had been testing minicomputer prototypes nearly as long as the Apple II had been on the market. To be sure, the antitrust suit brought in the waning hours of Lyndon Johnson's presidency had dragged on into the Reagan era (it would be dropped by the DOJ in 1982), and the battle had sapped IBM's energy and constrained its ability to enter new markets. But by the summer of 1980, as Apple mania seized Wall Street and VisiCalc demonstrated that personal-computer software could indeed have business applications, the IBM brass had agreed to go forward with a personal-computer product, christening it with a weighty name: The Manhattan Project. Needing to move fast and do things relatively cheaply, the managers in charge did exactly what Vannevar Bush had done forty years earlier. They outsourced, trekking out to all corners of the country in search of potential hardware and software partners.

Up in the evergreen upper-left, they found Microsoft.

BEYOND BASIC

After Altair went into its tailspin, Bill Gates and Paul Allen had moved Microsoft back home to Seattle in 1979 (dropping the hyphen in its name along the way). The founders had first contemplated going to the Valley, but Allen had argued against it. Bay Area engineers job-hopped too frequently. The cost of housing was too high. Seattle might have gloomy weather, but its rain would keep everyone in the office, working like maniacs.

Relocation northward coincided with a boom in sales, as more and more companies jostling for position in the personal-computer business started bundling Microsoft BASIC with their machines. Microsoft's enviable sales numbers soon attracted the attention of H. Ross Perot, who offered to buy the software company for a considerable sum (Perot later asserted it was upward of $40 million; Gates remembered it as significantly less). The young mogul had entertained the possibility seriously enough to go to the trouble of getting a haircut before the meeting with Perot, but he ultimately said no.[1]

By the time IBM knocked on its door that June, Microsoft was on its way to selling over $7 million in software and had forty employees. Like so many other high-tech founders, Gates and Allen had plunged everything into their company, making its success a 24/7 pursuit. Bill Gates had the same approach to vacations as Fred Terman: the only break he took from Microsoft's total immersion was one week per year at an intensely competitive tennis camp.[2]

Gates was still owl-eyed and skinny, but his self-assurance had grown along with his sales numbers. He wasn't surprised that IBM was coming to call—his software was becoming the global standard, after all—but he was excited about the possibilities of scale that a partnership with Big Blue might bring. Only a few weeks earlier, he'd persuaded a friend of his from Harvard, an exuberant Detroiter named Steve Ballmer, to drop out of Stanford's MBA program and take on some of the operational duties of his growing company. Microsoft was a land of sweatshirts and rumpled khakis; Ballmer was, as Gates put it, "a good suit-type guy." Gates asked his more presentable colleague to join him at the table for the IBM pitch.[3]

Ballmer may have had the suit, but Gates ruled the room. The twenty-four-year-old reveled in the technical details, and IBM didn't awe him. So many of the early tech billionaires had grown up in families without connections or wealth. Gates was different. He was the grandson of a banker and the son of a successful lawyer. His mother, Mary, was currently chair of the board of the national United Way—alongside none other than IBM's new CEO, John Opel. (When an IBM executive mentioned the Microsoft deal to Opel at a later stage of the negotiations, the CEO responded, "Oh, that's run by Bill Gates, Mary Gates's son.")[4]

Gates also remained steadfast in the conviction that had propelled him to write his famous letter to the Homebrew Computer Club several years earlier: that computer software should be a stand-alone intellectual and commercial product, as valuable as the hardware itself. The computer industry that IBM had defined and dominated for so long was a full-stack business, with hardware and software developed and sold as one. Apple had carried on that model as well. Now, in their haste to get a competitive microcomputer to market, IBM executives were willing to break up the stack, to do things another way—Bill Gates's way.

In that Microsoft conference room on that June day, the conversation soon expanded well beyond BASIC. Big Blue was interested in Microsoft's whole line of computer languages, and it also wanted Gates's help in finding the right operating system for it. In a chain of events that ultimately became legendary in computer circles (and whose disputed details became another proof point for Silicon Valley's later dislike of all things Microsoft), the IBM team tried and failed to make a deal with California-based developer Gary Kildall, designer of the operating system, CP/M, that seemed poised to become the market standard.

As the whole plan teetered, Gates swooped in, adapting an OS from another Seattle-based company into something he called MS-DOS. Kildall called it a clone of CP/M. Many years and lawsuits later, Kildall struck his own deal with IBM for his operating system, but by that time Microsoft had cemented its lock on the IBM personal-computer universe. The river of money to be made in software flowed northward to Seattle.[5]

SILICON VALLEY NORTH

Even as Microsoft and IBM continued their negotiations, investors had started vying for Bill Gates's attention. Microsoft was still a partnership between Gates and Allen; it hadn't even incorporated. As Burt McMurtry remembered it, by the autumn of 1980 Bill Gates was "having his door beat down by investment bankers" and others who hoped to get an early stake in the company. By that time,

McMurtry and Reid Dennis had parted ways, and the Texan had a new firm—Technology Venture Investors, or TVI—and a new set of partners. It was the youngest of them, a thirty-year-old rookie VC named David Marquardt, who managed to squeeze through Microsoft's door.

Gates wasn't dazzled by the wing-tipped Wall Street types with their term sheets and promises of spectacular stock valuations. Getting rich didn't interest him; he had an indifference to net worth that was a privilege of people who had grown up in affluent families. He cared about the technology, and he liked people who were "smart"—and by that he meant "technically smart." Plenty of his parents' friends were eager to invest, but hometown money came from timber and airplanes—not tech. Gates wanted technical smarts along with that cash, and that only could come through connections into the Valley's high-tech ecosystem.[6]

This is where Dave Marquardt stood out. He was a tech geek too: a mechanical engineer, a former Homebrewer, who understood compiler design as well as capital markets. Over many months of courtship, which sometimes involved accompanying Gates and his parents to University of Washington football games, Marquardt and TVI put in $1 million to secure a 5 percent stake in Microsoft. Marquardt became a Microsoft board member, staying on for more than thirty years.[7]

Marquardt's investment and mentorship became only one of several important threads—both collaborative and antagonistic—connecting Microsoft and the Valley. Not too long after Marquardt showed up in Bellevue, he introduced Gates to an intense and energetic computer scientist from Xerox PARC named Charles Simonyi.

Like Intel's Andy Grove, Simonyi was a Hungarian émigré, the son of a professor of electrical engineering in Budapest. He had escaped the oppressive regime of his native land as a teenager, first to Denmark and then to Berkeley and to Stanford, then to PARC. Despite Xerox's continued difficulties in commercializing PARC's marvels—"like one of those poor lottery winners that blew their millions," Simonyi said ruefully—the lab had continued to be a mothership of extraordinary invention, and Simonyi had racked up a string of important accomplishments during his time there. He had developed the "killer app" for the legendary Alto that allowed the characters typed into a computer to appear on the screen just as they might on paper. Further refinements using icons and drop-down menus turned a laborious editing process into an easy point-and-click. The graphical user interface—or GUI—was a huge breakthrough, and everyone in the industry knew it. The previous year, Simonyi's team had given a little demonstration of the software for an intensely interested Steve Jobs (who had persuaded Xerox to, in his words, "open the kimono" by allowing the photocopying giant to make a $1 million pre-IPO investment in Apple).[8]

Simonyi had become restless at PARC, and he'd heard interesting things about what one friend called "this crazy guy in Seattle." It only took one conversation for the Hungarian to realize that he'd found a high-tech soulmate, someone who could get just as animated and utterly consumed by the technological future. By Christmas, Simonyi had joined the Microsoft payroll, becoming what he later described as "the messenger RNA of the PARC virus" and bringing a whole new universe of software possibilities to Microsoft.[9]

Another piece of connective tissue between north and south: Intel, whose chips IBM had chosen to power its new PC machines in another disruption of the full-stack model. Intel was far more well-established and deep-pocketed than Microsoft at the time, of course. Yet for a company then locked in fierce fraternal battles with its fellow chipmakers and buffeted by a seemingly unstoppable Japan, this stake in the future of the market was transformative. Thanks to the chips inside all those PCs, Intel's revenue nearly doubled between 1980 and 1984, to more than $1.6 billion. By 1990, revenues were close to $4 billion, and rose from there as the "Wintel" empire of Microsoft Windows–platformed, Intel-powered machines took over the desktop market.[10]

One of the most enduring—and fractious—partnerships between Seattle and the Bay Area was, of course, that between Microsoft and Apple, and Bill Gates and Steve Jobs. As their companies and personal fortunes soared, the two mop-haired boy wonders became entrepreneurial celebrities like none seen since the era of Henry Ford, and their intense rivalry became fodder for countless magazine articles, books, and movies-of-the-week. Steve Jobs was a master storyteller, enrapturing customers and investors alike with his grand ideas and gloriously sophisticated computers. Bill Gates was the nerd's nerd, who didn't care what the computers looked like, or what sound bites you used to talk about them, as long as decent-enough software was inside.

With all that transpired later between the two boy wonders of tech, it is easy to forget that—even as Microsoft was agreeing to provide software for an IBM personal computer that would go head-to-head with Apple's products—these men and their companies had a close, mutually beneficial relationship for the first critical years of the personal-computer revolution. Apple had been an important Microsoft customer ever since Woz hadn't gotten around to finishing writing his own software for the II in 1977. Gates readily provided testimonials about the technical superiority of Apple's machines. The two companies produced different things, and they needed each other's products to grow.

Seattle was 800 miles up the road from the Valley, but it was compatible in history, in spirit, in the personal and professional connections flowing between north and south. It too was a former gold rush town turned Cold War region, transformed by the defense boom and advantaged by the postwar migratory rush westward. It too was home to a major research university and lots of knowledge-economy jobs, its tech scene a strange mix of straight-arrow aerospace types and Vietnam-era lefties, a place of early adopters and ambitious technophiles.

Yet Seattle wasn't an isolated Galapagos. During those formative postwar decades it was Boeing's company town, busy and connected. Even after its early-'70s bust, it didn't develop the Valley's teeming petri dish of VCs and lawyers and PR flacks. It didn't need them. Only a two-hour hop on a Boeing 737 would take a person like Charles Simonyi or Dave Marquardt from one place to the other, able to make a day trip to close a deal or broker a new partnership. Little surprise that Microsoft would become only the first of several market-upending technology companies to come out of the Emerald City in the decades to come.

Silicon Valley and Seattle might have felt like they were rivals, but they were actually two parts of a larger whole. The growth of one enabled the growth of the other. Of course, it wasn't too long before the Valley people were worrying whether that software outfit in Seattle might topple them all.

THE PC

To the disbelief of those familiar with Big Blue's glacial product development cycles, the IBM PC made its debut on August 12, 1981, a little more than a year after this Manhattan Project began. The microcomputer market changed forever. Across Silicon Valley, panicked executives piled into early-morning strategy meetings to figure how they were going to respond to the day's news. Some looked on the bright side. "The presence of IBM will take away the sting of saying that personal computers are just a fad," remarked an HP executive. Most quaked as they realized how IBM's deep pockets could shake up their young industry. With an enormous marketing budget and plans to sell everywhere from Computerland to Sears, "they'll sell these things by the thousands and thousands," predicted one analyst.[11]

Of all the machines out there, the IBM was most directly an assault on Apple, which held 23 percent of the market. Jobs and Co. already were struggling to make a hit out of the expensive and still-buggy Apple III, and IBM's new entry—at a similar price point and offering comparable features—wasn't going to make things any easier. But if Apple was nervous, its spin didn't show it. Instead, the company's marketers doubled down on their story. The personal computer was changing everything, and Apple was a transformative company with a revolutionary product. A late-coming machine out of stodgy old IBM wouldn't be much of a threat. In fact, asserted Mike Markkula, "their presence would stimulate demand for a product that Apple originated."[12]

To put an exclamation point on this messaging, the Apple marketing team placed another full-page ad in *The Wall Street Journal* in the wake of IBM's product announcement. It was classic Apple, all intensely competitive iron fist wrapped in a Garamond-font velvet glove. "Welcome, IBM," it read. "Seriously." Following that blaring headline came copy saturated with prim disdain. "Welcome to the most exciting and important marketplace since the computer revolution began 35 years ago," it read. "We look forward to responsible competition in the massive effort to distribute this American technology to the world. And we appreciate the magnitude of your commitment."[13]

Apple's cocky confidence came in part from its company DNA; the bureaucratic and rigid IBM, with its distinctly *un*friendly big computers, was the antithesis of everything Apple's executive team believed their company stood for. But it also reflected the fact that Steve Jobs was hard at work on another product—one that he believed would be as much of a market disruptor as the Apple II had been four years earlier. It was called the Macintosh.

The Mac's origin story—and how Jobs swooped in to make the project his own after his insufferable behavior got him kicked off his original pet project, the elegant and expensive Lisa—is another oft-told tale, and deservedly so. The project was unlike anything the computer industry had seen before, and its roaring debut remains a landmark in product marketing and advertising. The Mac took the sophistication of the Lisa and the Apple III and combined it with the Volkswagen-like utility of the Apple II. It boasted the user-friendliness of the Alto and its commercial offspring, the Xerox Star workstation, without the eye-popping price. Several PARC engineers had moved over to Apple to work on the Mac team, bringing their knowledge of the GUI that Xerox had created but failed to successfully commercialize.[14]

The Mac team had started as a tiny skunkworks project within the company and ultimately grew to a team of 100. Like most everyone else at Apple, nearly everyone was under thirty, and had put everything to one side to put in eighty-hour weeks on the project. But the sense of mission was strong. The people of Apple in those glorious early 1980s, already blessed by great stock-option fortune, believed they worked in a truly special place. "Apple is human-oriented," said Jo Kellner, who staffed Apple's on-site customer support center. "We're free to be individuals. And that freedom breeds creativity." Coming to work was a pleasure, said programmer Rilla Reynolds, because "almost everyone here is playing." "There's a spirit at Apple," chimed in product manager Pat Marriott. "I think it grows out of the conviction that we are doing the right things, making a quality product—making a dent in the universe."[15]

The Macintosh operation took these kinds of sentiments into overdrive. Apple spent a staggering $21 million on R&D in 1981, most of it for the Mac, and built a state-of-the-art automated factory to build the new machines. "Steve was so insistent on perfection," remembered Floyd Kvamme, who had moved over from the chip world to become Apple's marketing chief. "There was no way that computer could live up to what was in his head." Although the company was awash in money, a maverick attitude prevailed. At one off-site, Jobs set the tone by reminding the group that one of the key "Sayings of Chairman Jobs" was: "It's better to be a pirate than join the navy." Thus inspired, members of the team sewed together a custom-made skull-and-crossbones flag—with an eyepatch featuring Apple's familiar rainbow-colored logo—and hoisted it proudly above its building on the company's growing campus in Cupertino.[16]

PIRATES VERSUS BATTLESHIPS

As Apple's pirate flag waved bravely, battleship IBM was eating the market. Although the design-obsessed Jobs was dismissive of the homely and squat PC, some of the more pragmatic heads inside Apple understood the magnitude of the marketing threat. Big Blue had been in American offices for more than eight decades. Apple was struggling to get into the office-machine business; IBM *was* the office machine business, and they knew how to give enterprise customers what they wanted and needed. It wasn't just a matter of brand loyalty to a particular device. It was a matter of an entire ecosystem around how large companies had gotten accustomed to buying and using computers.[17]

If electronic data processing became a religion in the 1960s, then Management Information Systems (MIS) managers were its high priests. The job category didn't even exist prior to the coming of all those electronic brains at the start of the 1950s, but two decades later corporate information-technology executives had become immensely powerful. They bought the computers, ran the computers, and controlled the information. The MIS manager's mantra: standardization. There was no way to manage giant business computer systems if the different parts couldn't talk to one another. That was the genius of the System/360, the market-dominating machine IBM released in the mid-1960s and the apex of the full-stack approach: a collection of computers and peripherals that meshed together seamlessly, supported by compatible software, with good support and training, able to scale up or down as the MIS manager needed it.

As minicomputers and, later, personal computers surged into the American office in the 1970s, the MIS manager lost some of his iron grip. Devices with their own keyboards, operated by the user rather than by a punch-card-wielding intermediary, proliferated by the tens of thousands. When it came to microcomputers, the encroachment initially was minimal, advanced by Ben Rosen types who sidestepped corporate IT offices and simply bought their own personal computers to use at work. But the breakthrough of VisiCalc had been followed in short order by other office software applications. The spread of transistorized word processors like those produced by Wang further fueled corporate interest in desktop computing, and a recognition that a keyboard could be more than a typewriter and secretary.

Yet mainframes—many of them made by IBM—remained the thrumming heart of corporate computing environments. To make personal computers actually useful in processing core office tasks, they needed to link into the big computers that were already there. The Silicon Valley companies that had only recently emerged from their garages now had a daunting challenge before them. IBM may have been late to the personal-computer market, but its people understood office machines—and the psychology of those who made corporate purchasing decisions—better than anyone.[18]

The market onslaught by IBM was immediate, and devastating. Business customers had at first feared sinking several thousands of dollars into a personal computer, but IBM's familiar logo prompted many to take the plunge. For those less persuaded by those three blue letters (especially the baby boomers who started humming the soundtrack to *2001: A Space Odyssey* and thinking of HAL's blinking red eye every time they saw them), the company launched a full-bore ad campaign.

Beginning in the fall of 1981 and continuing for six more years, American popular magazines and television airwaves were saturated with IBM ads that featured an actor dressed like Charlie Chaplin, extracting himself from comic dilemmas with the help of an IBM PC and peripherals. All white space and clean typeface in print, all genial familiarity on television, the look and feel of the ads was strikingly similar to those early pitches from Apple. Here was another friendly little device to make your life better. The message bore the ghost of Tom Watson placing those perky young women in IBM's windows more than three decades earlier: the PC was so easy to use, even the Little Tramp could do it.

Between 1982 and 1983, the personal-computer market more than doubled, with over 11 million units shipped in 1983 alone. IBM owned 26 percent of it. Apple had fallen behind, to 21 percent, despite a relaunch of a new-and-improved Apple III and the debut of the Lisa. While the IBM PC sales chugged away, the Apple III went begging for customers and the Lisa floundered amid scathing reviews. "If an executive has loads of time, doesn't need to use the computer much, but wants some sort of machine to sit in his office," sighed a reviewer at *Fortune*, "Lisa is probably the best bet."[19]

Even though the computer-market pie was so much larger, the new pecking order wasn't what Apple had expected. Instead of adding new staff, 1983 brought layoffs. Jobs recruited a new president from the outside, Pepsi executive John Sculley. The hire brought consumer marketing expertise to the operation and would perhaps, as *Time* characterized it, "induce a streak of humility in an organization where confidence bordered on arrogance." Sculley brought other non-technical East Coast types along with him, including a bearishly friendly former college football coach named Bill Campbell, who'd most recently worked at Kodak. Microcomputer purists looked at the suits and ties on the new arrivals with bafflement. What could sugar-water and camera salesmen do for a company that was supposed to change the world? IBM "definitely won round one," Jobs acknowledged testily that October. When asked how he felt about it, Jobs replied: "I'm ticked."[20]

The air had been sucked out of the pirate-ship sails. Personal computing was a revolution; how could the revolutionaries possibly agree to use a machine built by the old guard? How could they fight back? Was Apple going to go the way of the UNIVAC? Other personal-computing start-ups started dropping like flies. Adam Osborne packed up his very heavy laptop and declared bankruptcy in October 1983, ducking reporters like a guilty defendant as he scurried into his office the day the news broke. Lore Harp's and Carole Ely's Vector Graphic was blindsided by IBM's onslaught and plagued with management woes. In a matter of months, their pioneering Southern California company plummeted from market rock star to struggling turnaround operation. It wasn't just the little guys: Texas Instruments announced it was giving up on the home computer market after taking a more than $220 million bath in the first nine months of 1983. Even worse—new competitors emerged, bringing expertise and cash reserves to the party.[21]

Hewlett Packard was Exhibit A in that regard. The company that began with audio oscillators and radar jammers already had made successful leaps into new markets before, first with minicomputers in 1966 (it became #2 in the market) and then with hand-held calculators in 1972 (it became the *entire* market). By late 1983, it had a staggering $700 million cash on hand, and it was ready to conquer anew. When it came to building a personal computer, "we can afford to experiment until we get it right—with no danger of heading for the poorhouse," said John Young, who had taken over from Hewlett and Packard in 1978. And HP saw which way the winds were blowing. "If personal computers are not HP's primary business," by the end of the 1980s, said HP computer-group chief Paul Ely, "then we won't be a successful computer company."[22]

It still was a big jump. HP had always sold products to scientists and engineers; this would be its first pass at selling to business customers. Mirroring the way Apple and Microsoft structured their work around intensely competitive product development teams, Young set a massive reorganization in motion at HP—one with lasting effects on company culture. The HP Way endured on paper, but its flat organizational structure and "management by wandering" gave way to a much more centralized system. Young was determined to be a leader, not an imitator, and the HP personal computers featured something no other competitor had: touch screens. Going after consumer markets for the very first time, the company launched television ads that fall. "Even though Hewlett-Packard's technology has produced a number of firsts," said one, "some of you don't even know who we are. Maybe now you will." With such humble messaging, it seemed that Silicon Valley's original start-up hadn't picked up the new generation's storytelling talents.[23]

Companies that had never been in the computer business got in the game. The U.S. Government had just forced AT&T to break up its telecommunications monopoly, but the phone company still had more than $34 billion of assets to play with. Ma Bell went big, announcing its own lines of both minicomputers and desktops in early 1984. "Count Us In," its ad tagline proclaimed. While AT&T's computers were too expensive to truly give IBM a run for its money, the entry of a communications company into the arena brought new ways to get computers to talk to one another. AT&T proceeded to do what it had so vigorously blocked competitors from doing ever since the days of Bunker-Ramo: it built local area network (or LAN) systems that could hook up machines up to a third of a mile apart.

All the frenetic competition left Valley techies burned out and worried about what might happen next, their personal lives limping along in a world of eighty-hour weeks and round-the-clock obsession with technology. The industry's gender gap remained staggering, looming largest after working hours. One woman who worked at Stanford recounted a particularly disastrous double date with two male engineers: "It was like the two of us [women] did not exist. Men used to get to know each other through football. Here that's been translated into computers." Divorce rates spiked higher than the rest of California, which already had a higher rate than the rest of the nation. "Wives are often very frustrated," said one Silicon Valley psychologist. "They feel like their husband is married to a chip."[24]

A local family therapist saw so much angst coming through her offices that she published a self-help book titled *The Silicon Syndrome*, brimming with tips for women married to antisocial computer nerds. "A dynamic new book for wives, mothers, daughters, lovers, colleagues and secretaries of the engineer/scientist," cheerfully burbled the cover. "If you long to better understand how to communicate and interact with your scientist/engineer partner, boss or mate this book is for you!" Silicon Valley's problems weren't with its intense and overwhelmingly male tech culture, assessments like these implied. It was with the women who couldn't figure out how to live with it.[25]

Companies, cueing in to an overstretched workforce and anxious to retain valuable employees, attempted a little behavioral modification. AMD paid for multiple psychotherapy sessions for any employee who wanted them. Hard-charging Intel had a moment of concern for its employees' work-life balance and timed all the lights in the building to switch off at 7:00 p.m., forcing everyone to go home for dinner. Yet the lines between professional and personal lives continued to blur as each twelve-hour workday passed. "You are terribly afraid to stop working," noted another local therapist, "because somebody will get ahead of you."[26]

As if its big blue shadow over Silicon Valley didn't already loom long enough, in December 1982 IBM bought a 12 percent stake in Intel, whose 8088 chips powered the PC. Six months later, it bought up 15 percent of ROLM, signaling its ambitions to make the IBM desktop a node in a fully wired "electronic office" of the future. The Valley soared higher than ever before, but it seemed to many that it had flown too close to the sun.[27]

THE PLATFORM

Yet the thing that really made IBM's PC into the ultimate disrupter wasn't the hardware. It was Microsoft's software, coupled with Intel's chips, that broke apart the full-stack regime and upended the economics of the computer hardware and software business.

The "PC platform"—an entirely different OS from the ones used by Apple and the other early microcomputer makers—rapidly became the industry standard, shoving nearly every other contender out of the way, because, unlike Apple's closed system, the MS-DOS used by IBM wasn't restricted to IBM machines. The Intel chips that powered the IBM PC could power other computers too. From Texas to Tokyo, companies sprang up building "clone" machines that offered the same platform and apps as the IBM, for a fraction of the price. It was a boon for consumer and small-business markets, where customers were reluctant to drop thousands of dollars on fancy machines.

As the PC platform scaled up, software companies finally found their market opportunity. The market now teemed with little operations who had been busily trying to monetize something that computer people long had thought should just come for free. Now they had an obvious destination for their products. After so many years of drought, the programs started multiplying like rabbits: video games, spreadsheets, word processors, educational software. The growth far outpaced the number of programs being created to run on Apples. Now, the big start-up success stories were software companies.

Silicon Valley might have been the cradle of the first microcomputers, but now some of the biggest hits started happening in other places. The ever-astute Ben Rosen, now a venture capitalist in Texas, sniffed out some of the best of the new breed, and he made a killing in the process. In 1982, Rosen backed an IBM-clone maker in Houston called Compaq, whose sales topped $100 million in its first year of business. The same year, Rosen took a bet on a young software developer in Boston, who was building programs for the PC platform. There were plenty of software entrepreneurs trying to do the same thing those days, but Rosen sensed something was different about this one. His name was Mitch Kapor, and his company was called Lotus Development Corporation.

Tech had been blowing Mitch Kapor's mind ever since he'd picked up a copy of Ted Nelson's *Computer Lib* soon after graduating from college in the early 1970s. It wasn't the programming that drew him in—by his own admission, he was only an "ok" programmer—but something largely unappreciated at the time: software design. "The software designer leads a guerilla existence," he would write several years later, "formally unrecognized and often unappreciated." Yet good design—a lack of bugs, ease of use, and an interface that delighted the user—was fundamental to good software. "One of the main reasons most computer software is so abysmal is that it's not designed at all, but merely engineered." Throughout his varied career, Kapor was determined to change that.

After a few years of post-college drift that included a stint as a DJ and a six-month-long advanced training in Transcendental Meditation, he scraped up money to buy his first personal computer. It changed his entire trajectory. Soon he was writing and selling software for Apple IIs out of his Boston-area apartment. Before long, Kapor moved over to Personal Software, publisher of VisiCalc, the app that turned the Apple into a business machine. There he kept writing more programs and, because he was a consultant, keeping a nice chunk of the royalties.

Drawn in by the "micro revolution" buzz about the Valley, he transferred out to Personal's Sunnyvale office in 1980. He hated it. "It was a monoculture," he remembered, where people "could only think one big thought at any time." Amid the suburban blandness and California aridity, he ached to return to the bookstores of Harvard Square. After six months, he did.

Kapor's time in Sunnyvale may not have been an enlivening one, but it implanted the start-up bug within. Everyone out there was starting companies; why not him? The IBM PC needed software; why not build some apps as good as VisiCalc for the new platform? Venture capitalists intimidated Kapor—he'd once met the steely Arthur Rock, and found him "very scary"—but he knew Ben Rosen, who had been a user of Kapor's first software product. So he wrote Rosen a seventeen-page letter asking for money to launch his software business, which he had named Lotus. Rosen had been a VC for mere months by that point, but he said yes. It turned out to be a great bet. Kapor's company had $53 million in sales in 1983, its first year of operation. The next year, revenues tripled.[28]

Lotus Software was on its way to becoming one of the biggest success stories of the 1980s, the crest of a new wave of software companies building apps for the ever-expanding empire of the PC. The micro's spiritual home may have been the Bay Area, but many of these software upstarts—Lotus in Boston, WordPerfect in Utah, Ashton-Tate in Los Angeles, Aldus in Seattle—were just like Microsoft: outsiders, ready to give the Valley a run for its money.[29]

1984

As 1984 began, the men and women of Apple were standing outside looking in at the growing empire of the PC. Education had become a huge success story for Apple, aided greatly by California's adoption of "Kids Can't Wait." Yet the education numbers were minuscule compared to the massive office market. Same for home computing. The market was vastly larger than it had been, but the number of North American households owning a personal computer still stood at about 8 percent. The percentage of people using computers at work was three times that, and it was spiking up rapidly.[30]

Chinks appeared in Steve Jobs's supremely confident armor. "IBM wants to wipe us off the face of the Earth," he admitted to a reporter. Apple—and Jobs—needed a very big win. The man who embodied Apple in the media sensed that his celebrity was slipping along with the company's market share, and it annoyed him no end that the business press didn't take him as seriously as they did IBM's John Opel. When a member of Regis McKenna's PR team informed Jobs that he wouldn't be appearing on the cover of *Fortune*, the Apple chief became so enraged that he took the glass of water he had been sipping and threw it in the associate's face. Dripping and equally furious, the latest victim of Jobsian wrath left the meeting and drove the twenty minutes to RMI. "Look what your 'son' did to me," the associate told McKenna in disbelief. "Regis, the guy is insane." A chastened Jobs called with an apology soon after. The dampened exec suspected that Regis had cajoled Steve into it, but agreed to return. After all, the pressure to beat IBM could make anyone go nuts.[31]

John Sculley's years in the cola wars made him more bullish. Just like Coke could exist alongside Pepsi, Apple could survive, and thrive, in the same universe as IBM. After all, IBM's big Christmas 1983 release had been a dud: an adaptation of the PC for the home market called the PCjr. The dumpy little brother of the mighty PC hadn't been developed via another Manhattan Project, but through IBM's regular development channels—and it showed. The PCjr reminded the market that "Apple is definitely the people's computer," observed one software distributor. "This is the year Apple fights back," announced Sculley in January 1984. "We are betting the entire company." And the big bet was the Macintosh.[32]

After a couple of years of being outspent by its bigger competitors, Apple beefed up its marketing budget and hired Los Angeles–based ad agency Chiat/Day, a longtime computer-company favorite that had recently made a splash with its ads for Honda motorcycles and Nike running shoes. Psychographic research from Arnold Mitchell at SRI helped Apple home in on its target market: "achievers" who "would rather be individuals, not part of a group." But the Mac, for all its pirate-flag cred, "*must unequivocally* be positioned/featured as a *business* product," cautioned Chiat/Day's planners.[33]

The Mac had a killer market advantage: you didn't need to know much about computers to use it. "People are intimidated by choosing" the right computer, Chiat/Day reminded Apple executives, and worried about whether what they buy might become obsolete too quickly. IBM's Charlie Chaplin ads had successfully deployed "easy-to-use" messaging. Apple's Macintosh was *actually* easy to use. It was stripped-down and simple compared to the Alto and the Lisa, but it still had graphics, icons, and a friendly little mouse. Yet Sculley kept reminding everyone of his Coke-versus-Pepsi model: Apple shouldn't imitate, it should present itself as something completely different. Steve Jobs agreed. "We need ads that hit you in the face," he said. "It's like it's so good we don't have to show photographs of computers." "Macintosh advertising," the agency brass concluded, "must be distinctive and mirror the radical, revolutionary nature of the product."[34]

And distinctive it was. On January 22, 1984, the Mac debuted on the world stage in a $1.3 million television commercial aired during the Super Bowl. Directed by Hollywood science-fiction auteur Ridley Scott, the ad was sixty seconds of jaw-dropping visuals riffing off George Orwell's *1984* and all the computational allusions made to it ever since. The tagline: "On January 24, Apple Computer will introduce the Macintosh. And you'll see why 1984 won't be like '1984.'" The computer itself never appeared.

Macintosh ads blanketed the nation's television airwaves in the months that followed, through the Winter and Summer Olympics, through a presidential campaign season of ads celebrating the Republicans' "Morning in America" and the Democrats' first female vice-presidential nominee. No other Mac ad was quite as memorable as that Super Bowl spot—few ads in history *ever* matched the buzz it created—but the message of all of them was the same. We are not IBM. We are not the establishment. Our computers will set you free. Over video of an elegantly manicured female finger clicking on a mouse, the tagline for all of them read: "Macintosh. The computer for the rest of us."

MORNING IN AMERICA

By October, as the economy sped up to a gallop and the Mondale campaign limped toward an expected drubbing on Election Day, Regis McKenna had gotten philosophical. "The good news is that Mondale is going to lose and we'll see the end of the traditional Democratic Party we have known," he told Haynes Johnson of the *Washington Post*. "I see that as good because there's a whole generation of young Democratic politicians coming up that are different. The bad news is that Reagan is terrifying, and I really mean that."[35]

But McKenna's worries couldn't dim the fact that the Valley was enjoying very good times in the autumn of 1984. Johnson's election-eve piece on the Valley bore the title "Silicon Valley's Satisfied Society," and there was plenty of evidence of that satisfaction. Japan's onslaught of cheap chips and electronics continued, but the personal-computer boom had vastly enlarged and diversified the market. The name of Silicon Valley's game was no longer only semiconductors, but computer hardware and software too. And while plenty of IBM clones rolled off East Asian assembly lines, America was #1 when it came to building new generations of personal computers and building the software to run on them. Worries about global competition hadn't disappeared, but they became softened by billowing blankets of money.

That Christmas, Apple threw nineteen holiday parties, including one with a $110,000 price tag that featured the odd pairing of a Dickensian village theme—complete with thirty strolling performers in period costume—and a concert by Chuck Berry. Jerry Sanders went one better with AMD's $700,000 black-tie bash featuring performances by a boys' choir, a full orchestra, and arena-rock heroes Chicago. "Sure, I get letters about the starving children in Ethiopia," said Sanders breezily. "But our people worked hard for this, and they earned it."[36]

The over-the-top company parties marked a new high-water mark of conspicuous consumption in the Valley. But this was not the last time that outrageous holiday shindigs would signal bumpier times just ahead. As 1985 began, Wall Street's mania for personal computers cooled. Even Microsoft hung back for another year before having its IPO. The young market shook out further, matured, and some of its pioneers found themselves rather unceremoniously out of work. Others became more powerful than ever. The semiconductor makers continued to battle for market share, their fortunes finally stabilizing only when Japan's economic miracle proved to be not so miraculous after all. New technologies and influential new players emerged in fields that not only built thinking machines but *connected* them to others: workstations and relational database software and computer networking.

And amid all of this disruption was something that had been there from the start, had never really gone away, and that in the 1980s had become more influential—and, to some in the Valley, much more ominous—than it had been in several decades. It was the tech counterculture's original Big Brother: the computer-powered federal government, and its very high-tech ways of making war.

CHAPTER 17

War Games

"SHALL WE PLAY A GAME?" the computer asked David Lightman. "Love to," the teenage geek responded. "How about Global Thermonuclear War?" With two text commands, the 1983 summer blockbuster *WarGames* shifted into high gear.

Armed with nothing but an IMSAI computer and a modem in his bedroom, a kid in suburban Seattle inadvertently hacks into a top secret Defense Department mainframe. Before long, the protagonist and his female companion find themselves deep in the Air Command's mountain bunker, frenetically trying to reprogram a supercomputer that is determined to launch thousands of nuclear warheads. At the last moment, Lightman's programming skills save the world from mutually assured destruction. The story was pure Hollywood, but for the popcorn-chomping millions watching in the cool dark of American movie theaters that summer, *WarGames* wasn't all that far from reality.

After decades of test ban treaties and détente, America was once again ramping up its nuclear arsenal and employing the highest of high technology to do it. The post-Vietnam American military could no longer keep up with the Soviets in terms of size—the U.S. had abolished the draft, while the U.S.S.R. had forced conscription—but it had a huge advantage when it came to technology. America was the capital of microelectronics, and even if Japan was nipping at the U.S.'s computing heels, the Soviet Union would have to spend mightily to get anywhere close. Defense Secretary Caspar Weinberger wasn't particularly bullish on tech, but he believed in the power of an economic strategy to drain the Soviet treasury. So, too, did Secretary of State George Shultz, who had been a Northern Californian ever since the Nixon years, and who had close ties to Stanford and the Valley tech community. Just as it had in the early days of the Eisenhower Administration, the defense agenda once again took a high-tech turn.[1]

Just three months before *WarGames*' release, Ronald Reagan had announced an audacious new program to create a sophisticated missile shield in space using satellites, lasers, and all kinds of computer-controlled technology. Northern Californian fingerprints were all over the proposal: the laser-based system had been championed loudly and early by Berkeley's Edward Teller, father of the H-bomb and director of the Lawrence Livermore Laboratories. David Packard endorsed the idea as well. Called the Strategic Defense Initiative, or SDI, the program pushed the outer edge of the technologically possible. And although the press reports included illustrations of lasers and satellites and pow-pow action in the upper atmosphere, SDI was really all about computers. Thus the arrival of SDI, and its accompanying political controversies, became inextricably intertwined with the other major DARPA initiative announced in the summer of 1983: Strategic Computing.[2]

Reagan already was battling a reputation of being a warmonger—he famously proclaimed the USSR an "evil empire" mere weeks before the SDI announcement—and despite presidential assurances that the new program was about nuclear deterrence, many of the nation's most prominent scientific names lined up to decry the program as a dangerous boondoggle. Technophilic Democrats in Congress howled too. Making the shield work would be "like hitting a bullet with a bullet," one aide informed Senator Paul Tsongas, who had become a particularly vocal critic. The program quickly acquired a not-so-charitable nickname: "Star Wars."[3]

The prospect of space battles between the light and dark sides of The Force might have been far-fetched. The possibility of a *WarGames*-like scenario wasn't. The program's deep dependence on computer software raised the ominous possibility of bad code triggering accidental global annihilation. Worst of all, in the minds of Americans now conditioned to think of government as an inherently bad thing, it was going to be built and controlled by bureaucrats. "Technology is not a panacea for our ills," explained *WarGames* director John Badham, when asked about the message of his film, "and bureaucracy is something that will surely get you in deep, deep trouble every time if left to run untrammeled." Sam Ervin couldn't have said it better himself.[4]

Unlike in the privacy wars of the 1960s and early 1970s, however, the computing world was no longer solely the domain of big government and big corporations. Now it was the playground of the gamers and hackers, masters of the microcomputer and modem, just like the young hero played by Matthew Broderick in Badham's film. The personal computer had triumphantly moved in, particularly for American children and teenagers.

Computer nerds had become familiar, sympathetic pop-culture characters, whether they were fictional figures from movies or television, or real-life multimillionaires like Jobs and Gates. Right at the same time that the SDI battles were brewing, journalist Steven Levy was immortalizing the history of this rebellious breed in *Hackers*. (The Valley tech community had been so delighted by their heroic portrayal in the book that they reclaimed the label as an honorific, and Stewart Brand began holding an annual "Hackers Conference" to celebrate the movement they had forged.) An outlaw programmer was the hero of William Gibson's sci-fi novel *Neuromancer*, a cult bestseller published right around the same time. Even run-of-the-mill geeks got the girl in teen hits *Sixteen Candles* and *Revenge of the Nerds*. As the Doomsday Clock ticked closer to midnight, it wasn't surprising that so many dreamed of a Hollywood ending where these hackers used tech to make peace instead of war.[5]

STAR WARS

Silicon Valley was ground zero for the Jedi Knights of the computer world, and not just because Apple had portrayed it that way during the 1984 Super Bowl. In the academic precincts of the Valley—Stanford, SRI, PARC—the antiwar sentiment of the Vietnam era never fully subsided, and by the middle of the 1980s these places had become hubs for the nuclear disarmament movement.

In the 1970s, Congress had restricted defense grants going to academic projects with a direct military application, and NASA and Department of Energy funding had further declined. The arrival of Reagan's team shifted those priorities: military spending at universities soared up from less than $500 million in 1980 to $930 million in 1985. The surge was particularly noticeable in computer science, which by two years into the Reagan era had nearly 60 percent of its federal funds for basic research coming from the Pentagon. With the arrival of SDI, that percentage promised to shoot up even more.[6]

The groundswell against the new defense dependence started soon after Reagan arrived in office. Throughout 1981, PARC's internal mail network buzzed with discussion of the bellicose new regime and its plans to build up the nuclear arsenal. By the spring of 1982, a petition signed by 150 PARC employees persuaded Xerox to sponsor a nationally broadcast television special titled *Facing Up to the Bomb.* By the fall, the group had a name: Computer Professionals for Social Responsibility, or CPSR. "We believe," they proclaimed, "neither that the path to national security lies in military superiority, nor that superiority can be achieved through the use of computers." CPSR's founding ranks contained the Valley's boldfaced names in AI and human-computer interaction, including a newly tenured Stanford computer science professor named Terry Winograd.[7]

A baby boomer with a liberal conscience and a PhD from MIT, Winograd had made a decision early on not to accept military research money himself. Seeing the rising tide of resources rushing into the world of academic computer science, he urged his colleagues to just say no as well. "Once the university has become dependent on military funds for its survival," wrote Winograd in an early CPSR newsletter, "it is very hard to take a stand on some 'minor' issue which could jeopardize everything. A few minor issues soon add up to a lot of control."[8]

The uneasiness percolated all the way to the top of the university administration. Then-president of Stanford Donald Kennedy was a biologist who had served as director of the U.S. Food and Drug Administration under Jimmy Carter. He wasn't going to say no to research dollars, but having spent more than twenty years at Stanford, he understood keenly how politics might mess with independent research agendas. The new defense windfall, he worried, "throws the balance of science research all out of whack."[9]

It wasn't just the ethics that troubled the academics. It was the science. The idea of an AI-enabled computer governing the nation's nuclear arsenal was *WarGames* brought to life, a nightmare that wasn't out of the realm of possibility. AI could do a lot, but the state of the art wasn't anywhere close to making a computer-advised battlefield possible. Those closest to the heart of the machine understood that better than anyone else. In fact, the leaders of DARPA were so uncomfortable with the technical and ethical aims of SDI that the Reagan White House soon moved the program's budget line to elsewhere in the Pentagon.[10]

THE SUSTAINED OPPOSITION TO NUCLEAR BUILDUP AND SDI WAS ONLY ONE slice of the Valley's Reagan-era story, however. Computer moguls grabbed the headlines, but the world of military electronics had never left the Valley of Heart's Delight. In the late 1970s and early 1980s, Santa Clara County had more defense spending per capita than any other place in the country. One-fifth of the Valley's economic output still came from aerospace and defense. The defense buildup that these technologists opposed so passionately was, quite literally, happening in their backyards.

ROLM, the company that had made Burt McMurtry's venture career, earned 99 percent of its $10 million revenue in 1975 from "mil spec" sales of rugged, impact-resistant minicomputers designed for soldiers' use in the field. A rising star of the 1980s, the relational database company Oracle got its start with a CIA contract that Larry Ellison and his co-founders completed while working at Ampex in the late 1970s. As military spending soared, this ever-present defense sector grew larger, even if it stayed partially submerged from view.[11]

The beating heart remained the place that started it all: Lockheed Missiles and Space, whose 24,000-strong 1980s workforce was nearly twice as large as Intel and five times larger than Apple. Over the three decades since the company first landed in Sunnyvale, Lockheed's 175-acre campus had grown so large that the company had its own fire department. Yet the security perimeter that surrounded it remained airtight. Defense contracts continued to make up the bulk of its business, perpetuating a secrecy and stealth that was in sharp contrast to the increasingly casual, collaborative culture of the rest of the Valley.

Lockheed people didn't chit-chat at the Wagon Wheel or on the sidelines of Little League games about their days at the office. They tended to be lifers, not job-hoppers. Many were military men or veterans, with the close ties to the defense establishment that had largely faded away in the new generation of Valley firms. Their hair stayed short; their shirts stayed pressed. They didn't worry about Japan. They weren't networked in to Valley venture capital firms, local lawyers, or public relations maestros—because they didn't need them. They had the Department of Defense.

Given that Sunnyvale had specialized in antimissile defense since the heyday of the space race, it was little surprise when Lockheed quickly emerged as a prime contractor for the SDI system. It was big money—$100 million here, $200 million there—spurring fresh waves of hiring and an economic impact that rippled across the region as Lockheed hired other firms as service providers and vendors. "People are on an absolute spending spree," marveled Sunnyvale City Manager Tom Lewcock, and he pointedly noted defense spending as a major cause of it. "It's just another paved-with-gold kind of period of time that we're going through." By early 1985, Lockheed's balance sheet was so flush and its future so bright that the company announced it would spend $5 billion over five years to modernize its facilities, including Sunnyvale. Star Wars might be a miracle, or it might be a money pit. One thing it was for sure: a windfall for Silicon Valley.[12]

The boom was as controversial as it was profitable. The prominence and sheer size of its SDI contracts made Lockheed a ripe target for protesters, who regularly clustered with banners and chants at the facility's gates to register their objections to the defense buildup. Twenty-one people were arrested at one protest in April 1986. More than twice as many were arrested that October, after a Halloween-themed event that included activists dropping pumpkins off an overpass on Highway 101 and onto incoming traffic. Lockheed management brushed off the demonstrations as a pesky nuisance and vowed that business would proceed as usual.[13]

It wasn't just the pumpkin-throwers who disliked the new order. Ed Zschau was keenly aware of how much defense contracts were helping his district, but he worried that the flood of money was diverting research resources—and research talent—from the private sector. "By virtue of our devotion to defense research and development," he fretted to a reporter, "we're making it more difficult to be competitive in global markets." What's more, the defense buildup had turned the procurement system into a runaway train of $640 toilet seats and other "gold plated" equipment, attracting vehement and bipartisan criticism on Capitol Hill. DOD cost overruns had garnered the president so much bad press that by the middle of Reagan's second term he was forced to establish a "Blue-Ribbon Commission on Defense Management." The person he tapped to run it: the ever-loyal and ever-thrifty David Packard.[14]

CANON WARS

Silicon Valley's Reagan-era battles between hawk and dove, Right and Left, weren't just about how much the U.S. Treasury was spending on defense, or on the ethics and science of high-tech weaponry. They extended even wider. And the place they raged the fiercest was the two thousand bucolic acres that always had been at the center of the Valley's story, its intellectual hub and town square: the campus of Stanford University.

And the center of it all was the 285-foot sandstone battlement of the Hoover Tower. Ronald Reagan's ascension to office had cemented the Hoover Institution's reputation as the nation's premier conservative think tank. The young economist whom Herbert Hoover had tapped all those years ago to run it, W. Glenn Campbell, had weathered the protests of pipe-smoking humanities professors and rock-throwing students in the 1960s, so unperturbed by all their wailing that he expanded the institution's research agenda beyond foreign affairs and into domestic policy soon after. The move into live-wire conservative topics like economic monetarism, deregulation, and welfare reform further amped up the visibility and the controversy, attracting generous private donations from billionaire right-wing backers. The expansion of policy scope also attracted high-profile "visiting fellows" like Reagan, who joined Hoover's roster after leaving the California governor's office, bringing along with him 1,700 boxes of gubernatorial papers and the tapes for eight rare episodes of his old TV Western, *Death Valley Days*. He donated all of it to the Hoover archives.[15]

Little surprise, then, that, within months of Reagan's inauguration, Campbell proposed that the Californian's future presidential library and museum be located on Stanford's campus. After quietly percolating for a couple of years, the library proposal began to get a serious review by the White House in 1983—and all hell broke loose back on The Farm.

When longtime Hoover fellow and current Secretary of State George Shultz gave the 1983 commencement address, protesters gathered to condemn the Reagan Administration's policies in general and Hoover's complicity in particular. A petition signed by 1,500 faculty and students demanded that the institution leave campus. When Gloria Steinem visited Stanford for a book signing that autumn, she couldn't resist weighing in. "You have my deepest sympathies," America's most famous feminist remarked. "The campus in the Midwest that is responsible for [Reagan's] lack of education should be responsible for his papers." A student group arose calling itself "Stanford Community Against Reagan University." Faculty living nearby objected to the likely traffic and noise the library would generate; the august Center for Advanced Study in the Behavioral Sciences, which occupied the oak-studded hillside next to the proposed library site, threatened to move away from Stanford altogether.[16]

The battle over the Reagan library raged on for four more years, until the president's team at last decided to abandon Palo Alto for the ideologically friendlier climate of Southern California. "The joggers, environmentalists, and free spirits with tenure who had opposed the library held self-congratulatory parties in the rolling foothills," wrote one Hoover fellow acerbically in *National Review*, "free at last from the terrors of Reaganism." Glenn Campbell retired two years later, claiming Don Kennedy forced him to, because the Stanford president "doesn't like me." The accusation of partisanship was deeply unfair, said an unrepentant Campbell. "Hoover doesn't lean to the right—it's an optical illusion because the rest of the campus leans so far to the left. The Hoover Tower points straight up."[17]

The high-powered, high-profile Campbell gave his exit interview to a little student newspaper that was barely two years old and only published a few issues a year. *The Stanford Review*, however, was already making its mark as a self-styled voice of reason for Stanford's conservative students.

Founded in the late spring of 1987, the *Review* was the brainchild of a sophomore philosophy major named Peter Thiel. German-born and California-bred, a regional chess champion and J. R. R. Tolkien devotee, Thiel had arrived on campus as the battle of the Reagan library raged. For the remainder of his undergraduate years and as a Stanford law student immediately after, Thiel focused his considerable intellectual energies on the *Review*, making its libertarian-conservative views an inescapable feature of campus life as a fresh, even more polarizing battle erupted: the war over the undergraduate curriculum.

The "canon wars" blazed hotly on many elite American campuses in the mid-1980s, as students and faculty demanded—and won—a more inclusive, multicultural approach to humanistic education. Civil rights and affirmative action victories of the 1960s had resulted in far more diversity on campus; people of color now made up one-third of the Stanford student body. But as access to college enlarged, so did what one national periodical primly called "public concern about student ignorance." Bestsellers like E. D. Hirsch's *Cultural Literacy* and Allan Bloom's *The Closing of the American Mind* bemoaned the state of American higher education, turning academic debates about what went on the syllabus into a flashpoint of the 1980s culture wars.[18]

This swirl was escalating on Stanford's campus by the time Thiel founded what he called "a forum for rational debate" in the *Review*, and its eventual heat and velocity—coming right on the heels of the headline-making fight over the Reagan library—made Stanford's canon wars national news. Where many students saw an overdue turn away from an emphasis on Western civilization and the work of dead white men, Thiel and his fellow conservatives saw an effort to "restrict the academic freedom of professors." Soon joining him on the masthead was another campus contrarian, law student Keith Rabois, whose love of Ronald Reagan was matched by his sharp distaste for liberal orthodoxy. Rabois (who later came out as gay) later gained notoriety for yelling a gay slur outside the house of a Stanford professor, following up with "I hope you die of AIDS!" He declared that his actions were simply a protest against the university's restrictive speech codes and left Stanford shortly afterward.[19]

Other *Review* stalwarts displayed a similar mix of ideological rigor and misplaced performative politics. Another local chess champion, David Sacks, arrived as a freshman in 1990 and found that his social awkwardness and his delight in pummeling liberal egos left him an odd man out everywhere except the *Review*. Sacks later disavowed many of the things he wrote in the newspaper's pages, but soon after he graduated he produced more of it as a published author, collaborating with Thiel on a book-length treatise titled *The Diversity Myth*. Multicultural education was "the intellectual equivalent of junk food," the two declared, and it was just the tip of the iceberg. America had become "a kingdom of victims," and university administrators like Provost Condoleezza Rice were encouraging it (ironically, the hawkish Rice was a Hoover fellow and later would serve as national security advisor and secretary of state for President George W. Bush). "Individuals must strike out and set their own destinies, free from both the historical cultures of the past and the newer multiculture."[20]

Thiel, Rabois, and Sacks all did exactly that: striking out into tech, and ultimately striking it very, very rich. They did not do so as individuals, however, but as an immensely powerful network of men Thiel had brought together in his many years at the *Review* and who, armed with their Stanford degrees and their well-worn copies of Ayn Rand, set out together to remake the brave new world of the Valley's Internet economy. Within a decade, the core group would be multimillionaires known as the PayPal Mafia, after the online-payment company they founded was sold to eBay for $1.5 billion in 2002. Nearly all went on to found and invest in other major tech hits.

Stanford's campus wars burned hot for only a few years, but they had lasting resonance. For the young people zooming on their bicycles through the sun-dappled quads in the late 1980s and early 1990s were the same people who would found and build some of the richest and most influential companies of the Valley's late 1990s and 2000s. The faculty who agonized over whether to take money from the Pentagon became advisors and mentors to the Valley's next generation of CEOs. The academic administrators steering the place in this era went on to become influential political advisors in the next. Stanford University's stormy Reagan years became the stage on which the politics of the next-generation Valley were formed.

As sharply polarized as Left and Right were on Stanford's campus during this time, some common threads connected them. Both Peter Thiel and Terry Winograd were concerned about freedom of speech on campus. Both Glenn Campbell and Don Kennedy believed that Stanford scholars had an opportunity, and a responsibility, to contribute to politics and policy. Both the students crying out for a new, multicultural curriculum and the conservative elders who tut-tutted at the closing of the American mind agreed that the college years shaped a person's trajectory for the rest of their lives.

This was especially true at Stanford. Fred Terman's university was no longer a dusty little outpost; it was now the center of a large and rich high-tech universe. It was the place to be if you wanted to push the edge of the scientific envelope, or if you wanted to build an entirely new kind of business. You had easy access to people and resources; you had the eyes of the world upon you. You were networked into the most tightly networked place on earth, perfectly poised for a brilliant career.

And while a few of these perfectly poised young men and women took their Stanford experience and continued the political fight on the Left and the Right, many others took away a different lesson from the wars over libraries and canons. Politics was exhausting, messy, hurtful, inconclusive. Business, in contrast, had the power to change the world. Just look at all the companies that had spilled out of Stanford, out of Homebrew, out of the VC firms over on Sand Hill Road. Tech didn't dictate a canon or tell you what to think; it created a neutral platform for creativity and free expression. Perhaps it was just easier to go heads down, focus on the code or the business plan in front of you, and get busy building the future.

AMERICA GOES ONLINE

As Reagan's defense budgets swelled, local computer hardware and software companies became attractive targets for large defense contractors in search of a high-tech edge. One of those was Tymshare.

Given how many time-sharing and networking companies had bitten the dust in the early 1970s, the company had enjoyed a remarkable run. Ann Hardy was still there, a company vice president at last. Tymnet became an early online platform for business applications that would later become ubiquitous: banking, bill paying, travel reservations. All of these online services were operating out of Hardy's division, and she and her team became among the first to try out these new technologies. She traveled a lot for work, and she taught her elementary-age children how to use e-mail so they could communicate with her while she was on the road. By the early 1980s, other time-sharing companies like CompuServe and Prodigy were branching out into lucrative consumer markets—offering e-mail, electronic newspapers, even some early online catalogs—but Tymnet was still sticking to its corporate customers. CEO Tom O'Rourke was ready to retire, and he didn't want to spend the money it would take to make Tymnet a competitor.[21]

Hardy wished her boss would be a little more ambitious about building out the business, for she saw how the market was beginning to spike. Larger companies already were circling around Tymshare for a potential acquisition. They saw value in the venerable, well-designed system and appreciated its $300 million valuation. In 1984, one of the biggest—McDonnell Douglas—snapped up Tymshare.

The acquisition made Hardy the big defense contractor's only female VP, and it was something her new bosses didn't quite know how to handle. At the first company meeting she attended, "every speaker started up his presentation with some kind of off-color joke," she remembered. At executive meetings, company chief John McDonnell began following Hardy around anxiously to make sure no one said anything insulting. No one had imagined there'd be a woman in the room. After that, "they spent a whole year trying to figure out how to get rid of me without having me sue them." Defense work wasn't any fun at all. Yet she didn't quite know what to do next. Go out on her own? "I didn't know anything about starting a company," she demurred. She was nice, not tough, and "nice women don't do that." But, boy oh boy, life at the new gig was awful.

Out of desperation as much as gumption, Hardy at last did what so many men around her had been doing for years. She started her own business. Its name was KeyLogic, and it made transaction-processing security software for mainframes. It was based on the secure OS designed so long ago at Tymshare; something, it turned out, that would become increasingly important in the years to come. Two decades in the Valley had given her great connections; she secured venture financing, assembled a strong team, and won promising early customers.

At long last, Ann Hardy had become an entrepreneur. She was a focused, unflashy one, a serious middle-aged executive in a time of computer-nerd cover boys. But her kind of business—big machines, big corporate clients, big government contracts—was still as much a part of the Valley's high-tech 1980s as the personal computer had become. "There's so much to Silicon Valley," she reflected. Hardy had been around long enough to understand that success didn't come from technical talent alone. "It's timing. It's luck." And she'd gotten lucky once again.[22]

COMPUTER NETWORKING HAD BEEN AROUND NEARLY AS LONG AS COMPUTERS themselves—Ann Hardy would be the first one to tell you that—but things that people were using the networks *for* began to change in the 1980s. If the fictional David Lightman's near-brush with thermonuclear war reflected America's Reagan-era reality, so did the mayhem he was able to cause via the modem hookup in his bedroom. For the explosion of the personal computer market had allowed computer networking to become personalized too.

At first, the arrival of the microcomputer created a royal headache for the time-sharing companies. They were in the business of distributing computing power to companies that couldn't afford their own machine; now, most could. CompuServe, the Ohio-based network founded in 1969, realized that survival would now depend on hooking the new personal computer users into the networked life. Also seeing a business opportunity were the makers of modems. The machines that made a phone handset into a portal for computer communication had been around since the start of the 1950s, too, but it was only in the post-*Carterfone* world that lots of different manufacturers could get in on the action. Modems were phone equipment, after all. As people started scooping up their Apple IIs and TRS-80s in the late 1970s, modem makers swooped in to make a sale.

It was a perfect target market. Lee Felsenstein helped build the Community Memory bulletin board before he was part of Homebrew, after all. This Whole Earth generation wanted "access to tools" so that they could communicate with one another. It was no fun to be at home with a computer without anyone to talk to. With early adopters settling micros on their desks, their modems awaiting the first screech of connection, the network companies built software to get new customers on line. By the summer of 1978 CompuServe had more than 1,000 users using its new service, MicroNET. By the summer of 1979, a new service called The Source launched out of Northern Virginia. Although it was a start-up, its spendthrift founders threw a launch event at New York's Plaza Hotel, rolling out the sci-fi star power with a special appearance by Isaac Asimov. "The information utility the world's been waiting for!" exclaimed the corporate press release. "This is the beginning of the Information Age!" Asimov chimed in.[23]

The profit-seeking online companies weren't the true children of Community Memory, though; nor were they like the ARPANET, which continued to purr away as the nerdy domain of academic computer scientists and government contractors. The thing that spread the anarchic, decentralized essence of these two original noncommercial networks was something entirely separate: Usenet. To enter this online world, you didn't subscribe to a content-creating service but instead tapped into a newsgroup whose *participants* created the content: a discussion board geared to a particular topic—microcomputing, gaming, gardening—containing threaded discussion of posts and responses.

Usenet's "poor man's ARPANET," launched in North Carolina in 1980, allowed users to transfer files (at glacial dial-up speeds) and exchange e-mail. In the early days the exchange of information would be batch-processed at certain times of the day, a throwback to the Stone Age before time-sharing. You could send a message to Europe via Usenet, but it would take two days to get a reply. Nonetheless, access to a decentralized, user-driven, specialized communication network was a marvel to its thousands of users at a time when ARPANET use was highly restricted. And the different Usenet groups—over 900 by 1984—turned the alien world of computer communication into something personal, something that tapped into the passions and enthusiasms people had in real life. Yet it went one better, providing anonymized connection through the veil of the username, making it a place for free expression that the dreamers and doers of Community Memory had always hoped the electronic bulletin board might be.

The Usenet group was just one flavor of bulletin board service, or BBS, that proliferated in the pre-Internet online world of the 1980s. Most of the people going online were high-income men in their thirties—no surprise, given the fact that this was the demographic targeted by the micro makers—and discussion threads reflected their priorities. In 1985 came the most famous of the early BBSs: The WELL, or Whole Earth 'Lectronic Link, started by Stewart Brand and his merry band of hackers up in Marin County. The WELL's fame came from the Silicon Valley celebrities who made it their first online hangout, including Grateful Dead lyricist John Perry Barlow, journalist Steven Levy, Lotus founder Mitch Kapor, and of course Brand himself. The bland Ohioans running CompuServe (now owned by even blander tax preparer H&R Block) couldn't compete with The WELL's glamour and dash.

The WELL's pedigree was decidedly countercultural, as it hired a clutch of its founding staff from the legendary Tennessee commune The Farm, and devoted considerable discussion-thread and file-swapping bandwidth to the Dead. But like the original countercultural computer guys, the 1980s WELL left 1960s debates over gender equity behind as it created its new frontier of online communication. Women on the WELL were few in number, and some early users split off to form more female-friendly boards of their own. The mostly white and male network that remained on the WELL would go on to play an immensely important role in the development and early evangelism of the World Wide Web; the boards for second-wave feminists and Gen X riot grrls didn't have the same amplitude.[24]

And for every BBS devoted to high-minded social causes—and there were quite a few—there were ones devoted to darker pursuits. A burgeoning white supremacist movement found BBSs to be fertile ground for recruitment and retention;

Ku Klux Klan grand dragons and leaders of the "White Aryan resistance" became early movers in the BBS world, using its reach and quasi-anonymity to rally new followers. Less nefarious, but equally determined to operate in the shadows, were the Cypherpunks, a kind of punk anarchist collective in cyberspace, whose members employed the intricacies of cryptography—technologies all developed within the military-industrial complex—to wall out potential electronic eavesdropping by government snoops.[25]

Tribalism set in early as the like-minded found one another online, an early signal of the immense power these networks would one day have as platforms for all kinds of political activism, propaganda, and rumor-mongering. By 1995, just before the spread of the Internet and its browsers would make the world of Usenet and CompuServe and BBSs fade into the sunset, more than 70,000 BBSs existed in the U.S. alone.[26]

PERESTROIKA

Ronald Reagan left office in the early days of 1989, less than a year after his first trip to Moscow and the speech where he rhapsodized about the glories of the microchip revolution before the computer science students of Moscow State, an event orchestrated by Stanford's own hometown hero, Secretary of State George Shultz.

No sooner had Reagan departed office for retirement on his Santa Barbara ranch than the George H. W. Bush Administration shifted toward a less costly, more agile SDI system that relied more heavily on miniaturized electronics and computerized control. Berkeley's Edward Teller became as much of a booster for the new approach as he had been for the original Star Wars plan, spending nearly the entire final year of Reagan's presidency lobbying the White House to go the miniaturized route. "For defense we do not need the big bang," Teller declaimed to a gathering of defense contractors in early 1989, "what we need is accuracy and more accuracy." Another bonus: each interceptor would now have the relatively modest price tag of $1 million each. Military planners dubbed the system Brilliant Pebbles. Although Lockheed got a prime contract, it was the first sign that the mighty boost that the Reagan years brought to Sunnyvale's fortunes would soon fade away.[27]

Yet the Valley did not leave the foreign-policy spotlight as the Reagan era came to a close. In the early summer of 1990, Soviet president Mikhail Gorbachev came to Stanford for an event that sounded like a call-and-response to Reagan's speech in Moscow two years earlier. The Berlin Wall had fallen; the West had won the Cold War. But "let us not wrangle over who won it," Gorbachev told the large campus crowd gathered in the cavernous Memorial Auditorium. Let us instead look toward the future, for "the ideas and technologies of tomorrow are born here in California." Gorbachev was as smitten with the Bay Area as Charles de Gaulle had been thirty years before. "I always wanted to come here, and I never had the chance," he told a clutch of waiting reporters. "Fantastic!"[28]

But future-tense California had never been without the government's invisible hand. The big money that flowed in via SCI and SDI contracts in the 1980s was a reminder that defense remained the big-government engine hidden under the hood of the Valley's shiny new entrepreneurial sports car, flying largely under the radar screen of the saturation media coverage of hackers and capitalists. Contracts for missiles and lasers and interceptors didn't get much airtime in the many studies considering the race to build "the next Silicon Valley." When it did merit a mention, defense spending appeared as something past tense, something foundational, an opening act for the entrepreneurial headliners of the 1980s and beyond.

New industries like microchips, personal computers, and video games won most of the press attention. But they also were buffeted by overseas competition, domestic challengers, and wrongheaded business decisions. In the hawkish, high-spending last days of the Cold War, the defense contracting business became insulated from much of that boom and bust. As they had before, these federal contracts subsidized the development of bleeding-edge projects that otherwise wouldn't have seen the light of day, within and beyond academic research, seeding industries and companies that would keep the Valley a leader into the next generation.

CHAPTER 18

Built on Sand

On the eve of Election Day 1988, *Washington Post* reporter Haynes Johnson made a return visit to the Valley to gauge the region's temperature, checking in with now-pessimistic Democratic power broker Larry Stone. "This place has changed and with that comes the awareness that the community's not near what it ought to be or [what] we thought it was," Stone glumly observed.[1]

Stone had plenty of data points for his dour assessment. The perpetually white-hot real estate sector had cooled, and people who had paid top dollar for ranch houses and bungalows a few years earlier now found themselves underwater on their mortgages. The maturation of the personal-computer market compounded the pain. No longer high-priced novelties, PCs were now an affordable mass commodity. With a computer on nearly every desk and in any tech-savvy household, sales had flattened and profits dipped. Engineers traded Maseratis for minivans, as worried locals wondered aloud if the boom times were gone for good. The Valley's silicon mountain looked like it might just be a hill of sand after all.

On top of all that, the tech guys had lost their congressman. Ed Zschau made a bid in 1986 to unseat incumbent U.S. Senator Alan Cranston, beating out a crowded primary field but losing by a heartbreakingly thin margin in the general election.

The bitter loss had been given an extra shove by Regis McKenna, who had joined Cranston at a packed press event to declare that Zschau's highly valued Silicon Valley business credentials were bunk. McKenna had taken a board seat at Zschau's System Industries after the congressman left for Washington, and he announced that he had found it disastrously managed. "If it were not for the people who followed Ed," McKenna said crisply, "it would not even be around today." Zschau shot back: "I think Regis is wrong. Obviously, he is making a political statement, trying to cast aspersions on my company." Revenues had doubled from 1980 to 1981, Zschau pointed out, and he had led a second IPO to raise additional capital before he left for Congress. McKenna stood his ground. His comments were well supported by others in the company's leadership, he noted. "It was not just my opinion." Several members of System Industries' board took out full-page ads in the local papers with the trumpeting headline, "Regis, you should be ashamed of yourself!" It was a political brawl unlike anything the convivial Valley had witnessed before.[2]

Two years after that, the Reagan era ended, and the men who had brought a little bit of Stanford and the Hoover Institution to Washington were heading home. The semiconductor crowd finally had their next-generation chipmaking operation, Sematech, but it wasn't even going to be in California. Instead, it was going to Austin, Texas. But the Valley got to leave its imprint in the end: after no one immediately stepped up to run the operation, Bob Noyce agreed with some reluctance to relocate to Texas to become its director.

Texas was riding high, too, with the arrival of incoming President George H. W. Bush, and the Valley was unsure of what changes might result. While the Bushies were certainly business-friendly, the new president's economic team approached the industry at an unfamiliar distance. "Potato chips, computer chips, what's the difference?" one Bush advisor was rumored to have asked; the quote was apocryphal and never firmly sourced, but it confirmed everything the Valley chipmakers suspected about the new regime, and it drove them crazy.[3]

HOW GRAY WAS MY VALLEY

Part of the loss of faith had to do with a realization that, when it came to old-economy woes like pollution and labor costs, the Valley's golden new economy was not all that exceptional after all. Ever since HP started making oscillators and Ampex began crafting magnetic tape, factories had stretched along Highway 101 from San Jose to San Mateo. Tens of thousands of employees (disproportionately female, Asian, and Latina) worked the assembly lines and fabrication plants during the semiconductor boom years, far outnumbering the white, male, professional workers who were the faces the tech industry presented to the world.

This hidden Silicon Valley had quite literally spilled over into the public consciousness earlier in the 1980s, when news broke that highly toxic chemicals had seeped into well water underneath Fairchild Semiconductor's fabrication plant near the working-class Los Paseos neighborhood of San Jose. Residents had already started noticing alarming spikes in miscarriages, stillbirths, and health problems in both infants and adults; now they believed they had found their culprit. High-tech boosters in the Valley had been praising the "clean" and "smokeless" virtues of their industry ever since Fred Terman first cooked up Stanford Industrial Park. Toxic groundwater did *not* fit the Silicon Valley narrative.

This was a shoe that had been waiting to drop for a while. From the chemical baths that produced microchips to the toxic metals embedded inside every piece of computer equipment, Silicon Valley had been in the business of particularly dirty manufacturing since the start. Compounding the danger to human health was the fact that this manufacturing boomtown had been built on farmland where water and sewer infrastructure had been built cheaply and quickly—if it had been built in the first place. Pollution wasn't just a problem for working-class neighborhoods like Los Paseos, where drinking water reservoirs had been dug alarmingly close to Fairchild's leaky holding tanks. In tonier subdivisions tucked away behind research parks and corporate campuses, tens of thousands of homes relied on private wells for their drinking water. Shallow and unregulated, these were even more susceptible to toxins leaking in, and not being caught in time.[4]

As poisons seeped further, the Valley's problems became impossible even for Ronald Reagan's regulation-averse Environmental Protection Agency to ignore. "It has become obvious," concluded EPA regional chief Judith Ayres, "that the absence of smokestacks does not mean the absence of environmental problems." The events shook the faith of local officials. "There was no doubt in my mind that this was a clean industry," lamented San Jose Mayor Janet Hayes. But the growing revelations set off by the Los Paseos leak upended her assumptions. "We now know we are in the midst of a chemical revolution." The EPA eventually designated twenty-three places in the Valley as exceptionally polluted Superfund sites, including a cluster of sites in the bucolic Stanford Research Park—the development that had set an international standard for "no smokestacks" manufacturing since the 1950s. A new, lucrative specialty cropped up in the Valley commercial real estate development business: environmental remediation.[5]

Bruised by bad PR and chafing under newly strict environmental codes in towns like Palo Alto, tech companies began looking around for other places to manufacture their products. Some chose a local escape hatch, simply moving a few miles east to the cheap flatlands across the Bay, territory that already was recognizably industrial, its water and soil long sullied by oil refineries and salt ponds, auto assemblies and chemical processing plants. In some of these East Bay plants, robots did an increasing share of the work, like they did in Steve Jobs's high-priced Mac factory in Fremont. Where sharp-eyed human oversight and fine motor skills were needed, companies increasingly found they could cut costs by sending factories overseas.[6]

Now American tech brands expanded beyond Singapore and Taiwan to southern China and India, where economic liberalization and privatization were just starting to create huge new opportunities for foreign companies. The plants themselves were usually owned and operated by an eager and sometimes mercenary cadre of subcontractors, putting the physical work of high-tech production at an even further remove from the Bay Area sunshine. (A later generation of Apple customers saw this economic geography etched on the back of every iPhone or iPad: "Designed in California by Apple, Assembled in China.")

Back in the Valley, the low-paid assembly line workers of the electronics industry that remained could no longer afford to live remotely close to where they worked. Labor organizers upped their efforts to unionize plants from California to Massachusetts and continued to find a stony wall of opposition. "The high-tech industry has working conditions and sensibilities that are sophisticated enough so that unionization is not an issue," scoffed one executive. "Unions exist only because the management mistreated their workers," Jimmy Treybig added. "People want to feel they're citizens of the company." Labor leaders would have none of it. Tech companies were pretending their blue-collar employees didn't even exist, protested labor organizer Rand Wilson. "It's not all it's cracked up to be. Many of these places are high-tech sweatshops."[7]

But voices like Wilson's were hard to hear over the groan of John Arrillaga's bulldozers, flattening aging fab plants to make way for tidily remediated office parks and corporate campuses. With Asian labor markets beckoning, it had become much easier to move the dirty business of high-tech manufacturing far away and out of sight.

Globalization of production allowed the mythos of a clean, white-collar high-tech world to stay alive and gain velocity. With every new silicon-chip-festooned magazine cover, it seemed like the world went crazier for high-tech research parks. Everywhere from Perth to Peoria still wanted to create a Silicon Something of their own. In their pursuit of an industry that promised white-collar jobs and abundant tax revenues, mayors and governors willfully ignored the tradeoffs that the original Valley had made as it became an industrial powerhouse. "Cities everywhere want to be home to companies with nonpolluting factories and campus-like offices," dutifully reported *U.S. News & World Report*, as if Los Paseos had never happened. "I don't think you'll find that there will be any pollution," Texas Governor Mark White declared as he announced yet another new high-tech facility, except from "the Japanese cars they drive to and from work to do it."[8]

WHAT'S NEXT

Appearances had been deceiving in the Valley business world too. 1984 had been a spectacular year for Apple and its charismatic co-founder. January had started with the grand-slam of the Super Bowl ad, followed by the main event of the Mac release itself. Through the spring, demand for the new computer exceeded supply. Steve Jobs had engineered his specially built Mac factory in Fremont to produce a million units a year—and he predicted that it would soon run at full capacity. It was, he proclaimed again and again, "insanely great!"[9]

The Mac was conquering a new market too: higher education. "These students are the knowledge workers of tomorrow," Jobs told *InfoWorld*. Students were smack in the middle of Arnold Mitchell's "achievers" psychographic; placing Macs on their desks at age nineteen would always associate the brand with their fondly remembered college days. (It also broadened the gender profile; college students were the only target market for the Mac that included women as well as men.) The purpose was to "get a generation growing" whose first loyalty was to Apple products, crowed Dan'l Lewin, the young marketing executive Floyd Kvamme assigned to the project. Lewin's hyperkinetic product marketing persuaded a number of achiever-filled elite universities to join a new "Apple University Consortium," receiving deliveries of tens of thousands of cut-priced Macs.[10]

Then, as Valley journalists Michael Swaine and Paul Freiberger later put it,

"Apple ran out of zealots." After a surge of early sales, the numbers for the Mac never met Jobs's stratospheric public projections. Early reviews of the new Apple praised its ease of use—at last, a thinking machine that didn't require a 100-page user manual and five hours of setup in order to use it—but knowledgeable computer users were quick to point out its limitations, particularly in business. The first Mac didn't have a hard disk drive. The graphical interface and built-in programs ate up most of its built-in memory, leaving little room for new software and storage. Because it was so sleek, it lacked plugs for peripheral equipment like printers. Bill Gates and Microsoft had worked closely with the Macintosh team since 1981 to develop a suite of software for the Mac, but the closed world that Jobs had created for his cherished new computer made it difficult for other developers to build software to run on its platform. The "1984" ad had made history—and Chiat/Day's reputation—but the same could not be said for its subject. Mused Regis McKenna, "The ad was more successful than the Mac itself."[11]

In the spring of 1985, Apple posted its first quarterly loss. By summer, it laid off 1,200 employees—over 20 percent of the entire company. By the end of September, the stock price had tumbled from a 1983 high of $62 per share to less than $17. In the mind of the board and John Sculley, the main source of all of Apple's woes was the mercurial, messianic, and megalomaniacal Steve Jobs. In one of the most celebrated firings in American business history, Sculley took away Jobs's operational authority and moved him into the ceremonial—and powerless—role of chairman. Four months later, Jobs sold all his Apple stock (except one, deeply symbolic share) and quit.

For a breathlessly watching business press, the showdown between Jobs and Sculley had reverberations far beyond Cupertino. The suit-and-tied Organization Man from the East had beat out the long-haired entrepreneurial visionary from the Golden State. Bold promises of being "insanely great" couldn't overcome the relentless metric of quarterly earnings. The personal-computer business was now *big* business, and quirky genius wouldn't cut it anymore.

In the post-firing postmortems, the qualities that had rocketed Jobs to business-world superstardom now became his greatest liabilities. "The very characteristics that lead entrepreneurs to start companies—independence innovation, and commitment to ideas—are the same ones that can cause their demise as managers," offered one business management expert. Wall Street immediately signaled its approval: Apple stock jumped up by a dollar as soon as Jobs left the building. But

Silicon Valley veterans weren't so sure. "Where is Apple's inspiration going to come from?" asked Nolan Bushnell. "Is Apple going to have all the romance of a new brand of Pepsi?"[12]

Stunned and vengeful, Jobs wasted little time in making bold moves that kept him in the headlines. Dumping in $7 million earned from his stock sale, he started a new company: NeXT Computer. Coolly informing Sculley and the Apple board that he wasn't going alone, he pulled in some of the top people from the Macintosh pirate crew as well as some favorites from the Regis McKenna universe. No more messing around with boring business machines. Jobs wanted to go after those college students—and their professors too. Sun Microsystems, founded in 1982 and roaringly profitable ever since, had taken a huge bite out of the Boston-dominated minicomputer business by marketing high-powered "workstations" offering both minicomputer power and the user-friendly features of a PC—all at an attractive price for business and academic use. Blending the Sun concept with sophisticated design and fresh software, Jobs called the NeXT "the scholar's workstation."

Once again, Jobs talked a big game. The new device would be "10 to 20 times more powerful than what we have today," he promised. Design and product marketing continued to be his obsession, and he commissioned a logo for NeXT even before he had designed the computer itself—paying top dollar to hire Paul Rand, legendary designer of corporate logos including, most notably, the blocky blue letters of IBM.[13]

From the earliest days of the Apple II, Jobs's evangelism about computers as engines for creative learning had won him admiration across the educational spectrum, even if not every teacher and educational expert bought into his bullish take. The roaring success of the Mac on campus strengthened Jobs's conviction that education was the next great frontier. The NeXT team courted universities' IT directors assiduously. "They just kept on going out and asking people what they wanted," said one. "We're fairly jaded and cynical, but these folks really did a good job."[14]

While the thirty-one-year-old resisted outside investment at first—he'd been beholden to the venture capitalists since the first days of Apple, and now he finally had the riches to shake them off—his pursuit of perfection had a staggering burn rate. "The honeymoon is over," he told his staff in the spring of 1986, only six months into NeXT's existence. To develop a product that met sky-high expectations, and ship it quickly, Steve Jobs needed a new infusion of cash.[15]

He got it from H. Ross Perot.

Relentless, fast-talking, and spectacularly good at sales, Ross Perot was a five-foot-six-inch business legend with a force of personality that rivaled that of Jobs. Perot had been successful at nearly everything he'd done in life: president-for-life of his U.S. Naval Academy class, a record-shattering sales career at IBM, and then the billion-dollar company he founded in Dallas in 1962, Electronic Data Systems, or EDS. Perot was "utterly self-assured," a later observer wrote, "the sort of person who walks into someone else's house and turns on the lights."[16]

With EDS in the early 1960s, Perot had pioneered a new and highly lucrative business model of selling software and consulting and IT services for mainframe computers. His early business success came chiefly from multimillion-dollar contracts to build and manage databases for the giant, newly created federal health insurance programs, Medicare and Medicaid. By the early 1970s, EDS processed more than 90 percent of the Medicare claims in the nation. The bounty of big-government business made this small-government conservative extraordinarily rich. Although Perot styled himself as a self-made man, the ultimate expression of free enterprise at work, one sardonic critic observed that he was in fact "America's first welfare billionaire."[17]

While the silicon boys of Fairchild Semiconductor wore shirtsleeves and the computer guys of Digital looked like graduate students after an all-night coding session, Perot's EDS battalions looked like IBM in miniature, with a dash of Texas ROTC thrown in. Everyone wore ties, sharp suits, and close-cropped hair. Employees had to sign ironclad non-compete agreements. Their stock options evaporated if they ever left the company. Later, Perot became so enraptured by a slim management how-to titled *Leadership Secrets of Attila the Hun* that he blurbed the paperback edition and bought 700 copies to distribute at company meetings.[18]

By the time he encountered Steve Jobs, the tenacious Texan was looking for a defining next act. In 1985, he had sold EDS to General Motors in a deal that netted him an enormous amount of money, but that had left him a very square peg in the round holes of GM's organizational hierarchy. Used to being in charge and being listened to, Perot couldn't resist loudly giving advice to GM head Roger Smith about how he could improve operations. After about a year of this yammering, an exasperated Smith forced Perot out.

Not too long after, Perot happened to catch a public-television documentary that featured a rapt profile of Jobs and NeXT. Here was another visionary, a maverick, someone who believed in pursuing bigger and higher ideas. And he needed money. Perot picked up the phone, and a few weeks and $20 million later, the squarest-of-the-squares computer billionaire had revived Jobs's considerable ambitions and secured a 16 percent stake in the company.

It was one of the tech world's most unlikely partnerships. Here was the spit-and-polish mainframe-era mogul teaming up with the once barefoot-and-bearded evangelist of California cool. Yet the egalitarian conceit behind Jobs's vision appealed to Perot's durable populism. "With these electronic tools," the Texan enthused, "you can bring the very finest 'courseware' by the very finest professors to even the smallest liberal arts school with no endowment." Jobs was similarly complimentary toward his new investor. "Even though I've never lived in Texas and he's never lived in Silicon Valley, it's become clear we've had similar experiences."[19]

One of those common experiences, of course, was getting ungracefully kicked out of your own company after butting heads with its top executives. It was possible that the high-octane combination of Perot and Jobs would be too volatile a mix. "If a guy like Roger Smith can't take Perot on his board," mused Wall Street analyst Richard Shaffer, "I don't see how Steve can." Esther Dyson, who was steadily building Ben Rosen's newsletter and conference business into a tech-forecasting empire, had a more optimistic take. "Perot brings to the party a lot of wisdom and a lot of real-world thinking," she said. "I think they're a good match."[20]

But the glories that Jobs and Perot predicted for NeXT never came to be. The company never made a profit, and it burned through every cent that the Texan had invested in it. And Perot and Jobs weren't all that compatible after all. The former Navy man keenly understood how lucrative government contracts could be, and he tried to persuade his new protégé to go aggressively after federal business. Jobs was uninterested. When Perot's people tried to chase after a lucrative contract with the National Security Agency, Jobs put his foot down. He wasn't going to let his company go into the spy business. An exasperated Perot immediately picked up the phone to call the young mogul, but he couldn't get a response. "On any given day I call the White House and get through to the President," Perot steamed to Dan'l Lewin. "I'm in business with you, so why can't I talk to Steve?" At the next board meeting, Perot stepped down from his seat. Lewin soon left the company as well.[21]

Jobs's grandest ambitions to become the workstation in every classroom fell far short. He pivoted into business markets, and then made a last-gasp flail into selling software. The Valley's original wunderkind remained as compelling a storyteller as ever, and the NeXT machines were certainly beautiful, but they were never as intuitive to use as the friendly little Mac. "Steve's trouble was that he tried to do another Apple," observed a colleague. "He is like a person who goes from marriage to marriage trying to get the same relationship."[22]

The only thing that saved NeXT was the prodigal son's returning to Apple. In 1997, after more than a decade of flailing against the relentless forward march of the PC platform, Apple fired its CEO and replaced him with Steve Jobs. Stepping in on an interim basis, and then later taking over entirely, Jobs embarked on a complete reinvention of the company's product lines that spawned a series of market-upending successes: the iMac in 1998, the iPod in 2001, and, biggest of all, the iPhone in 2006. NeXT became part of Apple as part of Jobs's reentry deal. The workstations soon were no more. Its Unix-based software, however, lived on—becoming the heart of the OS that powered these new devices and, eventually, turned Apple into the richest company on the planet.

THE CALIFORNIAN INVASION

Steve Jobs still might be getting all the press, but there were close to a million people working in the computer industry in California in the middle of the 1980s. The vast majority of them weren't working amid the glamour and flash of PC companies and software makers. They weren't making six-figure salaries, either. They were the programming army, the people who lived and breathed software languages and knew the ins and outs of every OS. They worked on mainframes, not micros. They were Californians like Trish Millines.

The basketball champ from New Jersey had gotten tired of Arizona's heat and small-city sleepiness after two years into her job at Hughes. She rented out her house, bought a motor home, and relocated to San Francisco. Yet if the micro world was booming when she arrived in early 1982, the bigger companies that employed programmers in bulk were still locked in the Reagan recession. Once they pulled up from it, contracting work was more common than actually being hired on full-time. Without a steady paycheck to cover San Francisco's steep cost of housing, she lived for a while out of her motor home, camping out with other high-rent refugees along Marina Green until the police swept through and shooed them out.

Employees at places like Apple might have been going on and on about all the fun they had at work, but for Millines a programming job was just a way to pay the bills. Her recreational rugby team was the thing that made the city fun. Even then, she found San Francisco to be not quite her style. One of her first jobs was teaching programming at a computer learning center—one of the many outfits that had sprung up those days to provide a few months of quickie training to would-be technologists. She got fired after only one term for being too tough a grader. "I was giving people the grade they deserved," she chuckled, "instead of the grade they wanted."

She eventually found a steadier programming job down in Redwood City, a little closer to the Valley action, but the transience of the Bay Area wore her out. People were hopping jobs and, as prices rose and employment slowed, they were leaving town. "I got tired of making friends all over again," she remembered. She was learning a lot, but the Bay Area didn't get any cheaper. By the end of 1984, she had decided to move again—this time to Seattle. Green and rainy, cheaper and friendlier, not many black folks but lots of jobs for software engineers. By January, Trish Millines was driving the 800 miles north, toward her next adventure. Silicon Valley might have contained a pot of gold for some, but the housecleaner's kid from the Jersey Shore never found it.

Millines wasn't the only one heading in that direction those days. A slowing economy and sky-high real estate prices were pushing people out of both Northern and Southern California, and a good number of them ended up in the Pacific Northwest. Portland, Oregon, was now home to a large manufacturing and design facility for Intel. Seattle had Microsoft and Boeing and more. Life was a little slower and considerably more affordable up there; it offered all of the Bay Area's culture and none of its traffic jams.

Seattle's metropolitan population spiked by nearly 400,000 over the course of the 1980s. While people came from all over, the conventional wisdom in town was that Seattle was under siege from "Californication." Seattle's self-appointed town curmudgeon, newspaper columnist Emmett Watson, likened the California migrants to house cats that had filled their own litter boxes and needed somewhere to do their business; transplants who didn't change their car license plates quickly enough would get honked off the road.[23]

But the Californians kept coming. Silicon Valley had turned into a place that didn't make *things* anymore. Semiconductors got made elsewhere. Personal computers had reached a sales ceiling. New-age heroes like Steve Jobs hadn't managed to fend off old-style competitors. The region's raw materials were no longer silicon and copper wire, but people and ideas. The product came as strings of software code, intangible other than the floppy disks on which they were written. Software was a good business to be in when capital flows were uncertain. You needed less capital upfront, and you didn't need to sell out most of your ownership to VCs to get started. You didn't need as large a workforce, and you could hire large numbers of workers as contractors. "Software," observed *Forbes*, "is where the future money is." It turned out that software was where Seattle's future money was as well.[24]

THE VELVET SWEATSHOP

When Trish Millines arrived in Seattle, Microsoft was still small. When she came to work at Microsoft as a contractor three years later, it was bigger. When she came on as a full-time employee two years after that, Microsoft had grown to 3,500 employees, inhabiting four buildings in a verdant corporate campus fifteen miles northeast of downtown. "People were there," she remembered, "because they liked the work."[25]

They also were there for the perks. As the *Seattle Times* detailed in a long 1989 feature on the company, life was very good at Microsoft. Break rooms bulged with free sodas. Nearly every office had a window. Two computers sat on every desk. Workdays were punctuated with midday soccer scrimmages or goofy contests where top executives dared one another to jump in a nearby lake. Making everything more delectable was the money. Nearly every hiring package at Microsoft included options to buy thousands of shares of Microsoft stock. The company had gone public in 1986, making its early employees instant multimillionaires. Young employees would strut around the office wearing lapel buttons reading F.Y.I.F.V.: "Fuck you, I'm fully vested."

The free sodas and stock options might have been an upgrade from the early

Future venture capitalist David Morgenthaler during his wartime deployment in North Africa and Italy, c. 1943.

Ann Hardy, programmer in pearls, at IBM in the late 1950s.

Families streamed into the Valley as the defense economy boomed. Palo Alto first graders, 1955.

MIT's Vannevar Bush, architect of the wartime research effort, at work on his "mechanical brain," the differential analyzer, 1930s.

David Packard (*left*) greets his graduate mentor, Stanford dean and provost Fred Terman, as William Hewlett looks on, 1952.

Charles de Gaulle tours Palo Alto, 1960.

The men and women of Fairchild Semiconductor, 1960.

Radcliffe students use a dormitory teletype to access a remote mainframe computer, mid-1960s.

An enterprising teacher persuaded the parents of Seattle's Lakeside School to raise money for its own computer lab. Paul Allen, class of 1971, and Bill Gates, class of 1973, became two of its most fervent users.

Lee Felsenstein with a bullhorn he designed for use in Berkeley campus antiwar protests, 1969.

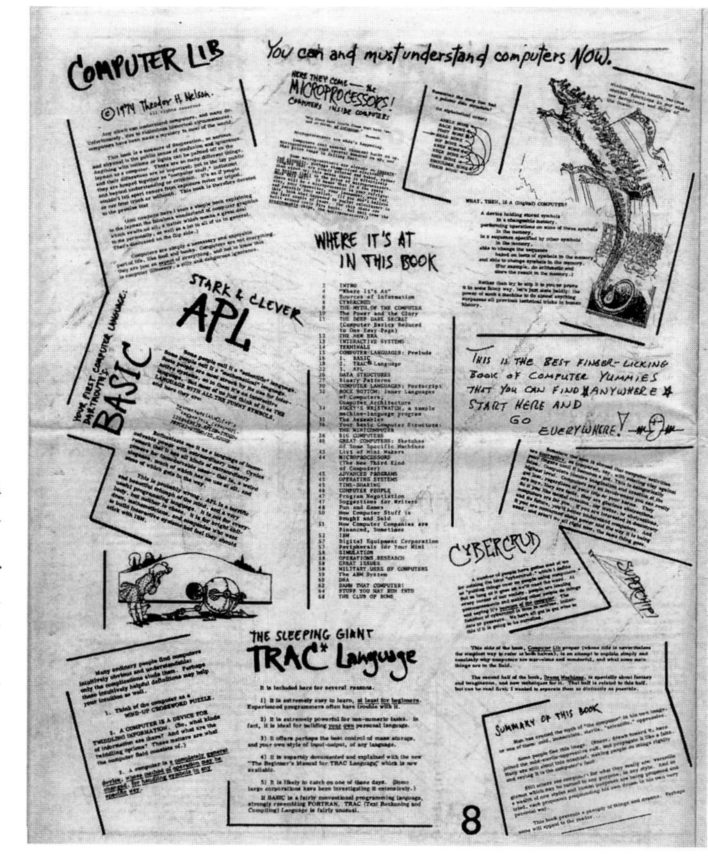

COMPUTER LIB

You can and must understand computers NOW.

© 1974 Theodor H. Nelson. All rights reserved

HERE THEY COME— the MICROPROCESSORS! COMPUTERS INSIDE COMPUTERS

WHAT, THEN, IS A (Digital) COMPUTER?

YOUR FIRST COMPUTER LANGUAGE: DARTMOUTH'S BASIC

STARK & CLEVER APL

WHERE IT'S AT IN THIS BOOK

THIS IS THE BEST FINGER-LICKING BOOK OF COMPUTER YUMMIES THAT YOU CAN FIND ANYWHERE START HERE AND GO EVERYWHERE!

CYBERCRUD

SUPERCHIP

THE SLEEPING GIANT TRAC* Language

It is included here for several reasons.

1) It is extremely easy to learn, at least for beginners. Experienced programmers often have trouble with it.

2) It is extremely powerful for non-numeric tasks. In fact, it is ideal for building your own personal language.

3) It offers perhaps the best control of mass storage, and your own style of input-output, of any language.

4) It is superbly documented and explained with the new "The Beginner's Manual for TRAC Language," which is now available.

5) It is likely to catch on one of these days. (Some large corporations have been investigating it extensively.)

If BASIC is a fairly conventional programming language, strongly resembling FORTRAN, TRAC (Text Reckoning and Compiling) Language is fairly unusual.

This side of the book, Computer Lib proper (whose title is nevertheless the simplest way to refer to both halves), is an attempt to explain simply and concisely why computers are marvelous and wonderful, and what some main things are in the field.

The second half of the book, Dream Machines, is specially about fantasy and imagination, and new techniques for it. That half is related to this half, but can be read first; I wanted to separate them as distinctly as possible.

SUMMARY OF THIS BOOK

8

Part underground tabloid, part technical manual, Menlo Park–based *People's Computer Company* launched in 1972 on its crusade to demystify computing. In its third volume, a two-page spread promoted Ted Nelson's newly released *Computer Lib.*

Ed Zschau (*left*) chats with Wisconsin Republican Rep. Bill Steiger (*right*), sponsor of 1978 legislation that slashed tax rates and encouraged venture investment.

Master marketer Regis McKenna talks to client and friend Steve Jobs, c. 1980.

Employees of Regis McKenna, Inc. pose for a team photo, early 1980s.

Toyko, 1983. Long the home of cheap transistorized electronics, Japan set out to move up the microchip value chain in the late 1970s, precipitating a crisis in the California semiconductor industry.

Atari home computer brochure, 1983.

Venture capitalist John Doerr at the 1984 PC Forum.

Unlikely business partners Steve Jobs and H. Ross Perot at NeXT, c. 1986.

Conference impresario and tech evangelist Esther Dyson holds forth with Bill Gates at the 1991 PC Forum, as John Perry Barlow looks on.

Microsoft's Bill Gates and Intel's Andy Grove, 1992. Their "Wintel" duopoly ruled the decade's computer market.

Vice President Al Gore and President Bill Clinton at the first Net Day in Concord, Calif., 1996.

The Internet era created new opportunities for Indian IT workers and companies.
Infosys training session, Bangalore, 2000.

Amazon.com founder and CEO Jeff Bezos poses before the Seattle skyline at company headquarters, 2001.

Young tech workers flocked to "pink slip parties" in the wake of the dot-com crash. A red dot on your name tag meant you were looking for a job. San Francisco, 2001.

Google cofounders Sergey Brin (*left*) and Larry Page (*right*) with CEO Eric Schmidt (*center*), shortly before the company's 2004 IPO.

Presidential candidate Barack Obama talks to a young journalist after a town hall meeting at Google, 2007.

Mark Zuckerberg launches the Facebook Platform, standing before logos of brands now connected into the growing social network, 2010.

Chamath Palihapitiya, new-style venture capitalist, 2013.

“He always gave us the best of himself,” Gary Morgenthaler said of his father, David, pictured here in his early 90s, “and brought out the best from us in return.”

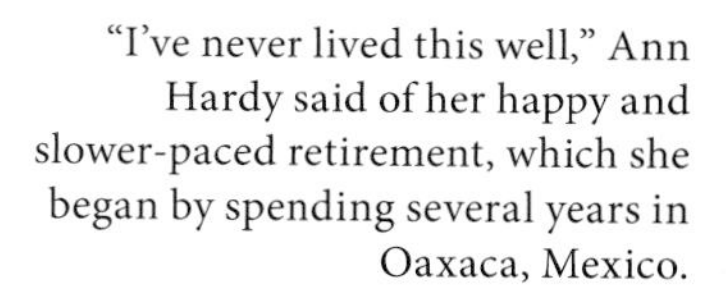

“I’ve never lived this well,” Ann Hardy said of her happy and slower-paced retirement, which she began by spending several years in Oaxaca, Mexico.

“It was very easy to think about giving back,” said Burt McMurtry of his philanthropy. Burt and Deedee McMurtry at the dedication of the Stanford Arts building, 2014.

Facebook headquarters, 2009.

days, but the crushing work routine still remained the norm. Gates rarely hesitated to angrily call people out if they missed a meeting, and he practiced the highly unusual habit of working on e-mail at home until late in the night. At Microsoft, "the company does many things to create a pleasant environment," one employee remarked. But "work definitely comes first. It's a velvet sweatshop." Many at Microsoft were not from the Pacific Northwest, but from far-flung places across the U.S. and around the world. But they had little time to experience the city around them. Another employee told the reporter: "There are programmers at Microsoft who after two years have never been to Seattle."[26]

Unlike the perpetual job-hoppers of Silicon Valley, many people at Microsoft had never worked anywhere else. Gates and Ballmer liked to hire straight out of college or graduate school; experience at other tech companies was irrelevant at best and distracting at worst. Wearing stock options' golden handcuffs, and with few other tech companies in Seattle of comparable size and wealth, the people of Microsoft tended to stay put. With a workforce that was mostly male and mostly under thirty, flush with the hypercompetitive and hyperactive vibe set by the two men at the top, the evergreen-ringed campus had "a corporate culture marked by devotion, self-flagellation, and a searing suspicion of the non-Microsoft world." Another chronicler later described Microsoft as "the frat house from another planet."[27]

Looking northward from California, the Seattle behemoth didn't just seem like it was on another planet. It was the mothership of a hostile alien invasion, relentless in its desire to gobble the software market whole. The friendly days of the early 1980s were long gone. Apple and Microsoft had moved from being business partners to opposing parties in both courts of law and courts of public opinion. For in 1983 Microsoft released the first version of Windows, its GUI operating system that sprang from the seeds planted by Charles Simonyi when he relocated from PARC three years before. It was a shot across the bow to Apple—never mind the fact that both companies had appropriated an idea first executed at PARC—and it was the beginning of an escalating war that crested on St. Patrick's Day 1988, when Apple sued Microsoft for copyright infringement.

It was a PR war as well. Microsoft had gotten better at corporate positioning, courtesy of a marketing chief who had once worked at Neutrogena, and what worked for face cream seemed to work pretty well for software too. Gone were clunky, hard-to-remember program names; now every program had "Microsoft" in its title: Microsoft Word, Microsoft Excel. The company mailed out free disks containing Word inside copies of *PCWorld*. Although Microsoft only sold to other computer equipment manufacturers, it began to blanket consumer electronics shows with announcements of upcoming products.

And along with the launch of Windows came the launch of the public persona of Bill Gates, the boy genius for the rest of us. Steve Jobs had Regis McKenna. Gates had Pam Edstrom, a petite bundle of laser-focused PR energy who made it her mission to turn the unpolished computer mogul into one of Middle America's most recognizable celebrities. Edstrom got Gates on the covers of *Fortune* and *Time*, and she also got him on *People* magazine's list of "25 Most Intriguing People." Taking a page from the Regis playbook, she invited reporters up to Seattle to meet with Gates and other executives, and once hosted a "pajama party" for twenty-five national reporters at the Gates family retreat on the shores of Puget Sound.[28]

Making it all the more infuriating, from the perspective of Valley engineering purists: Microsoft wasn't particularly innovative, and it routinely released buggy, clunky products. Gates had taken a page from the book of Big Blue, the company that had come late to the computer market and come to dominate it utterly: "Don't be the first to introduce the technology. Be second, and make money on it." The Seattleites used their first customers as product testers for a more functional second edition. That is, when they managed to release anything at all. And consumers might whine, but Microsoft's most important market was its enterprise customers. The business market didn't need something to be perfect. They just needed something to be good enough, and they needed it fast.

Gates was famous for promising to deliver software to computer manufacturers well before Microsoft had even designed it. In 1983, Valley venture capitalist Ann Winblad put a name on it: vaporware. The name stuck—even more so when Gates and Winblad began dating the following year. Love and loyalty led Winblad toward a more positive spin on Microsoft's release patterns. "Hey, talk about willingness to take risks!" she later remarked. "The company gets a new one out there, lets all this stuff come back, thinks about how crummy this stuff is, and then goes after the right stuff."[29]

With every new round of software and successive version of Windows, Microsoft edged further into Silicon Valley's territory. Then, in May of 1990, Gates & Co.

released Windows 3.0 and blew the doors off the barn. At last, here was a new operating system that was ready to live up to its promises. Proud mother Mary Gates called the day of its release "the happiest day of Bill's life." With Windows 3.0 and the updated versions released in quick succession in the early 1990s, Microsoft gained staggering market share and earned the eternal enmity of Apple and its fanboy universe. The blue screens of DOS gave way to WYSIWYG programs that offered all the user-friendliness of the Mac at a PC-platform price. And conveniently bundled up with this new OS were Microsoft products that riffed on the era's bestsellers. Lotus had Notes and 1-2-3; Microsoft had Word and Excel.

Now, the headlines about Microsoft weren't just about a new-generation CEO; they were about a company that was taking over. "Mighty Microsoft Breeds Fear, Envy," blared *PC Week*. *Business Month* called the company "The Silicon Bully." Around the cubicles of competitor tech companies, engineers grimly referred to Gates's company as "The Death Star." The destructive swath that Microsoft cut through the software industry, Lotus' Mitch Kapor concluded darkly, had turned early-1990s Silicon Valley into "the Kingdom of the Dead."[30]

THE NETWORK IS THE COMPUTER

As Bill Gates trained his big guns on Silicon Valley, Scott McNealy was spoiling for a fight. McNealy was one of four graduate students who had come together in 1982 to found Sun Microsystems, a company whose hockey-stick growth showed Silicon Valley how much market potential lay beyond the PC.

Here was another enormous success story made possible by the Valley's distinctive ecosystem: Scott McNealy and Vinod Khosla, Stanford MBAs; Andy Bechtolsheim, Stanford computer scientist; Bill Joy, Berkeley engineer. Added into the mix was another Berkeleyite, John Gage, who had spent the 1960s organizing antiwar marches and became Sun's adult supervision—first as its sales head, and then as its chief scientist. John Doerr of Kleiner Perkins backed the deal in one of his first investments after leaving Intel to become a venture capitalist. It was a very good bet. By 1988, Sun had annual sales of more than $1 billion and CEO McNealy was being celebrated in the pages of *The Wall Street Journal* as "the bad boy of the computer business." He relished the attention.

Like Bill Gates, McNealy was a child of privilege, who had grown up in the leafy suburbs of Detroit, the son of the man who ran marketing for American Motors in the 1960s. McNealy learned the value of a good sales pitch from his father, and he also learned tough lessons about holding on to market share as he watched his father's company get pummeled by Japanese competition and its mightier domestic rivals. Like Gates, he went to Harvard; unlike Gates, he finished his degree. From there he went to Stanford for business school, part of the wave of would-be entrepreneurs flocking west at the dawn of the microcomputer era to make their fortunes. The good-natured and no-nonsense McNealy was a workaholic, an unrepentant bachelor, and an ardent libertarian. When he finally tied the knot and started a family, he named his son Maverick. And he and his partners were utterly consumed with turning Sun into a major player in both the hardware and the software business.

While Sun's workstation made the company's reputation, it developed software for all kinds of computers that enabled file sharing across local networks—a market breakthrough in an era dominated by static, sit-on-the-desktop machines. The team announced an even bolder plan in 1987 to do for workstations what the IBM PC had done for personal computing: invite other companies to build clones that used the same hardware and software design. The gambit worked, creating a new ecosystem of workstations that, increasingly, were connected to one another via routers and networks.

Sun wasn't a computer company, it was a *systems* company—designing and developing its circuit boards and software, then subcontracting or purchasing all the rest. The future power of computing, McNealy and his colleagues could see, lay in a return to the networks created by time-sharing systems of old. Lone computers on desktops could only do a fraction of what was possible if you connected computers to one another and shared their power. "The network is the computer," proclaimed Sun's marketing tagline.[31]

By the early 1990s, Sun was raking in about $1 million every ninety minutes, and it was attracting the Valley's star engineers and highest-octane managers. Yes, the pace was punishing there, too, but it also included a particularly hypercompetitive strain of goofing around that was possible only in Silicon Valley. Along with the Friday afternoon beer busts and team-building events aplenty, there were annual April Fools' Day pranks that workers played on the company's senior executives, with the execs gamely playing along. Pranksters frequently resorted to sabotage of the top dogs' personal possessions, including very expensive cars. One year, workers put Bill Joy's car in a pond. A couple of years later, another group moved Andy Bechtolsheim's Porsche 911 into his office and set a massive fish tank inside it. These were jokes that were funny only if the victims were multimillionaires.[32]

Sometimes company "fun" veered into alarmingly misogynistic territory. Each morning, the engineering staff would launch the in-house computer systems that powered the company's operation—each of which had been given women's names by their male creators. In keeping with the highly gendered computer lingo, the verb used to describe bringing a computer system on line was "mounting." One female engineer angrily reported the results to a supervisor: "Every morning I have to listen to my male colleagues yelling down the hallways, 'I am mounting Cathy now!' or 'I mounted Judy a few minutes ago and she is purring!'" As another female employee put it, a career at Sun was like living "in the boy's dorm."[33]

Like Fairchild and Apple before it, Sun spawned an entire ecosystem—both of related products and connected people. There was Silicon Graphics, developer of workstations and software that brought 3-D modeling and sophisticated graphics capabilities to architecture firms and Hollywood studios. There was Cisco, a company that sprang out of Stanford's Artificial Intelligence Lab in 1984 and whose hardware made Sun's computers into networks, and whose routers would soon power much of the Internet age to come. There was MIPS, maker of microprocessors and computer systems designed for the workstation platform.

Executives at Sun and the other companies of the turn-of-the-1990s Valley ecosystem—both workstation and PC—went on to leadership roles in the companies of the next generation. Others, like the semiconductor generation before them, became venture capitalists. People moved on, but they stayed in the game. Such talent churn was distinctive to the Valley, and the other high-tech planet up in Seattle couldn't yet replicate it.[34]

YET FATTER BALANCE SHEETS DID NOT MEAN THAT THE VALLEY'S GRAY CLOUDS had dispersed. The shifts toward higher-priced hardware and infrastructure also shifted up the costs of founding a company. "There were days where $10 million would take a company from start-up to an initial public offering," said MIPS CEO

Bob Miller. "Today the average is $40 million to $50 million." The field winnowed; scrappy little start-ups were harder to find.[35]

The chipmakers were large and prosperous, but even they weren't getting what they wanted anymore. Going against the recommendations of a government semiconductor advisory committee that included Valley luminaries such as Jim Gibbons (the assistant professor long ago detailed to Shockley Semiconductor was now Stanford's dean of engineering), the Bush White House announced at the end of 1989 that Sematech funding wouldn't increase in the future. More government consortia also weren't going to happen. The money "is not now available, and it is unlikely ever to be available," Bush's science advisor told Congress flatly. Back in the Valley, high-tech power players felt that the new president just didn't get it.[36]

Another, bigger jolt came only a few months later. Bob Noyce died suddenly in Austin in June 1990, felled by a heart attack at the age of 62. Sematech no longer had its dynamic chief executive. The bereaved Valley crowd lost one of their technological pioneers and their most reliable liaison to Washington's power brokers. His death signaled a generational change, the end of an era when Valley leaders had been men much like Bob Noyce: crew-cut engineers in shirtsleeves who were children of the Depression, molded by the Cold War, makers of tangible things like chips and computer terminals.[37]

As if to underscore the passage of the founding generation, the Cold War had come to an end. The defense programs that had buoyed the bumpy times in the 1980s now faced massive cutbacks, turning California's economy upside down. Lockheed's Sunnyvale plant cut more than one-fifth of its workforce as the company struggled to find non-defense customers. Overall, California lost 60,000 aerospace jobs by 1991. *Time*, an ever-reliable gauge of the zeitgeist, rolled out a cover story referring to the Valley as "Gloomy Gulch."[38]

Draw back from the defense-cut disruption and the Microsoft onslaught, however, and the bummed-out prognostications didn't seem quite as on point. Sun and its generational brethren were growing, and they were again *making* things. These weren't just computers and floppy disks of software, but networking infrastructure: equipment of metal and molded plastic that enabled computers to talk to one another, to share information, to increase their computing power exponentially.

PRELUDE TO A GOLD RUSH

Another thing happened on the way to the 1990s. Silicon Valley decisively lapped Route 128 both in the number of firms and in the number of tech jobs. By the time of the first Bush Administration, the Valley had twice as many companies with sales of $5 million or more. It had three times as many people working in tech. The two regions had developed largely in tandem for the first several decades of the high-tech age, propelled by their respective research universities and by the river of government funding that went disproportionately to the two regions. MIT remained a colossus in the world of academic computer science, and the rest of Boston's research labs couldn't be beat, but when graduates decided to head into the private sector, they usually hightailed it to California.

For Silicon Valley had changed, and Boston hadn't. Minicomputers remained the beating heart of its business; defense contracting mattered even more. The biggest venture funds, the most experienced operators, the most ambitious MBAs: they all were out West. Boston didn't have enough raw material for its next electronic generation, and some of the greatest Route 128 success stories had fallen far and fast by the time the 1980s drew to a close. Ken Olsen infamously missed the boat on microcomputers, and as the PC market ascended in 1983, Digital had nearly gone out of business. Retrenching and rising again, it found itself in a near-death spiral between 1989 and 1991, cutting over ten thousand jobs and having to slash billions from its operating budget. Olsen had hung on as the company's president throughout, but by 1991 had to give up and turn over the day-to-day management of his company to someone else.[39]

Wang, too, had fallen hard and fast, out of step with the rise of the business PC market and never moving swiftly enough to catch up with the rise of workstations. When in 1986 An Wang at last relinquished control of the company he had started thirty-five years earlier in a North End storefront, he turned over the reins to his son. Keeping things in the family turned out to be a terrible business decision, and the patriarch had to ease his heir out a few years later. Wang's beloved company filed for bankruptcy in 1992.[40]

AS BOSTON RECEDED, THE TECH WORLD'S ENERGY DECISIVELY SHIFTED TO THE West Coast. Silicon Valley was no longer just a place in Northern California. It was the command and control center of a network whose influence spread across the globe, center of a vast supply chain that stretched from Chinese fab plants to Israeli research laboratories to its rainy Pacific Coast competitor and doppelgänger, Seattle.

The 1980s might have been rocky, but boom and bust was the nature of the tech business. Other regions might be sunk by recession and technological obsolescence, but the Valley had its specialized ecosystem, its VCs and lawyers and real estate men and research labs. It had Fred Terman's Stanford, its steeples of excellence spiking higher than ever. That meant that the region's near-constant state of precarity was also a perpetual state of renewal, with a new cadre of technologists rising up to quickly take the place of the old.

David Morgenthaler had been reading these tea leaves for some time. Although he had long held out the hope that robotics and AI might revive the manufacturing economy of the Midwest, he realized soon after hitting the Apple mother lode that "California was going to be the big winner." He was already closer to seventy than sixty, but he decided that he'd move his venture business to the Bay Area as soon as he could find someone to help him manage it. That someone, fortuitously, was his son Gary, who had found great success as an operator and investor in a range of tech and biotech companies. Cleveland's Morgenthaler Ventures moved to Palo Alto in 1989. Exactly forty years after his first visit, David Morgenthaler finally got off that train.[41]

ACT FOUR

CHANGE THE WORLD

Never get high on your own supply.

CHRISTOPHER WALLACE (THE NOTORIOUS B.I.G.), 1997[1]

Arrivals

STANFORD, 1990

"Shoe." That was the only English word Jerry Yang knew when he arrived in California in 1978. Barely ten years old, he and his younger brother had landed in San Jose with their widowed mother, a college professor who had first fled mainland China, then Taiwan, in search of political freedom and economic opportunity for her two boys.[1]

The Yang family's journey had become an increasingly common one by the late 1970s. Barely more than a decade after the Hart-Celler Act tore down the quota system that had restricted Asian immigration for so long, the U.S. Asian-American population swelled to three million. Arriving in an era when middle-class housing and jobs were shifting from city to suburb, and Rust Belt to Sunbelt, a disproportionate number of the new arrivals gravitated to the suburban sprawl of the Pacific West—places like San Jose. The city had less than 15,000 residents of Asian descent in 1970. In 1990, it had more than ten times that number, far outpacing the overall growth of the city's population.[2]

Immigrants had been getting the job done for some time in Silicon Valley, from European-born refugees like Andy Grove and Charles Simonyi to the Asian-American and Latina women who assembled microchips in fab plant clean rooms. But the wave of immigration that began in the 1970s had a scale and impact that the Valley had never seen.

From San Mateo to Sunnyvale to Fremont, bedroom suburbs whose populations had been nearly entirely white now became dynamic and diverse communities of high-achieving, highly educated immigrants from India, China, Hong Kong, and Taiwan. They started newspapers, opened businesses, built houses of worship and schools and arts centers. They followed family members, college classmates, co-workers, creating communities of ethnic diversity rarely seen before in the United States. And they worked in—and became founders of—technology companies. By 1990, foreign-born engineers made up 35 percent of the Valley's engineering workforce. The numbers spiked higher still after the 1990 creation of the H-1B visa program, which allowed technical workers a path to permanent residency.[3]

This demographic earthquake had transformed the Valley by the time Jerry Yang became a teenager. The shy immigrant kid had transformed as well. He excelled at everything, vaulting from remedial to advanced English, acing math tests and winning tennis championships. He was high school valedictorian *and* student body president. He took so many advanced-placement classes that he completed the equivalent of his freshman year before he graduated. By the time college application season rolled around, he had his pick of where he wanted to go, and scholarships to boot.

Yang chose Stanford, even though the offer wasn't a full ride and he'd have to work part-time to pay for it. But it was close to home, and it was *Stanford*. He hadn't been a computer nut in high school, really, although he and his brother enjoyed playing games on their home Apple II. But life at the academic center of the Silicon Valley universe quickly turned him on to "real computer stuff," as he put it, so much so that he completed an electrical engineering B.S. and a master's degree in four years flat, sailing straight into Stanford's PhD program by age twenty-two. Yang was so much younger than the rest of the graduate students that they gave him the nickname "Doogie," after the hit television show about a teenage doctor.[4]

Jerry Yang was a standout, but he wasn't alone. The schools of the South Bay—and the classrooms of Stanford—now filled with all-American kids like him. They were high-achieving children of high-achieving immigrant parents, citizens of the world and of MTV-era America, who'd grown up with Apples in their bedrooms and Ataris in their living rooms. And they all were coming of age at a moment and in a place when the technology industry was about to blast their generation off into the economic stratosphere.

The rocket that took them there was the Internet.

CHAPTER 19

Information Means Empowerment

The racket woke up all the neighbors on Balderstone Drive. It was 6:00 a.m. on a weekday morning in May 1990 when the San Jose detectives raided the ranch house in this quiet corner of suburbia, hot on the trail of high-tech thievery. Their target: an eighteen-year-old college freshman who ran a BBS named "the Billionaire Boys Club" out of his bedroom, and who possessed such prodigious hacking talents that he'd built his own IBM clone, fully outfitted with copyrighted software. "He's a pretty clever kid," the officer who led the raid admitted. However, he might be committing federal crimes. The agents didn't arrest the teenager that day, but they did something nearly as devastating: they confiscated his computer and the towers of shoeboxes that contained his floppy disks.[1]

The maestro of the Billionaire Boys Club wasn't alone. Hackers were stealing software code from the big computer companies and long-distance dialing instructions from the phone companies, trading them online and mailing them to other programmers. It was phone phreaking 2.0, the latest iteration of the kind of pranks that computer titans like Jobs and Woz and Gates used to do to show off their programming smarts. But now the kids in suburban bedrooms had ways to connect with others, over the blooming network of message boards riding on the backbone of dial-up networking services. Now all of corporate America ran on chips and bits and bytes of electronic data, and it had millions to lose if information landed in outsiders' hands. And now the companies founded by Jobs and Woz and Gates were big businesses, too, whose closed-off, proprietary software systems represented an establishment that hackers deeply mistrusted.

The raid in San Jose was only one of fourteen that happened that day across the country, as a total of 150 agents fanned out in a high-profile sweep that the Secret Service's Arizona field office—which led the sting—dubbed "Operation Sun Devil." By day's end, the feds had made three arrests, seized forty-two computers and more than 23,000 floppy disks, and attracted a wave of publicity for a war they had been waging stealthily ever since Congress had passed a stringent computer crime law in 1986.[2]

Prosecutors were adamant that bold measures needed to be taken. "It is possible to transmit computer information for an illegal purpose in the blink of an eye," asserted U.S. Attorney Stephen McNamee as the raid made the front page of *The New York Times* that June. Advocates for computer freedom and civil liberties were equally sure that the raids violated the Constitution. "The Fourth Amendment provides strict limits on rummaging through people's property," cried San Jose's unapologetically liberal Congressman Don Edwards.[3]

Silicon Valley's "information wants to be free" crowd was incensed too. In the years after leaving Lotus Software, Mitch Kapor had found his next chapter online, at the WELL. There, Kapor had started to post extensively on the issues he cared most about: software design, privacy, and free speech. Operation Sun Devil suddenly made him realize how vulnerable any chat room could be to government snoops. "This could have been me," he realized.[4]

Flush with many millions from selling his Lotus stock, Kapor established a legal defense fund for the accused hackers and began raising money among his friends. One contributor was fellow WELL denizen John Perry Barlow, for whom the raid had prompted flashbacks to the 1960s and a government that was "a thing of monolithic and evil efficiency." By summer's end, Kapor, Barlow, and several other tech-industry insiders went even further, establishing a new group to fight government threats to the free flow of computerized information, and to establish some new rules for a medium that defied so many of the old ones. They named it the Electronic Frontier Foundation. *Upside* magazine, a short-lived but influential chronicler of the dot-com boom, later called the EFF "the ACLU of the infobahn."[5]

The tech world had always found frontier metaphors irresistible, and the growth of the online world had Silicon Valley throwing them around with abandon. "Like the Old West," *InfoWorld* columnist Rachel Parker cheerfully explained to her readers announcing the EFF's formation, "the electronic frontier is uncharted territory—no one can stake any claims to it, and boundaries are rare." Illegal hacks were unfortunate, but unavoidable. "Corporate databases are something akin to Wells Fargo ponies running valued goods across the new territory. Someone, somewhere is bound to shoot at the riders." Barlow also talked about the cyberworld in terms steeped in plenty of John Wayne myth and very little Native American history. "It is vast, unmapped, culturally and legally ambiguous, verbally terse . . . hard to get around in, and up for grabs."[6]

Operation Sun Devil faded into the desert sunset. But the questions it raised about the laws of cyberspace became ever more urgent. For the same May that the feds made that predawn raid on a San Jose subdivision, a British computer scientist named Tim Berners-Lee began to circulate a modest proposal to adapt Ted Nelson's thirty-year-old notion of "hypertext" to organize the sprawling surge of information on the Internet. He called it the World Wide Web.[7]

INTERNETTING

The Internet was more than thirty years old by the start of the 1990s, and it still had the academic and proudly noncommercial spirit it started with in 1969. The sharp push toward deregulation of communications and information industries during the Carter and Reagan years had turned dial-up networking into a sharp-elbowed marketplace. Prodigy and CompuServe had user bases in the millions, Usenet groups and BBSs sprouted by the hundreds, but they all mainly connected over telephone wires, just as time-sharing companies had done for decades. Even the Valley-insider group on the WELL relied on dial-up networks.

The reach and user base of the Internet, however, was expanding significantly. In the mid-1980s, to the great delight of academic computer scientists who had bemoaned the ARPANET's slow rate of growth under Defense Department supervision, the National Science Foundation took over the network and renamed it the NSFNET. Along with the name change came a renewed focus on serving academic researchers and on wiring as many campuses as possible, as DARPA's push to compete with the Japanese on supercomputers had yielded fruit, and now the feds had five brand-new supercomputer centers in need of high-speed and high-capacity network connections. NSFNET did the job. One serendipitous offshoot: because of the expanded number of universities in the network, thousands of college students now had Internet access in their dorm rooms, fueling the growth of online communities among a tech-savvy younger generation.[8]

Along with chat-room-dwelling collegians, the NSF opened up the Internet to private-sector companies too—but only if they adhered to the agency's "acceptable use policy" and used the network for communication only, and not for commercial transactions. Despite these strictures, business eagerly jumped onto the Internet backbone as soon as it could. The network was faster than anything else out there, and it could handle large amounts of data that the phone lines couldn't. By 1990, for-profit carriers were on their way to linking more than three thousand business customers into the network. The twenty largest computer companies delivered hardware, software, and consulting services over its wires. Networking conferences that previously had been the domain of only the most die-hard computer science types now teemed with men and women in business suits as well.[9]

One of the firms that used the Internet in this way was the French conglomerate Schlumberger, where Marty Tenenbaum led a Valley-based AI lab for the better part of the 1980s. Tenenbaum—the guy who'd bought a year's worth of books before moving out from MIT in the 1960s—saw the growing online traffic, and he became increasingly frustrated at the missed commercial opportunities. "Stuff was going back and forth," he mused, "and no one was making any money." Departing Schlumberger after the 1987 stock market crash prompted company-wide consolidation, Tenenbaum finagled an office on the Stanford campus and started thinking hard about how to build a system for electronic commerce.

Over long walks with colleagues and long runs in the hills, Tenenbaum cooked up his first idea: convert old-fashioned mail-order catalogues into electronic files and send them via e-mail, which by that time had about twenty million users. In 1990, he decamped from Stanford, secured a couple of DARPA grants, and started what became the world's first e-commerce company, a tiny little outfit called Enterprise Integration Technologies, or EIT. Never mind that sharing such files over dial-up would be a glacial process as long as commercial Internet transactions remained illegal. Tenenbaum believed the NSF would budge from its noncommercial stance eventually.[10]

Significantly, the innovation that changed everything was not a product of DARPA, NSF, or one of their American academic grantees. It emerged from outside the U.S. altogether, from the mind of a British scientist employed by the European Organization for Nuclear Research, or CERN, in Geneva.

Nonetheless, American hacker-and-homebrew culture provided soul and inspiration for Tim Berners-Lee. He wanted information to be organized, but he also wanted it to flow freely and transparently. Working on a NeXT workstation (just like any self-respecting member of his scientific tribe), Berners-Lee and his CERN team created many of the building blocks of the online future. Hypertext markup language, or HTML, provided a common tongue for all the information now riding atop the Internet, both textual and visual. Hypertext transfer protocol (HTTP) was a platform on which to share the new language. A mailing-address standard (the uniform resource locator, or URL) got it to the right place. And a browser, "WorldWideWeb," became a portal through which the user could access it all.[11]

When Berners-Lee posted the server and browser programs to an online newsgroup on August 6, 1991, he unsealed powerful channels of Internet communication to non-experts much in the same way that the Apple II had opened up the microcomputing world to non-technical users fourteen years earlier. But unlike an Apple, the Web wasn't something you bought at your neighborhood Computerland. It didn't require paying for a pile of floppy disks. Despite the raging Mac-vs.-Windows war, the Web remained as neutral as its Swiss origins. It was free for anyone to download, and deliberately designed to work with everything and transcend what Berners-Lee politely referred to as "the technical and political battles of the data formats."[12]

Meanwhile, back across the Atlantic, the Internet was experiencing serious capacity problems. As more and more users joined the online universe, the middle-aged network didn't have the speed and bandwidth to keep up. In the words of an all-star group of computer scientists convened by the National Academy of Sciences, the existing Internet networks were "fragmentary, overloaded, and poorly functioning." The United States long had been at the technological forefront, but now Japan and Western Europe threatened to jump ahead.

The drumbeat grew in the tech community to open up the Internet to buying and selling, spurring the commercial carriers on the NSFNET to form a trade association to lobby Congress for a change in policy. Pressure was on for additional common, global standards that would permit true "Internetting" to occur. Downloads of Berners-Lee's Web client racked up into the thousands, but other browsers and protocols were being downloaded a lot too. NSFNET had adopted some common rules and classification systems; this included, significantly, the TCP/IP software protocol developed by Vinton Cerf at Stanford close to two decades before, which allowed different kinds of computers to communicate with one another. But it still was not enough to accommodate the pace of Internet growth, and certainly not enough for a full-blown commercial Internet. Some rule-minded grownup needed to bring order to this house party.[13]

THE GENTLEMAN FROM TENNESSEE

Enter Al Gore. The onetime House backbencher was now a U.S. Senator. Unlike many of his fellow Atari Democrats, Gore long had been a computer *user* as well as computer advocate. When the personal-computer market was still in its infancy, he declared that micros presented "dramatic new opportunities for optimizing our resources and exploring new frontiers" and organized computer classes for his fellow legislators. He regularly invited top computer scientists to his office to explain emerging trends in hardware and software. He had three home computers. He was typing a future bestseller, *Earth in the Balance*, on an early laptop. He went to computer-industry conferences, wrote articles for *Scientific American*, and fluently spoke the language of VLSI and AI, RAM and ROM. On top of all that, he was a baby boomer who invoked E. F. Schumacher and listened to the Grateful Dead.[14]

The Tennessean's worldview was centrist, market-focused, and technophilic. He believed that access to powerful, networked computers would open up marvelous new horizons in education, economic opportunity, and democratic communication. Tech wasn't just a policy, it was a *solution* to all sorts of other kinds of policy problems. Gore already had made one bid for the presidency in 1988 and was a fair bet to try again in 1992. For a man with presidential ambitions, beefing up and commercializing the Internet could be a signature issue—an "Information Superhighway" to provide a shot in the arm for an American economy limping along in an early-1990s recession.[15]

Gore might have been an early adopter, but he was hardly a lone wolf. It was clear by the start of the 1990s that a new wave was brewing in tech, and that government action now would make or break the Internet's ability to realize its promise later. Bipartisan support enabled the Gore-sponsored High Performance Computing Act to become law in December 1991, only five months after Tim Berners-Lee released his Internet browser. President Bush endorsed it, and so did House Minority Whip Newt Gingrich. The Act ushered in an era of better standards and faster, higher-capacity connections, while still keeping the decentralized, democratic structure of the Internet intact. Soon after, as Marty Tenenbaum had hoped, the

NSF started to pull down the walls of its online garden, changing the terms of its "acceptable use policy" to permit business transactions online.[16]

Mitch Kapor ran with a crowd that made no bones about its antipathy for government. Along with Barlow, one of his EFF co-founders was John Gilmore. An early Sun Microsystems employee with a net worth in the many millions, Gilmore was a key figure behind the Cypherpunks, a libertarian hacker collective devoted to the pursuit of building cryptography-based monetary and communications systems. The future pursued by the Cypherpunks was an exalted state of "crypto-anarchy," unbounded by government control.[17]

Perhaps because he'd always been bouncing back and forth between East and West Coasts, Kapor understood that making change wasn't always about storming the barricades. It also came from changing the establishment from within. He'd given $100,000 to Michael Dukakis's bid for president in 1988, and he was close with Massachusetts Representative Ed Markey, who chaired the committee responsible for setting telecommunications policy. Along with leading the EFF, Kapor took on the chairmanship of the new lobbying group for commercial Internet providers formed in 1991.

Wearing both those hats, Kapor came to Capitol Hill in the spring of 1992 to underscore the importance of keeping the Internet free, and to warn against letting the structure and management of the network become dominated by just a few, powerful private-sector gatekeepers. Big telecoms and cable companies already were eyeing the commercial Internet hungrily, and he worried that the system's decentralized and neutral spirit was in jeopardy. "The new information infrastructure will not be created in a single step," he warned, "neither by a massive infusion of public funds, nor with the private capital of a few tycoons." For decades, the Internet had functioned largely as an academic sandbox of ideas, for iteration and collaboration unencumbered by the demands of the stock market and quarterly earnings reports. Government's role was to ensure commercial competition, expand access, allow for free expression, and keep it a system where no single entity was in control.[18]

The hearing rooms were nearly empty—only Markey, Gore, and a few committee staff people understood enough to pay attention—yet Kapor's message resonated deeply with his listeners. At a time when few lawmakers understood what the Internet was, much less had any way to experience it directly, the small band of tech-industry advocates and mostly Democratic political staffers began to lay out the future blueprint of the online world.[19]

REINVENTION

Right about this time, the world of technology crashed headlong into the 1992 presidential election. The same week that tech mogul Kapor was giving thoughtful, nuanced commentary on Internet policy on Capitol Hill, candidates Bill Clinton and Jerry Brown were brawling like undisciplined second-graders. The Arkansas governor already had slogged through several near-death moments in the 1992 primary season, many of them self-inflicted—sordid allegations of a longtime affair, revelations of a youthful effort to avoid the draft, a protestation that he once smoked marijuana but "didn't inhale." He'd fended off an early and strong challenge from another familiar high-tech name, Paul Tsongas. Now, in the heat of the New York State primary and with the Democratic nomination in his sights, Clinton had to fend off Brown, who'd entered the race making bold promises and with few bridges left to burn.

After losing his 1982 bid for Senate, the former California governor had retreated into the cushy world of corporate law and into the political wilderness. Now Brown was coming out swinging, going after Clinton's wife, Hillary, for *her* corporate law-firm ties and condemning the governor for his "scandal a week" campaign. Clinton could barely keep his explosive temper in check, scorning Brown's popular proposal for a super-simple flat tax as something that would benefit only the rich, and demanding that Brown reveal his considerable net worth by releasing his tax returns. The Californian was the longest of long shots, but his scrappy outsider message had appeal in a recession-dampened election year. As he glad-handed morning commuters on the Staten Island Ferry, Clinton had to put up with passersby yelling "Go Jerry!" and "Brown, all the way!"[20]

Underscoring the hunger for a different sort of politician was the entry into the race of the unlikeliest of presidential contenders: Ross Perot.

Perot may have been the first tech mogul to run for president, but he didn't spend much time talking about the computer hardware and software that had made him rich (much less the Medicare and Medicaid contracts that helped him do it). Instead, he put himself forth as the ultimate self-made businessman, a straight talker who'd fix the mess in Washington and fix the economy along with it. Perot had left old Republican loyalties behind (as well as any lingering friendliness between him and his fellow Texan George H. W. Bush) and ran as an independent. Scorning politics as usual, he announced his candidacy on CNN's wildly popular *Larry King Live* and pledged his loyalty to the "everyday folks" who'd written in, urging him to run. Bush's people considered him a kook, and Clinton's people dismissed him as a cable-TV gadfly, but by June, Perot was neck and neck in the polls with both of his rivals.[21]

When it came to winning the hearts and minds of many in the tech community, however, there wasn't much of a contest. Jerry Brown might have had a long history in Silicon Valley, and Ross Perot had been a computer business giant for more than three decades, but Bill Clinton had been waging a two-year-long charm offensive to persuade the titans of the new economy that he was going to be their man. And it was starting to bear fruit.

A couple of years earlier, at about the same time that the agents of Operation Sun Devil were busting up Balderstone Drive, Larry Stone had received a phone call. After two terms as mayor, Stone was back on the Sunnyvale city council, where one of his colleagues was Regis McKenna's wife, Dianne (she had run for office partly after Stone's encouragement). Although the Valley had twice gone for Ronald Reagan and many of its CEOs were Republican red, the rank and file of the tech world leaned slightly to the left. When it came to presidential politics, all the Valley needed was the right candidate. So when Bill Clinton's closest political aide Craig Smith rang his office one spring day, Stone picked up the receiver. The governor was coming to town. Would Stone help host a fund-raiser? "We're having a difficult time getting people to this event," Smith explained apologetically.

Eyeing the date on the calendar, Stone's heart sank. He had baseball tickets that day. Oakland A's tickets, no less. It was a great time to be an A's fan—the team had won the World Series in 1989 and were on their way to winning the pennant again that year—and Larry Stone hated to miss a great game for some long-shot Southern governor. But he was a good Democratic foot soldier, and he reluctantly said yes.

Getting his friends to come was like pulling teeth, Stone remembered, but when the night of the event rolled around, Bill Clinton exceeded everyone's expectations. The young governor didn't know much about tech, but he wanted to learn.

Everyone in the room got the full Clinton treatment: close listening, lots of questions. Clinton's preternatural talent for making his audience members feel that they were the most significant people in the universe—even if ever so briefly—hit the mark in Silicon Valley, a self-important place with a chip on its shoulder about whether outsiders could ever properly recognize its greatness.[22]

The event was the beginning. Shortly before Clinton formally announced his candidacy, he had a one-on-one meeting with Dianne and Regis McKenna amid the grand trappings of San Francisco's Fairmount Hotel. The three talked tech, and they talked politics. "Were there any Willie Hortons lurking out there?" Dianne McKenna asked Clinton, referring to the race-baiting wedge that Republicans had used to derail Mike Dukakis's 1988 campaign. After being assured otherwise, the McKennas became early and ardent supporters.[23]

Silicon Valley people weren't just rich sources of campaign cash. They were smart, they were powerful, and they were building the future. Clinton was a son of the Sunbelt, a fan of self-made entrepreneurs, public-sector poor but drawn to private-sector riches. Silicon Valley tech people were exactly the kind of company he liked to keep. "He was so alert, in touch and technologically literate," enthused one CEO. "He *listened* to us."[24]

Dave Barram was one of those convinced that night. Like so many others, Barram came west as a young man after his discharge from the Navy, lured by a place where, he once said mistily, "dreams were being fulfilled every day." He started his career at Hewlett-Packard in 1969, the same year that Dave Packard went to Washington as Nixon's Deputy Secretary of Defense. While Barram's politics skewed left, not right, the boss's move showed him that it was possible to be both a tech leader and a political statesman. At the end of the 1970s, he ran for Sunnyvale city council along with Regis McKenna's wife, Dianne. She'd won, and Barram hadn't, but he and the McKennas remained good friends, bonded by the lonely distinction of being among the few outspoken Democrats in tech. Eager to share what he knew with national party leaders, Barram had sent bullish position papers on tech policy to every Democratic presidential nominee from Jimmy Carter to Michael Dukakis. He never got a reply.

As Silicon Valley's star had risen, Barram had found a far more willing audience among policy wonks for his ideas about education and the tech economy, including Hillary Rodham Clinton, whom he first had met in 1987. Five years later, Barram had even greater wealth and industry connections—he'd been an early hire at the high-performance computing superstar Silicon Graphics, and then moved into a senior role at Apple—but hadn't lost his desire to find a Democratic leader who shared the Valley's worldview. With Clinton, he believed he'd found his man. And given the moribund economy and the chilly relations with the Bush White House, the time was right to lure some Republicans over to Clinton's side as well.[25]

Fired up and ready to go, Barram took a leave of absence from Apple in the spring of 1992 to work full-time on Clinton's election effort. Drawing on decades of personal and professional connections, Barram organized a private meeting between Clinton and leading Valley CEOs, including lifelong Republicans John Sculley of Apple and John Young of HP. Young was perhaps the biggest fish, and the hardest sell—he had chaired Reagan's Council on Industrial Competitiveness, after all, and he remained close with many Bush insiders.

Clinton entered the room and started talking. Twenty-five rich and powerful men listened. Young took copious notes. Clinton sounded some of the same themes as Packard had in an earlier era, talking about a government that helped people help themselves, and that invested in education and research. He echoed the message that Gary Hart and others had put forth the decade before, about the need to build an economy around sunrise industries and to modernize trade policy. He emphasized how important technology—*Silicon Valley* technology—was to the nation's future. He was sensible, self-deprecating, and obviously smart. The group was impressed. Afterward, Clinton walked straight over to John Young, who, in the words of one observer, "gushed" with enthusiasm.[26]

Not too long after that, on a stickily hot Arkansas day in early July, Clinton made the move that sealed his tech credentials. He announced that Al Gore would be his running mate. The choice was surprising, given that Clinton and Gore were both in their mid-forties and from neighboring states; to maximize electoral appeal, vice presidential picks often presented a demographic contrast to the top of the ticket. But, Clinton promised, Gore would be a different kind of veep. He'd be an equal, not a subordinate, responsible for a broad policy portfolio that included being the nation's "technology czar."

At long last, Dave Barram had someone to read his position papers. What's more, he now was being asked by Gore himself to beef up the campaign's broad-brush technology policy. Barram, in turn, invited John Sculley, Mitch Kapor, and others to chime in. Kapor took politicians' bold promises with a grain of salt, but he sensed that this was just the kind of thing that could shake the political and business worlds out of their Internet slumber. "It's psychological," he told a reporter. "If people believe there is genuine national leadership, they will be prepared to move."[27]

By September, a Silicon Valley that was usually quiet about its politics loudly declared its loyalty. Young, Sculley, and thirty other tech executives gave a public endorsement of the Democratic ticket. "I am still a Republican," Sculley said, "but I am voting for Bill Clinton because I don't believe America's industries can survive four more years of President Bush." The ghosts of the chip wars loomed large in John Young's mind as he made the endorsement. All his work in Washington hadn't resulted in much policy change. Clinton "understands business and technology and the need for a highly skilled workforce," said Young. Added Xerox chief Paul Allaire, "Bill Clinton has indicated his willingness" to work with high-tech business in fighting overseas competition. "Bush has not."[28]

The endorsements rocked the tightly-knit industry. Dave Packard, still the gray eminence of the Silicon Valley Republicans, released a testy letter in response that chided his old friends for being "caught in the updraft of Bill Clinton's hot air balloon." Packard then rallied a second group of tech executives who signed a letter in support of George Bush. But by the time that letter came out in mid-October, it was clear to political insiders on both sides that the Democrats were going to win.[29]

While both endorsements came too late to sway the basic electoral math, the 1992 campaign became a watershed moment for both industry and party. Silicon Valley was no longer a business-sector sideshow, but a power player. And the Democrats were in its corner. Clinton's campaign theme song had been "Don't Stop Thinking About Tomorrow." Silicon Valley *was* tomorrow—both in the products it made and in how it ran its companies—and his endorsers reinforced the message that Clinton was not, in the words of one aide, "a traditional tax-and-spend Democrat."[30]

GEEKS-IN-CHIEF

The Clinton win thrust some of his biggest tech supporters into the political spotlight. Amid Beltway buzz about whether they might be up for Cabinet jobs, Sculley and Young both won coveted invitations to the president-elect's Little Rock economic summit in mid-December 1992. Cerebral, breezy, and telegenic, Sculley long had enjoyed making big-think statements about tech and the future. He relished the opportunity to play policy wonk on a big stage. "The biggest change in this decade is going to be the reorganization of work itself," he told the assembled luminaries. "In this new economy, the strategic resources are no longer just the ones that come out of the ground, like oil and wheat and coal, but they are ideas and information that come out of our mind."[31]

One month later, Sculley found himself sitting at the right hand of Hillary Clinton at the new President's first State of the Union address. Sitting on her left was another former Republican, Fed chief Alan Greenspan. The choice spoke volumes about the new economic order and the Clintons' aspirations to reinvent the business-government relationship: Silicon Valley on one side, Wall Street on the other, and an extraordinarily powerful First Lady in between.

Only days after John Sculley enjoyed the best seat in the house at the 1993 State of the Union, Clinton and Gore flew west to bask in Silicon Valley midwinter sunshine and engage in some heady talk about high technology. First came dinner at a bistro in the tiny and tony suburb of Los Gatos with a close group of their top tech supporters. Regis McKenna sat next to Clinton; Dave Barram sat next to Gore. The new president sipped on caffeine-free Diet Coke as his vice president "waxed poetic about the 'gestalt' of gigabits." Emerging from the restaurant nearly three hours later, Clinton and Gore were greeted by a buoyant crowd of more than a thousand locals who'd been waiting throughout the chilly evening. They proceeded to shake hands and coo over babies well into the night. Los Gatos had never seen anything like it.[32]

The main event came the next day. "The policy wonks met the computer nerds," wrote one local reporter, "and neither side blinked." The two leaders sailed into the headquarters of Silicon Graphics for a televised town hall meeting emceed by CEO Ed McCracken. Soft-spoken and buttoned-down, with a boyish flop of blond hair, McCracken was another recovering Republican as well as a HP veteran who'd been inspired by Dave Packard's model of entrepreneur-statesmanship. Silicon Graphics had continued its run as one of the Valley's highest Wall Street flyers, and the eye-popping special effects it created for Hollywood blockbusters made it significantly more glamorous than its beige-box-producing Valley peers. SGI made cool technology that ordinary people could *see*. McCracken felt it was a good time for this kind of engagement. "As a company approaches a billion dollars in sales," he reflected, "it ought to play some visible role in the community."[33]

After a demonstration of remote video chat (Gore wielded the mouse, Clinton looked on in gee-whiz amazement), the two leaders unveiled the federal technology initiative that included things so many Silicon Valley executives had been dreaming about for years: $17 billion of investment in new technological research, more Sematech-like alliances, a permanent corporate R&D tax credit, fiber-optic information infrastructure, and more. After all the political debates and reports and commissions of the past decade, the Democrats had come back to where they started a decade before. It appeared that high tech was getting a national industrial policy.[34]

Dave Barram spun the proposal to waiting reporters. "We've had industrial policy since the Civil War. We built railroads, canals, we had land grant colleges," he explained. "The government and private sector have always had a relationship—here's what we think is most effective." Not everyone in the Valley agreed. Flashy and splashy Cypress Semiconductor CEO T. J. Rodgers—an ardent libertarian and one of those who fiercely denounced the Noyce-led chipmakers as "crybabies" in the 1980s—scoffed at the Clintonistas' earnestness. "I'm adamantly opposed to picking winners and losers," he said. But to the engineers in the room that day in Silicon Graphics, Clinton's charm and Gore's tech smarts won them over. "I'd just like to say, I didn't vote for you," one employee ruefully admitted to the new president during the question-and-answer period, to laughter around the room. "I wish I had."[35]

THE TECHNOLOGY CZAR

Despite the hype and hope on display at Silicon Graphics that afternoon, and the tech luminaries now in the White House orbit, Gore's job as high-tech point man was harder than it seemed. For one thing, the expansive size of his overall policy portfolio kept staff resources somewhat limited. The vice president's policy A-team busied themselves with marquee issues like trade, the environment, and the massive "Reinventing Government" initiative, Gore's ambitious charge to streamline and modernize government bureaucracy. For another, the Clinton team had a staggeringly ambitious agenda that included financial stimulus and deficit reduction, health care reform, welfare reform, and free trade. In the Old Executive Office Building, policy "war rooms" sprouted like mushrooms after a spring rain.

Tech issues remained difficult for ordinary people—and non-technical policy people—to grasp in 1993. Everyone in the Clinton White House had an e-mail account (Clinton sent the first presidential e-mail with great fanfare that March), but only the most dedicated geeks used it to communicate with people outside the building, much less in other cities or countries. The network principles of the web were baffling even to some of the most senior people in the Clinton Administration, including Wall Street multimillionaire Robert Rubin, the man responsible for the nation's economic policy. "I remember the first time I showed Bob Rubin a web browser," said White House tech policy advisor Tom Kalil. "He asked me, 'Who owns this?'" No one, Kalil replied. The financier couldn't believe it.[36]

Thus, the Information Superhighway plan that the Clinton-Gore team developed in consultation with industry leaders didn't attract all that much attention from D.C. insiders. Now titled the National Information Infrastructure, or NII, its September 1993 rollout got largely drowned out by press coverage of the administration's health care reform legislation. Significantly, Clinton appointees from the Department of Commerce—not the FCC—led the policy effort, despite the fact that it was, at its core, about regulating telecommunications. Clinton's Commerce Secretary was Ron Brown, powerful political fixer and former Democratic Party chair, and one of the more prominent people to hold this Cabinet spot since Herbert Hoover way back in the days of Harding and Coolidge. Placing the NII in Commerce's bailiwick signaled that the Information Superhighway was an economic policy with technology, and high-tech industries, right at the center.[37]

"Information means empowerment—and employment," declared the report. NII would overcome "the constraints of geography and economic status, and give all Americans a fair opportunity to go as far as their talents and ambitions will take them." Bold language aside, however, the initiative was not going to be another government-funded moon shot. It wasn't even going to be another interstate highway program. The private sector was going to build, own, and operate it. The government's chief role was to regulate—and that involved a considerable amount of up-front *de*regulation, knocking down the barriers that long had divided telecom and television and tech as separate markets.[38]

The business-first approach of Gore's proposal delighted the telecommunications and cable-television industries, and alarmed tech's left wing. Within a matter of weeks, sixty interest groups announced they were forming a coalition to fight for an information superhighway free of big-business influence. Those arrayed against the White House included anti-SDI warriors Computer Professionals for Social Responsibility, who now received a healthy chunk of their funding from the EFF. A corporate-run information superhighway, CPSR warned, could be a cyberpunk novel come to life. "It doesn't take a novelist's imagination to recognize the rapid concentration of power and the potential danger in the merging of major corporations in the computer, cable, television, publishing, radio, consumer electronics, film, and other industries." No central authority should dictate what could or could not happen on the Internet; its self-policing chaos was part of what made it so powerful. "Life in cyberspace seems to be shaping up exactly like Thomas Jefferson would have wanted," wrote Mitch Kapor, "founded on the primacy of individual liberty and a commitment to pluralism, diversity, and community." The politicians shouldn't mess it up.[39]

But an even thornier issue was brewing as 1993 rolled into 1994—one that went right to the heart of the privacy questions that had brought about the EFF in the first place. As more data zinged around the Internet and illegal hacks rose, the intelligence community had developed a new, impenetrable encryption technology called the Clipper Chip. This kind of crypto-privacy was what computer makers long had been hungering for, as it allowed them to offer a higher degree of security to tech-wary users, but it came with a catch. The feds intended to keep the keys to the Clipper, allowing them to unlock any piece of software or hardware that carried it. The bad guys couldn't snoop on what people did online. But the government could.

The prospect of this untrammeled federal eavesdropping power sent Silicon Valley's free-speech crusaders over the edge. Under no circumstances should there be a back door where a third-party power could get into people's information, even "if Mother Teresa and the pope were the two agencies" holding the Clipper Chip keys, said a CPSR spokesman. The large tech companies hated it as well. Sticking a piece of U.S. government spyware on their products would damage their ability to sell to consumers at home and seriously hamper their ability to build markets abroad. Why would a Chinese company buy Microsoft Office if they knew that it would mean the CIA could see every spreadsheet or e-mail?[40]

The Vice President hedged. The plan was not "set in concrete," Gore assured his tech-industry friends in February 1994, as the Clipper Chip controversy swamped the information infrastructure discussion. Furious intelligence officials pushed back against a wavering White House, warning darkly that national security would be compromised, but the anti-Clipper forces ultimately won the day. Yes, there'd be encryption, decreed the Clinton team. But it wouldn't have to be one single standard, and the third party with the unscrambling keys didn't have to be the government. "Clipper is dead," rejoiced EFF executive director Jerry Berman. Gore had made "a big step, both for privacy and security."[41]

The Clipper Chip may have gone nowhere, but disagreements about law enforcement's ability to snoop into the digital world remained a sore point between the Valley and D.C. into the next millennium, intensifying with the advent of mobile devices that could potentially record a person's every move. Some of Al Gore's early fans never quite got over the fact that the Vice President had been associated with the dreaded chip in the first place. "That was when it became clear, that Gore's sympathies were more with the national-security state than with cyberspace," sighed John Perry Barlow. "Al Gore is a Deadhead; he's also an authoritarian."[42]

DURING THE FIRST FOUR YEARS OF THE 1990S—WHEN 97 PERCENT OF AMERICANS had no connection to the Internet, and when going online involved the whir of a dial-up connection and a text-only interface—policymakers and advocates shaped critical ground rules for the online universe to come. Perhaps as important as what Washington policymakers did is what they did *not* do: place the network under centralized control, either of the federal government or via a private monopsony, like a wired version of Ma Bell.

Instead, the Internet stayed true to its academic roots, as a decentralized, networked world where no one was in control. The telecoms couldn't ration, the spooks couldn't spy; instead, Mitch Kapor's small-scale, independent, Jeffersonian vision could bloom. After a century of everything getting bigger—government, business, systems of social organization—the small and the networked had won. The Vietnam generation that had questioned the notion of progress, looked for the soul in the machine, and made the computer personal now was in charge, and the commercial Internet was their marvelous legacy.

CHAPTER 20

Suits in the Valley

You knew Wall Street was visiting when you saw the black town cars. They rolled all over Silicon Valley in those dizzying days of the late 1990s, when the NASDAQ couldn't stop climbing and everyone was scrambling to get a piece of the action. If a town car was in the parking lot, it was a sign that the East Coast money guys had come courting, hoping for the chance to underwrite another dot-com IPO. They were doubly hard not to notice once they got inside, blaring beacons in $3,000 suits and silk ties amid a California-casual sea of T-shirts and cargo shorts. Over time, the visitors started to clue into the dress code. On the second trip, they ditched the tie. On the third, they switched to a sport coat and khakis. The bankers in their town cars never went so far as to adopt the full Silicon Valley uniform, but they tried.

The extraordinary Internet run-up of the last years of the twentieth century, outpacing even the most bullish of predictions, shone the spotlight onto Silicon Valley like never before. The Valley completed its metamorphosis from place to idea, shorthand for a technology industry whose products had altered nearly every aspect of how the world worked, learned, and played. The 1990s left the Lamborghini-driving-and-Chablis-swilling wealth creation of the 1980s in the dust. In the PC-happy summer of 1983, the NASDAQ crested at a little over 300. In March 2000, it surpassed 5,000. During the eight years Bill Clinton was president, the tech-heavy composite grew by nearly 600 percent.[1]

But the avalanche of shareholder value was just one measure of how the online era changed the American economy. Software vapor floated atop a very tangible hardware infrastructure: millions of miles of copper wires and broadband connections and routers and servers that flowed into offices and schools and homes. Old-economy corporations spent billions retrofitting their operations for the online age.

The business of buying and selling—to consumers, to other businesses—migrated online. From the presidential bully pulpit to Capitol Hill committee rooms, Washington, D.C., steered public investment to accelerate the Internet's spread. Free-trade fever spread across the globe, opening giant labor markets to American tech companies and swelling the customer bases for their products.

The cumulative investment made the American economy roar. GDP grew for ten consecutive years—the largest peacetime boom in U.S. history. The Internet allowed transactions to happen in milliseconds, erasing time zones and language barriers and making the Valley into the command-and-control center for a global network of commerce. The distinctive business culture of Valley companies, which had been slowly percolating into the public consciousness since the early days of the microchip, went from object of curiosity to a model to emulate. Business innovation meant taking the ties off, beefing up the software engineering staff, and building a volleyball court out back.

And it all happened mind-blowingly fast. *The New York Times*' technology reporter John Markoff filed his very first story about the World Wide Web in early December 1993. "In the next four years," he remembered, "I was run over by a Mack truck." The Internet became the story of the decade, the world-changer to close out the twentieth century and open the twenty-first. Scribes like Markoff—a Palo Alto native who had been living in and writing about the Valley since the Homebrew era—saw their bylines move from the back of the business section to the front pages as they tried to simultaneously explain the technological underpinnings of this brave new world and document its astounding ascent. Technology was the newest hot subject in the magazine trade, spawning a glossy flock of Bay Area–centric monthlies like *Wired*, *Red Herring*, and *The Industry Standard*. The tech economy and its kingpins were the subjects of no fewer than thirty-four cover stories in *Time* and *Newsweek* between 1994 and 1999.[2]

So much of the dot-com boom seemed extraordinary, not least in how quickly its irrational exuberance came crashing to the ground in 2001, taking much of the 1990s prosperity with it. Yet this chapter of the silicon age exhibited similar patterns to those that came before, albeit on a supersized scale. Out of university classrooms and laboratories came young men with novel ideas, met by venture capitalists who funded and advised and connected them into the industry's networks of expertise. Valley lawyers helped young entrepreneurs cross their t's and dot their i's. Real estate moguls plowed under the last generation's office parks to make way for the next. Evangelists both inside and outside the tech world spread the story of Silicon Valley's magic, boosting a new generation of boy wonders and convincing America's leadership class that this really might be growth without end.

And as there always had been, there were the suits: not just the Wall Street bankers, whose pinstriped swarms grew with every upward tick of the NASDAQ, but the politicians and policy wonks, who turned the well-worn footpath between Washington and the Valley into a roaring superhighway.

It all started with a browser.

THE WIZARD

John Doerr liked thinking five years ahead. When the St. Louis native and Rice University graduate was a first-year business student at Harvard in 1974, he became so intrigued by the promise of Silicon Valley venture capital that he began cold-calling firms to see if any would offer him a job. But the market was at its lowest point. Forget about venture, the VCs told him. Go to Intel and work for Andy Grove. While at Intel, Doerr distinguished himself not only for his affability amid Grove's mercurial, hard-charging environment, but for being one of the first semiconductor guys to get bullish on microcomputers, pushing the company to make and market its own motherboard to hobbyists even before the two Steves had released the Apple I. As a new Kleiner Perkins partner in the early 1980s, as microcomputer mania seized the Valley, he was hanging out at academic conferences learning about the VLSI and 3-D technologies that would make chips faster and turn cheap desktops into powerful workstations.

So when Doerr saw his first demo of the Mosaic browser at the start of 1994, he already had been paying attention to the Internet for a long time. Eight years earlier, he had taken a leave of absence to spend several months in Tim Wirth's Senate office, absorbing all he could about the workings of the ARPANET. Since this stint as Capitol Hill's highest-net-worth summer intern, Doerr had continued to rack up enormous wins in computer hardware and software. All the while, he was increasingly convinced that the Internet would be the next big thing. And that's what he wanted to nurture. "We think in terms of building new businesses," he once explained. "In our more grandiose moments, we think of building industries." In the 1990s, Doerr and his partners did exactly that.[3]

Mosaic was their launching pad. If Tim Berners-Lee's Web client was the Apple II of the Internet, Mosaic was its Macintosh: the portal that opened up the online world for millions. Created by a group of graduate students at the University of Illinois supercomputer center who were tired of the text-only HTML environment of the earliest Web, Mosaic turned the Internet into an immersive, colorful, point-and-click experience. Within months of its 1993 release, the new browser had seized the imagination of Silicon Valley insiders. Its chief student inventor, Marc Andreessen, headed west to capitalize on the excitement. Andreessen found a first landing spot in the scrubby cubicles of EIT, where Marty Tenenbaum was busily building his e-commerce platform and knew that a good, graphical browser was essential to bringing customers online. But within months, the twenty-three-year-old Andreessen had been lured away from Tenenbaum's shoestring operation by a plump paycheck waved by Jim Clark, a former Stanford computer scientist who had founded Silicon Graphics.

Clark had been one of the people teaching John Doerr about the future in the early 1980s. Now, he had become annoyed with his company's cautious corporate direction (and CEO Ed McCracken's mooning around at the White House) and was casting around for something new to do. Andreessen was it. After signing on the dotted line, the two new partners promptly flew back to Champaign-Urbana and signed up the entire Mosaic engineering team. Weeks after that, Clark tendered his resignation at SGI and began pitching his old friend Doerr for help with his new venture. Kleiner Perkins put in $5 million, and Mosaic Communications was born.[4]

The rapid rise of the company that soon was renamed Netscape showed the tightly knit, decades-in-the-making Silicon Valley network in prime form. The leadership Doerr assembled for Netscape was a seasoned mix from tech and other venture-backed new industries, all grade-A men for a grade-A idea. Doerr recruited AT&T Wireless CEO Jim Barksdale, formerly of FedEx; his chosen VP, Mike Homer, was a veteran of Apple. Wilson Sonsini provided outside counsel. Kleiner Perkins itself was bursting at the seams in those days with high-tech venture partners: Sun's Vinod Khosla, Fairchild and Apple's Floyd Kvamme, and Regis McKenna, who at Doerr's urging had joined in 1986 as a general partner. Even Netscape's real estate had history: its first offices were sublet from HP, and it soon scaled up to grand headquarters on the newly-remediated site of Fairchild Semiconductor's Mountain View manufacturing plant. The Wagon Wheel was just around the corner.[5]

There was no better product around with which to create an entirely new industry than Navigator, the browser Netscape released in early 1995. Built to run on any type of computer—Mac, PC, Unix workstation—Navigator decisively left the desktop wars of the 1980s behind. Designed to run on high-speed broadband networks (like the NSF-funded one Andreessen had enjoyed in his U of I lab) and secure enough to host financial transactions, Navigator anticipated a dial-up–free future where nearly anything could be bought with a single click. The company further disrupted the usual order by giving its browser away for free. It was a startlingly efficient way to build a user base. One year into its existence, Netscape had six million customers.[6]

Netscape was just one star in the growing Internet constellation financed by John Doerr and Kleiner Perkins during those heady days of the mid-1990s. The firm put money in scrappy Northern Virginia–based America Online (AOL), the successor company to The Source, which was elbowing into CompuServe's and Prodigy's market share by inundating American households with free CD-ROMs pitching its wares. And KP incubated entirely new companies, like @Home, which promised to bring high-speed broadband into homes across America. (The venture later became part of one of the dot-com era's most celebrated failures. At the height of the late-'90s froth, @Home merged with the Internet portal Excite in a deal valued at $7.2 billion; two years later, the stock had lost 90 percent of its value and the company was bankrupt.)

And an earlier generation of Kleiner Perkins hits helped the online world build out and scale up. In March 1995, making good on its promises that "the network is the computer," Sun Microsystems introduced Java, the first programming language written with the Web in mind. Java product manager Kim Polese sensed that the new language was going to be a big deal after her team pitched a story about the launch of what they then called "Hot Java" to Silicon Valley's paper of record, the *San Jose Mercury News*. When the big day came, Polese frantically flipped through the business pages to see if they'd picked up the story. Not finding it, she cast down the paper in frustration. Then she saw the front page. There it was: "Why Sun Thinks Hot Java Will Give You a Lift."[7]

Java wasn't just a lift; it was a game changer. It brought the Web alive, allowing programmers to build applications with graphics and animations as robust as any found on current desktop software. And Java ran through a browser, not an operating system. The Web was not going to be balkanized like the PC platform. Programmers could build applications on Java and have them run on any kind of machine. Websites went from static and clunky to animated and agile. As Sun CTO Eric Schmidt put it, "It's the difference between a telegraph and a telephone." Java made Sun a jazzy Internet-era brand and turned Polese into a Web-era superstar (she soon left to start Web multimedia company Marimba). By 1996, Java so ruled the world that KP opened a $100 million fund solely to invest in companies building Java-based software.[8]

Doerr, who had emerged from the 1980s chip wars as a faithful disciple of Japanese management principles, called all of this "the Kleiner *keiretsu*." It was a collection of different firms sharing resources and expertise, operating symbiotically to create an entire market ecosystem—from the networks and routers to the browsers and portals to the programming languages and software applications. It was how Japan's electronics consortia had ruled the world in the 1970s and 1980s. But it also fairly characterized the web of personal relationships and specialized expertise and tacit knowledge that made Silicon Valley soar. Other Valley VCs, both veteran and new, had giant Internet wins, of course. But Kleiner Perkins came to symbolize the dot-com generation more than any other. The firm was the MITI of the Internet era, and John Doerr was the wizard behind the curtain.[9]

FARM TO FACTORY

Gordon Moore liked to observe that the most important thing Fred Terman's university did for Silicon Valley was to graduate 800 masters and PhD students per year, replenishing the region's intellectual pool. The other Moore's Law certainly held true in the case of the young people who came out of Stanford's computer science and engineering programs in the 1990s. Four decades of innovation in AI, software design, and networking had turned The Farm into the home for many of the world's best computer scientists, and a magnet for many of the very best graduate students. Stanford's encouragement of tech transfer meant that the very best of them already had well-formed dreams of commercializing whatever they built. If you wanted to study computing, you went somewhere else. If you wanted to study computing *and* found a company, you came to Stanford.[10]

The grad student dubbed "Doogie" became the Internet era's first brilliant case in point. After steaming through his first two Stanford degrees, Jerry Yang had found himself stalled as he worked on his doctorate. When he entered the program, he'd opted for computer-aided design as a focus, which in 1990 seemed like a great bet at a moment when Silicon Graphics was becoming the darling of the Valley. A few years later, the landscape looked a little different. Thanks to being at Stanford, he and his closest friend in the PhD program, David Filo, had cottoned on early to the wonders of Mosaic. When their advisor went on sabbatical, the two hunkered down in a pizza-box-strewn trailer at the fringe of the engineering quad and started spending all their time building websites: a homepage for a fantasy basketball league, a tribute to sumo wrestling.

The two spent so much time Web surfing that Filo put together a list of his favorite sites so that he could better navigate the swelling sea of online information. Yang—whose undergraduate work-study job shelving books in the Stanford library had introduced him to the joys of the Dewey decimal system—realized they could go one better by turning the list into HTML and putting it online for others to enjoy. The attention-shy Filo didn't really want his name on the thing (they were supposed to be writing dissertations, after all), so Yang simply dubbed it "Jerry's Guide to the World Wide Web." It was late 1993. Within a matter of months, the site had ballooned into a real business, as more people downloaded Mosaic, piled online, and tried to navigate through the Internet's dizzying bounty.[11]

By the start of 1995, the guide had gotten a million hits, Web-related start-ups were brewing across the Valley, and the two grad students realized that their time playing hooky in that trailer had generated a serious business opportunity. Yang and Filo left Stanford, taking with them a number of their campus friends, who now became full-time "Web surfers"—human beings who read the web and attempted to thoughtfully sort its contents into various categories. In contrast to the overwhelmingly male engineering teams that populated the new generation of start-ups, a good number of these early surfers were women.[12]

Now courted by the Valley's leading investors, the grad-school dropouts opted for $3 million from journalist-turned-VC Mike Moritz (now at Don Valentine's

Sequoia Capital) and named their new venture Yahoo! "Their greatest strength," elder statesman Valentine observed sensibly about Yang and Filo, "was the recognition of their weaknesses and their lack of experience." Once again, the Silicon Valley mentorship network, new and old, got to work. Moritz and Valentine connected Yahoo! to lawyers, PR experts, and Internet providers. They found a seasoned CEO, Tim Koogle, a Stanford graduate who'd been working in tech companies big and small for two decades. Marc Andreessen offered Yang and Filo server space at Netscape so they could move off Stanford's overburdened network.[13]

The corporate exclamation point and goofball job titles—Yang was "Chief Yahoo," and it got more fanciful from there—weren't the only ways in which the operation reflected new trends in the Internet-age Valley. Yahoo! ran on programming smarts, but it wasn't software or hardware that could be sold at a retail store or bundled into a desktop. Its product was *content*: a distinctive classification system for the Web's vast information landscape, a job first performed by the human Web surfers, then later—as Yahoo! scaled up to undreamed-of heights—by an algorithm. Intel sold microchips; Microsoft sold packaged software. Yahoo! and its dot-com brethren gave away their product for free. The only way to make money was through advertising. It was a profoundly new model for Silicon Valley.

After four decades of perfecting its art, the Valley's business ecosystem knew exactly how to nurture funny little companies that built chips or wrote code. It had little inkling of what might happen when those companies became information platforms.

THE COMMERCIAL SUPERHIGHWAY

The viral spread of the Netscape browser and the ease with which Yahoo! opened up the Web to surfing brought more and more people onto the Internet by the middle of the 1990s. They were browsing, searching, chatting, and e-mailing. But they were only doing a minimal amount of buying and selling. The act that later became a mundane part of Americans' everyday consumer life—go to a site, point and click, enter your credit card number, and hit "buy"—was an alien, anxiety-inducing idea in the early dot-com era. There was no guarantee that a credit card number wouldn't get stolen. There wasn't an obvious way to pay for shipping, or even to choose how quickly you'd receive your purchase. The solution arrived in the Valley via an under-the-radar federal grant, clocking in at a relatively modest $2.5 million, that at last answered Marty Tenenbaum's dreams of making the Internet into a bustling marketplace.

As often happens in politics, the money from the U.S. Commerce Department hadn't been designed with the Internet in mind at all. Instead, it was part of the blandly named "Technology Reinvestment Program," or TRP, created at the tail end of the Bush Administration to alleviate the economic pain felt by defense-dependent regions—like California—as the Cold War wound down. In the Valley, however, the money that the Clinton Administration awarded in early 1994 went toward creating CommerceNet, a new, Tenenbaum-run industry association dedicated to e-commerce. "Capitalism is coming to the Internet," declared *The Wall Street Journal*. Under Tenenbaum's management, CommerceNet developed software to protect credit card data and ensure safe shipping. It helped companies and engineers develop new e-commerce security tools. Most important, the enterprise got lumbering old-economy companies comfortable with conducting transactions online. At one point the association had 800 corporate members. In those early days, "everyone who did e-commerce was part of it," said Tenenbaum.[14]

By the end of the year, a host of other ventures were competing to develop their own electronic-payment software, while new companies were coming into the world that gave users their first taste of buying and selling on the Internet. "Right behind sex, commerce on the Internet seems to excite people the most," remarked Jim Bidzos, whose data-encryption technology was at the core of CommerceNet's software. In 1995, Bidzos commercialized his technology via the Web authentication company VeriSign—whose seal of approval became a familiar sight on many a payment page, assuring online consumers that they could enter their credit card numbers without hackers snooping in.

That same year, an Iranian-American engineer named Pierre Omidyar launched online auction service eBay out of his one-bedroom Silicon Valley apartment. eBay wasn't fully electronic in its commerce at the very first—to buy auction items, most customers opted to put a paper check in the mail—but that changed as more technologies spun out into the world and new users flocked to these portals. Omidyar's innovation was to blend the community-building power of the BBS or the Usenet group with the platform of e-commerce. He wasn't selling things; he was creating a community of people who sold things to one another. And with the auction model, he created an addictive product, with buyers returning to the site again and again to see if they could make the winning bid. eBay grew by 40 percent per month for fourteen straight months. It and other e-commerce portals developed and bought their own proprietary technology, but many of the basics of doing credit-card transactions and shipping online came from CommerceNet. Given the trillions of dollars eventually generated by online retailers, those TRP dollars invested at the start of 1994 may have had the greatest return on investment in Valley history.[15]

Marty Tenenbaum and Jim Bidzos had solved the problem of secure and reliable Internet shopping. Pierre Omidyar had shown it was possible to get people obsessively buying and selling with one another online. But all the VeriSigns in the world couldn't fix Internet retail's other big problem: the reluctance of consumers to buy something from a retailer sight unseen, at a possibly higher price, with no obvious way to get their money back if they didn't like it.

That was exactly the kind of thorny problem Jeff Bezos loved to solve.

REGRET MINIMIZATION FRAMEWORK

Jeffrey Preston Bezos's path to the Internet economy started when David E. Shaw decided to use computers to make money on Wall Street. Shaw had a PhD in computer science from Stanford and a comfortable faculty job at Columbia when he was first lured to the Street for a high-paying job at Morgan Stanley at the height of the high-rolling Reagan years. He lasted eighteen months before he decided to leave to start a hedge fund of his own in 1988—a different kind of fund, one that applied the most sophisticated of algorithms to snap up the best deals and trade at a scale and speed only possible by computer. The trading floor of D. E. Shaw & Co. had more than four times as many Sun workstations as it had employees, machine and man working around the clock to extract more power and profit from the market. Massive returns had Shaw inundated with résumés; he only hired 1 percent of applicants. In December of 1990, he hired Jeff Bezos.[16]

Born in New Mexico, raised in Texas and Florida, Bezos was another future titan who'd displayed early giftedness, from his aptitude for building electronic gadgets in elementary school to his photographic memory of every play sequence when he was defensive captain of the football team. A major influence on his young life was maternal grandfather Preston Gise, a missile specialist who had been part of the founding team at ARPA and later ran the entire Western region for the U.S. Atomic Energy Commission (his vast domain included the thousands of scientists of Los Alamos, Sandia, and Livermore Labs—the then employer of Ann Hardy, LaRoy Tymes, and many more).[17]

After retirement, Gise returned to his West Texas ranch, where his grandson spent every summer. By day, Bezos learned the tough work of ranching—laying irrigation pipes, fixing machinery, vaccinating cattle—and gained an appreciation for what could be accomplished through resourcefulness and hard labor. By night, he would marvel at the vast constellations and galaxies carpeting the Southwestern sky, kindling his passion for space exploration. He briefly considered becoming an astronaut before heading to Princeton to study electrical engineering and computer science, and his fascination with outer space never diminished. Slipping the bounds of earth, shooting into the final frontier: that was the ultimate in self-reliance, the grandest exercise in long-range planning. His grandfather had been part of that great flurry of shoot-the-moon optimism in the early 1960s; the grandson wanted to recapture some of that hope and grand vision as he embarked on his own career.

After graduation, Bezos had his pick of prestigious big-company tech jobs, but the emerging opportunities for computer scientists on Wall Street sounded more audacious and interesting. The people he admired most in the tech world were those who hadn't played it safe, who'd been confident enough to push boundaries early on: Bill Gates at Microsoft, Alan Kay at PARC. A few years later, Bezos affixed a (mis)quote from Kay to his e-mail signature: "It's easier to invent the future than to predict it."[18]

Although he presented an affable front, with his ever-present bark of a laugh and his eagerness to describe himself as a "nerd," Jeff Bezos was a meticulous analyst, so much so that Shaw tasked him with assessing the Internet's prospects early in 1994. Bezos was a big believer in finding the white spaces and jumping in, and the Internet clearly was a massive white space. Someone was going to make an immense amount of money selling things online. A few months into the research he started to ask: Why not him?

Of course, it was a crazy idea. Bezos was raking in money in his current job, not to mention that it was the middle of the year. Walking away meant he'd lose his 1994 bonus. Was he really going to throw it away for the off chance that an Internet-based retail start-up might actually work? Bezos's story of what happened next was one that he told again and again to curious reporters during those early years: urged by Shaw to take some time to think about the decision, Bezos decided to sketch out what he called a "regret minimization framework." ("Only a nerd would call it that," he'd chuckle.) If I got to age eighty and looked back on this decision, would I regret having tried this? Bezos remembered asking himself. The answer was a firm no. He knew he wanted to sell things on the Internet, and he also knew that he couldn't do it in New York. For he'd figured out what he wanted to sell online—something that you didn't necessarily have to touch and see first, something that wasn't that expensive, and something that wasn't breakable—a book. He would become an online bookseller, and he would do it from Seattle.

As much as Seattleites might have liked to dream otherwise, and as much as reporters' spin might have framed it as another heroic young man heading west to find his destiny, Bezos's decision to move to the Emerald City was the unsentimental choice of a detail-driven Wall Street analyst. Most of the Internet early adopters were in California. But the rules of the Internet road were that customers in the same state as a firm's headquarters would have to pay state sales tax on their online purchases. Scratch that; he wanted to be somewhere with a smaller population, and fewer taxpayers. The fact that Washington State didn't have an income tax was an added bonus.

Then there was the matter of logistics. The West Coast's biggest book distributor was close enough to Seattle that his team wouldn't need a warehouse of their own at first; they simply could order up the books and mail them from there. But there was another Seattle factor, one that signaled that Jeff Bezos had far more than books on his mind. The region had Microsoft and its thousands of software engineers, some of whom might be restless enough to leave the mothership and join a start-up. If this was going to be what he planned, he'd need a lot of them. "Life's too short to hang out with people who aren't resourceful," Bezos once said. And he was the most resourceful of them all.[19]

It took him 60 meetings and considerable powers of persuasion to raise his first $1 million from 22 investors—"anyone who knew anything about the book business did *not* invest," Bezos remembered—but by the summer of 1995, Amazon .com was open for business. (Bezos toyed with calling it Relentless.com, but opted for something a little softer.) The early clientele were computer nerds, too, and computer manuals dominated the bestseller list, but that didn't last long. It was immediately clear that selling products over the Internet fixed the chronic inventory problem that had sunk many a brick-and-mortar operator. Amazon didn't have to keep things on the shelves in anticipation of customer demand, but it could fill that demand for any title, from the celebrated to the obscure. Consumers flocked to the service, and Michael Crichton page-turners quickly lapped the computer books on Amazon's sales charts. "If it's in print, it's in stock," was Amazon's boastful motto, and it was right on point.[20]

While Amazon was in Seattle, its ties into the Valley were so tight that it might have well been in Sunnyvale. The local presence of Microsoft, of Boeing, of the University of Washington—that was nice, but not yet particularly relevant for a garage-based operation that only had a handful of employees. Bezos needed the kind of start-up support that only the Valley could provide. Amazon became an early member of CommerceNet, adopting the software and security protocols so that nervous book buyers would stop putting checks in snail mail to pay for their new books. Then came the coup: an early funding round from John Doerr, who put $8 million and a management team into Amazon, ending up with 15 percent of what would one day become one of the world's largest and richest companies. Little surprise that Bezos and Doerr would immediately hit it off, for Bezos, too, was a five-years-in-the-future sort of guy, "willing to plant seeds and wait a long time for them to turn into trees."[21]

Doerr didn't have to wait too long for those trees, however. Between 1996 and 1997, sales jumped nearly tenfold, from $15 million to close to $150 million. Forty percent of business was from repeat customers. Amazon went public that year, with Bezos keeping control of 41 percent of the company. By the end of 1998, its customers were in the millions and its market valuation was a gobsmacking $30 billion—more than the century-old emperor of American retail, Sears.[22]

The gee-whiz press stories about the nerdy guy in Seattle focused on the books he was selling and the frenzy he was creating on Wall Street. They paid less attention to the thing that was attracting so many customers and keeping them coming back for more: the data. For Amazon wasn't just able to give its visitors the books they came to buy, but the books that they didn't even realize they wanted. Carefully mining data from every transaction, Amazon was able to track its customers' tastes with eerie accuracy. Bought a Stephen King thriller? Here are five other books you might like, and here is some music, too. "You might also like" became an addictive feature precisely because it was so on-target. Bezos might have seemed like a goofy book lover, but he was pure quant, and his work on Wall Street had shown him the extraordinary things computer models could do. The humming engine behind Amazon lay in its vast and tightly guarded rooms of blinking black server blades. From the very beginning, Amazon wasn't a bookstore. It was a data platform.

Jeff Bezos also gave Steve Jobs a run for his money as the most relentlessly on-message CEO the tech world had ever seen. As Amazon's valuation climbed and Bezos's net worth soared into the billions, his persona remained as earnest and humble as ever. He still drove his beat-up Honda Accord and sat at a desk made of an old door and blocky, nailed-on two-by-fours. Amazon rented grungy office space in marginal neighborhoods, their insides as messy and makeshift as a freshman dorm during finals week. It didn't produce TV commercials, host splashy launch events, or even put a big sign out front.

All was in service of the message that Bezos pounded home to every reporter, in every press release, and in every annual letter to his shareholders: the customer came first. Valley dot-coms might have outrageously luxe offices and holiday bacchanals. Amazon.com had door desks and pizza parties. "Everything connected with the company," noted *Adweek* admiringly, "is carefully scripted to create the image of a scrappy underdog that cares more about people than profit."[23]

THE QUEEN

As all this percolated out West, something happened back East that became one of the luckiest of the tech industry's many lucky accidents: the stock market flooded with money. Many things triggered the flow of investment capital. The early 1990s recession ended; the pain of post–Cold War defense cuts faded. The Clinton economic program cut government spending and hacked away at the deficit, recapitalizing banks. Telecom deregulation and NAFTA opened new markets and lowered labor costs, increasing corporate profits. Bucking tradition and economic laws of gravity, Fed Chair Alan Greenspan kept interest rates low, creating incentives for investors and individuals alike to borrow, borrow, borrow.

While money sloshed around the markets, the Valley's PR mavens filled the business pages with stories about Internet companies and the whiz kids who ran them. As they read about the latest generation of endearing geeks—Andreessen, Yang, AOL's Steve Case, and others—Wall Street investors got Internet fever. By April 1995, AOL's market capitalization was close to $1.3 billion, or $640 per subscriber. A company's actual capacity to earn a profit didn't seem to matter in this wild world: AOL's market cap was 7 times revenue; Netcom, an Internet service provider that went public in late 1994, had capitalization that was 14 times revenue. The market was still extremely volatile—nearly every article about Internet stocks included a worried quote from an expert who doubted their staying power—but people kept buying.[24]

Then, on August 9, 1995, Netscape went public and Wall Street went nuts. The company was little more than a year old. It hadn't made a dime of profit. But the excitement among investors was running at such a fever pitch that NASDAQ delayed the opening bell by ninety minutes that morning at the request of Netscape's underwriters. When the market opened, the shares zoomed up to more than two and a half times their offering price, peaking at $75. Netscape came out of the day with a staggering valuation of $2.3 billion. On paper, Marc Andreessen now had a net worth of $80 million. Jim Clark was a billionaire. "People started drinking my Kool-Aid," Clark gleefully told his biographer, Michael Lewis. "What the IPO did was give anarchy credibility."[25]

It also gave credibility, and celebrity, to a new generation of tech-industry analysts. The leader of the pack—the woman whose unerring instinct for Internet picks gave her the title "Queen of the Net"—was a thirty-five-year-old Morgan Stanley analyst named Mary Meeker.

Just like Ben Rosen and his microcomputers one generation earlier, the Indiana native had been doing deep-dive research on the Internet while no one on Wall Street was paying much attention, and she was bullish about what she saw. She also had forged connections with key people in the Valley's Internet revolution, including Doerr, whom she had met as Morgan Stanley took business and financial software maker Intuit public in 1993, and worked with again as the firm took Netscape public two years later. Meeker understood that the market was going to be big, and that it was going to turn tech into a fundamentally different business. "It's a media market," she told the *San Jose Mercury News* in late 1994, "and the winners will be companies that do the best job of editing and presenting information."[26]

Even after the Netscape earthquake, the tech market remained distressingly unpredictable, with stocks spiking up and down with alarming velocity. Meeker sailed above it all. "If I believe in the company," she said. "I buy the stock." And she believed in AOL, in Netscape, in Amazon, and in eBay. When she rated something as a "buy," she stuck with it for the long haul. Meeker's unflappability and her instinct for picking tech winners quickly generated a large and eager audience for her research. Morgan Stanley's annual Internet report featuring Meeker's insights had been released earlier in 1995 without making much of a stir beyond Wall Street. After Netscape's IPO, the bank got so many requests for copies from unlikely places—schools, small investors—that they struck a deal with HarperCollins to publish it as a book the following spring.[27]

Meeker wasn't the only Internet-era star at Morgan Stanley. King of the dealmakers was mustachioed and assertive Frank Quattrone, who had been a technology banker in the firm's Bay Area office since graduating with a Stanford MBA in 1981. He'd led the deals for Silicon Graphics and MIPS and Cisco, and he leveraged his Valley connections to land some of the juiciest deals of the boom, starting with Netscape in 1995. Quattrone left the bank in 1998 for Deutsche Bank and then Credit Suisse, earning upwards of $100 million per year.

Then there was Morgan Stanley banker and close Meeker ally Ruth Porat, a Palo Alto native with deep connections to the tech world. Her physicist father, Dan, had worked at SLAC; her brother, Marc, was CEO of General Magic, an already legendary company that tried and failed in the early 1990s to build a pocket-sized computer—an iPhone before its time. Together, Ruth Porat and Meeker vetted nearly every Internet start-up that came down the chute in the late 1990s, and the bank became the primary manager on fifty of them.[28]

Some observers wondered whether it was healthy to have tech's most bullish analyst and its most connected bankers working together so closely; some sharp-elbowed bankers (including Quattrone) already had a reputation for pressuring their researchers to talk up a stock. But such qualms quickly got lost amid the enthusiastic din of a market that was going wild for all things Internet.[29]

Soon it was hard to pass a newsstand without seeing magazine covers emblazoned with the smiling mugs of the newest generation of high-tech cowboys and cowgirls. Freshly-minted MBAs opted for jobs at tiny Silicon Valley dot-coms instead of the Fortune 500, willing to make the risky bet that stock options would be a path to fame and fortune. The Wall Street IPO became a rite of passage for nearly any Silicon Valley start-up, no matter how new or untested. "Why do they do IPOs?" parried one broker. "Why do rock stars marry models? In part, it's because they can." Wall Street was clamoring for all things Internet, Meeker observed. "These companies would be silly not to take access to capital."[30]

CREATIVE ACCOUNTING

The storming of Wall Street by new, mostly profitless dot-coms wasn't embraced by everyone. Business columnists tut-tutted about "Netscape fever" as soon as Clark and Andreessen's company shot out of the gate. "For someone who looks at the fundamentals, this really represents a dangerous sign of overspeculation," warned one analyst.[31]

There was reason to be concerned if you knew much about the fundamentals underneath the dot-com flash and pop. Not only were most of these firms profitless, but they used a particularly creative—and, frankly, deceptive—kind of accounting when reporting to Wall Street. As Valley firms had done ever since the first tilt-ups appeared amid the fruit orchards, the new crop of Internet companies gave employees stock options, luring in talent that start-up–stage enterprises otherwise couldn't afford. This made the payroll line smaller than it otherwise might have been on the balance sheet, and built loyalty and drive among employees, whose job-hopping inclinations were tempered by the golden handcuffs of stock options that had yet to fully vest. Like any trait developed in relative isolation, the Valley's use of stock-as-compensation became more noticeable and distinctive over time.

The practice that made eminent sense within the high-tech Galapagos became problematic when these firms pushed to center stage in Wall Street's 1990s bull market. Once public, companies had to start filing earnings reports with an obscure but powerful entity called the Financial Accounting Standards Board, or FASB. Essentially, they maintained two sets of books: one for the IRS, and one for FASB. And where a firm's incentive with the tax man was to express as little profit and as much loss as possible, with FASB you wanted to arrive at high-earnings numbers that would encourage Wall Street to buy. While regular payroll had to be charged against earnings, there was no rule that required tech companies to count stock options the same way. If you counted these fat options packages as compensation—which, in fact, was their main purpose—then the profit-and-loss statements looked considerably less rosy, particularly as stock prices climbed.

The Valley's ability to sustain such unusual accounting practices was the result of an important political victory it had achieved in the early years of the dot-com boom, another sign of the growing clout that its Internet-era companies had gained in D.C. In the summer of 1993, just as the Mosaic browser was bursting onto the scene, FASB regulators had proposed doing away with the stock-option exemption. The corporate outcry was immediate, and sustained, spurring the Valley into a fresh flurry of political activism. Things came to a full boil by the spring of 1994, as software engineers packed the San Jose Convention Center to rally against the measure. From the podium, tech executives were eager to frame the whole matter as yet another David-vs.-Goliath battle between earnest and scrappy Silicon Valley and the backward-looking bureaucrats back East. "FASB is a bunch of accountants who sit in a vacuum in Connecticut," T. J. Rodgers thundered to the crowd. "The accounting in this instance has a face," exhorted 3Com's Katherine Wells. "This is not about debits, it's about dreams."[32]

The whole business was a giant headache for Clinton's newly appointed Securities and Exchange Commission Chair, Arthur Levitt. A seasoned Wall Streeter of centrist politics and an impatience with corporate bloviation, Levitt's sympathies lay with the individual investor—and in providing such investors with the most full and transparent accounting possible of company financials. Stock-option accounting was a shell game, and he didn't like it one bit.

Yet once the tech-sector outcry reached such a volume that lawmakers of both parties on Capitol Hill began making moves toward legislation that would keep the options free, Levitt decided that resistance was futile. He encouraged FASB to drop the proposal, which it did. Only one thing changed (and it didn't even go into effect until 1997): companies had to add a footnote to their filings that showed the earnings number with stock options charged against it. The difference was dramatic—using the alternative calculation, Netscape's earnings dropped by nearly 300 percent—but as many in the Valley knew, few people read the fine print. Levitt later regretted having given in. "It was probably the single biggest mistake I made in my years at the SEC."[33]

THE CHARMED CIRCLE

The Valley folk were right about one thing: those stock options could change lives. Trish Millines could hardly believe it when her Microsoft stock portfolio hit $1 million. She'd first started at Microsoft as a contractor in 1988, three years after arriving in Seattle and two after Microsoft had gone public. She came on as a full-time employee in 1990 and stayed on through six years and five stock splits. Over the time she'd been there, the massive revenues generated by MS-DOS had been dwarfed by the stupendous profits driven by the Microsoft Windows operating system and its cheerful, icon-filled software applications like Word and Excel. Millines—soon to go by her married name, Millines Dziko—finally had enough money to do something different.

Through her two decades working in a massively expanding and staggeringly wealthy industry, Millines Dziko had seen very little change in the makeup of who worked there. Nearly all her colleagues, and especially her bosses, were white and male; when she and another black manager at Microsoft once found themselves running a meeting together, it was as memorable as a unicorn sighting. At one point, realizing that only forty African Americans worked at *the entire company*, she co-founded a group called Blacks at Microsoft to provide some of the same kind of professional support that so many other engineers took for granted.

She moved from a technical role to become a diversity supervisor for the company, working to recruit and retain more women and minorities. But while she appreciated Microsoft's recognition of the problem, she realized that its source wasn't just in how hiring happened. It was a matter of who was in the hiring pool in the first place. Tech had a pipeline problem, the product of decades of exclusively male engineering programs and often-blatant sexism and racism in the workplace, exacerbated by the popularization of a tech culture whose freaky and geeky face was almost always white and male. Many in the industry got defensive when someone like her pointed out these dismal inequities. Tech was a meritocracy! You got ahead because you were a smart engineer! Plus, look at all the people who were first- or second-generation Indian and Chinese—didn't they count toward "diversity"? Even a Microsoft millionaire like Millines Dziko couldn't seem to make a dent in the problem.

If things were bad at Microsoft, they seemed even worse in the Internet-era Valley. The early online world of BBSs and cyberpunks had been mostly white guys, for sure, but there still had been plenty of women there in the beginning. Just as in the early days of micros and homebrewed motherboards, lots of people on the early Internet were self-trained programmers and participants who'd come online from a diverse range of backgrounds. It didn't seem far-fetched to imagine this might continue as the Internet commercialized. If the business of the Valley was no longer about things made by electrical engineers and software hackers, but about ideas and content that floated atop software platforms, shouldn't that open up the tech world to people who hadn't been part of it before?

Nope. Instead, as the Internet wealth machine accelerated, the old patterns intensified. The tech might be new, but the VCs and lawyers and marketers and senior operators weren't—and their attitudes about hiring were the same as ever. They encouraged employee referrals; they recruited at the very top programs. Sun filled 60 percent of its jobs through referral; Netscape hung a sign in its headquarters asking, "Who is the best person you've ever worked with? How can we hire him/her?"[34]

As the boom crested in the spring of 1998, the *San Francisco Chronicle* investigated the state of diversity in Silicon Valley and came back with some sorry findings. Their survey of thirty-three firms found a workforce that was 7 percent Latino and 4 percent black (at the time, the Latino population of the Bay Area was 14 percent and blacks were 8 percent). True, there was ample opportunity for immigrants, especially in engineering. Close to one in four employees were of Asian descent. The vast majority of them, however, were male. Tech bosses were adamant that this wasn't discrimination: "It wouldn't matter if you were green with white stripes, if you could code you will get a top job," said one HR director. At Cypress Semiconductor, where the employee base was 3 percent black and 6 percent Latino, CEO T. J. Rodgers declared, "We hire the best people for the job."

The Valley's racial minorities responded accordingly. Latinos Anglicized their last names. Black entrepreneurs spent their off hours bonding with white VCs on the golf course. South Asian immigrants, well-represented in the tech ranks but less so in the top jobs, banded together in entrepreneurial networks and hired one another en masse.

The charmed circle became so homogeneous by the late 1990s that Valley firms attracted the attention of federal officials for their failures to properly ensure diversity.

Anyone with a federal contract had to adhere to affirmative action guidelines, and tech was missing the mark. "Being the fastest-growing software company ever, we shot past the mark that the government sets down for putting an affirmative action plan in place," countered Bob Sundstrom, whom Netscape belatedly hired as its manager of diversity programs after it was rapped on the knuckles for its failures. Apple had to pay over $400,000 in back pay to fifteen black workers who were rejected for jobs. Oracle was fined for pay inequity toward female and minority employees.[35]

Trish Millines Dziko watched it all with resigned frustration. Software was transforming the world, and it was important to have a diverse set of minds shaping that software. She had weathered life in the extreme minority at Microsoft, an isolation made sharper by being in the overwhelmingly white Pacific Northwest. Over time, Blacks at Microsoft had morphed to being much more than just a place to compare notes on where to get a haircut, sponsoring events that brought minority high schoolers over from Seattle's Central District for a day at Microsoft. The experience had sparked new ideas about how the tech industry might change. "All we're doing," Millines Dziko mused, "is creating a bunch of consumers, and until kids begin creating technologies, the gap will always widen." The answer lay at the beginning—in teaching minorities and girls how to do it themselves.[36]

In 1996, many other newly minted Microsoft millionaires were leaving their jobs, and many were starting their own philanthropies. Bill Gates eventually would do the same thing, on a much, much larger scale. Millines Dziko's bank account was small stuff in this high-flying crowd, but she had enough to do something that might start moving the needle. That something was the Technology Access Foundation, an academy where low-income minority kids could go after school and learn how to design and program computers. If the big companies weren't going to fix their pipeline problem, then she'd try to fix it herself.

CHAPTER 21

Magna Carta

"The central event of the 20th century is the overthrow of matter." So began *Cyberspace and the American Dream: A Magna Carta for the Knowledge Age*, a declaration of cybernetic independence, rocket-launched in late summer 1994 out of a Washington think tank called the Progress & Freedom Foundation, or PFF. The outfit might have been obscure, but the essay's four authors were anything but.

Taking the lead was Esther Dyson, whose annual PC Forum and monthly newsletter *Release 1.0* had become the way the most powerful people in tech learned about the future. Then came "Doctor SDI," George (Jay) Keyworth, who as Reagan's science advisor had been one of the High Frontier's most bullish defenders. Another Reagan-era boldface name on the roster was former presidential speechwriter George Gilder, evangelist of the supply-side gospel and a pop-science gadfly whose musings about the dangers of feminism once prompted NOW to dub him "male chauvinist of the year." Now Gilder had turned techno-futurist in the mold of Alvin Toffler, the essay's fourth author, who supplied grandiose textural flourishes. Not a full-time Silicon Valley resident in the bunch, but all people whose ideas had left a deep imprint on the Valley's state of mind.[1]

Esther Dyson hadn't expected to be the Thomas Jefferson of this particular enterprise, but given her talent to end up at the center of everything, it wasn't surprising. The daughter of famed theoretical physicist Freeman Dyson, she had grown up among the giants of early digital computing at Princeton's Institute for Advanced Study and had honed her understanding of the tech ecosystem ever since arriving at *Forbes* magazine as a fact-checker in 1974. As a reporter there, she sniffed out one of the very first stories on the rising electronic might of Japan. Before too long she had moved into investment banking at Oppenheimer & Co., then joined Ben Rosen at Rosen Research.

By 1994, eleven years after she had joined and eventually taken over Rosen's newsletter and conference business when he became a venture capitalist, Dyson had become the most influential member of the newest generation of Valley storytellers. She had gone global, expanding her empire into Russia and Eastern Europe after the Iron Curtain fell. Esther Dyson combined Ben Rosen's keen industry antennae with Regis McKenna's ability to create media buzz and throw a great dinner party. On top, she added her own distinctive techno-futurist gloss—Tofflerism with a stock-picker's sensibility. It cost over $600 a year to subscribe to *Release 1.0*, and 1,500 of the tech industry's most powerful read its every elliptical word.[2]

Dyson leaned libertarian in her politics. She had never voted. Her work in the former Soviet bloc gave her firsthand knowledge of the damage wrought by authoritarian states. Yet she also had a nuanced understanding of the codependence of states and markets, and of the necessity of a productive working relationship between D.C. and tech. "Whether you like it or not," she reminded her audience at the 1993 PC Forum, "there are people in Washington who have more control over your future than Microsoft does." When the Clinton White House asked Dyson to join the NII advisory committee, she readily agreed. In 1995, she became the chair of the EFF and its well-oiled lobbying machine. The Internet economy was only starting to bubble (Amazon was one month old at the time of the essay's release), but government regulation would be critical to its growth.[3]

Although later critics called it a quintessential example of wide-eyed techno-libertarianism, the vision of *Cyberspace and the American Dream* was one of disruption rather than revolutionary overthrow. Yes, a Third Wave society needed and deserved a "vastly smaller" state. "But smaller government does not imply weak government, nor does arguing for smaller government require being 'against' government for narrowly ideological reasons." Squint, and you could mistake some of its passages for talking points from the Clinton-Gore National Performance Review, also known as "Reinventing Government," which was then attempting to reorganize an ossified federal bureaucracy for the digital age.

But the national politician whose philosophy it more closely resembled was someone closely affiliated with the PFF, a party compatriot of Keyworth and

Gilder, and someone who had spent years soaking in the futurist gospel of Alvin Toffler. That politician was Newt Gingrich.[4]

THE BOMB THROWER

In his fifteen years in Congress, the Georgia congressman and House Minority Whip had become the progenitor and master practitioner of a fire-breathing brand of sharply partisan, made-for-cable-TV intellectualism that decimated his rivals and horrified prim defenders of Congressional decorum. It also was terrifically effective politics.

In a staid Washington landscape, Gingrich stood out for his delight in sweeping historical analogies as well as an unapologetic futurism. He'd been a friend of Alvin and Heidi Toffler since the early 1970s, when as an unknown assistant history professor he had flown halfway across the country to hear Toffler give a speech. After he got to Congress, he invited the Tofflers to Capitol Hill to speak to the Conservative Opportunity Society, a group he'd helped to organize that was something of a GOP analogue to the tech-centric Atari Democrats. While the Tofflers didn't share Gingrich's conservatism on social issues—their take on sexual freedom and unconventional family structures always raised eyebrows on the right—they shared with him a conviction that, in Toffler's words, "a weakening of Washington and a dispersal of power downward is where we must go."[5]

By the summer of 1994, the Clinton Administration found itself careening from one crisis to another as the sunny optimism of its campaign promises met the drab reality of budget caps and legislative horse-trading. The GOP had adopted the bomb-throwing tactics of Newt Gingrich to successfully turn health care into a symbol of dangerous government overreach. Now, Gingrich was busily working with his lieutenants to formulate a campaign platform for the House midterm elections called "The Contract with America," a declaration of conservative priorities that the Republicans vowed to pursue if they won control that November.

All this was on Gingrich's mind as he showed up late in August to the small Atlanta conference thrown by the PFF to accompany the release of *Cyberspace and the American Dream*. The think tank was less than a year old by that point, but its alignment with conservatives like Gingrich gave "the only market-oriented institution focused on the digital revolution" special clout. Jay Keyworth was its president, and Republicans filled its ranks. (Founded at a moment when the EFF enjoyed high-level Democratic cosseting and generous corporate backing, the PFF's acronym was no accident.)

Dyson met Gingrich with guarded curiosity. "Everyone I knew seemed to think he was the antichrist," she recalled. Yet the Georgia congressman knew more about the Internet than she expected, and he was clued in to the fact that, as the manifesto put it, the rise of new information technology "spells the death of the central institutional paradigm of modern life, the bureaucratic organization." The Democrats in the White House understood the technology, too, but they didn't seem able to escape a Second Wave world of Information Superhighways and government micromanagement. With little fealty to governmental tradition, Gingrich sounded ready to do something really different, and something a little closer to the revolution that Silicon Valley was hoping for. "I left not convinced, but intrigued, and sure that he was hardly the one-dimensional figure my friends had supposed," Dyson later wrote.[6]

Five months later, after a Republican tsunami overtook Washington and Gingrich became the third-most-powerful politician in America, the new House Speaker devoted his first full day in office to all things tech. He unveiled Congress' new website, a service called Thomas, that put every bill, hearing, and report of legislation online. "This is going to lead to a dramatic shift in thinking and talking about ideas rather than personalities," he boldly predicted. "It's going to make for a dramatically healthier dialogue among Americans." Gingrich then moved from podium to committee room, to testify in favor of his new proposal to give tax credits so that every poor person in America could buy a laptop. The price tag for such a measure was potentially staggering, but Gingrich wanted to make it clear: he had arrived to carry the nation into the Internet Age.[7]

GINGRICH REVOLUTION

The second annual PFF conference that summer of 1995 showed how high the scrappy little think tank's standing had vaulted in Gingrich's Washington. The conference had left muggy Atlanta for the mountain air of Aspen, summertime playground of the powerful and rich. The halls overflowed with corporate lobbyists hoping to gauge what the Speaker might be thinking about the next big tech issue on the docket: *The Telecom Revolution—An American Opportunity.* So ran the title of that year's conference and its accompanying white paper, taking advantage of the fact that the telecom bill that the White House had brought to the Hill at the start of 1994 hadn't gone anywhere, shelved by recalcitrant Democrats who didn't like the idea of letting broadband Internet companies operate on a more open playing field. Now that the old lions had been knocked off their committee-chair pedestals, Gingrich planned to go much, much further. "We should be driving for as little regulation as possible," he declared. The FCC would turn into a dinky office in the White House. The many-layered governmental bodies controlling the rules of the Internet would go away.

Inspiration for Gingrich's approach, he said, came from his understanding of the computer hardware and software industries, a field of entrepreneurial dreams made possible because they developed free from burdensome federal regulation. Silicon Valley happened because Washington "got out of the way," Gingrich later explained matter-of-factly to *Wired* reporter John Heilemann. "I think it's pretty clear we're at a point where we ought to just liberate the market and let the technologies sort themselves out." A favorite proof point for Gingrich was the thirteen years and millions of dollars the U.S. government spent trying to prove IBM was a monopoly; all for naught, as the technology world sorted out the winners and losers on its own.[8]

The PFF conference vibe echoed Gingrich's "get out of the way" sentiment. Silicon Valley had been grumbling about the bureaucracy ever since Dave Packard's anti–Great Society jeremiads at the Palo Alto Rotary Club in the 1960s, and the Vietnam-era Homebrewers had taken the dyspeptic antigovernment rhetoric to an even higher plane. Yet the talk by tech emissaries in Aspen was more sharply pitched than it had been in a while. The federal government was "irretrievably clueless," said John Perry Barlow. The Internet would break all the rules of politics, of commerce, of everything. In the online marketplace, predicted Esther Dyson, "only the good stuff will survive."[9]

Silicon Valley's free-market wing was thrilled by the changeover in Washington. The founding generation of venture capitalists and chipmakers remained unwaveringly Republican, and some of the new stars were as well. "The U.S. has got to get back to privatization," said Scott McNealy, "to move the economy back into the private hands." Plus, McNealy smiled, he liked Gingrich's style, which "makes me look like a diplomat."[10]

The Silicon Valley's techno-liberals weren't any more enamored of government than the right-wing deregulators were, but they despaired at how corporate the entire business had become. A crisis developed within the EFF, which even before the Gingrich revolution was relying on hefty corporate donations from the likes of AT&T and IBM as well as Microsoft and Apple. "I've written them off," said Cypherpunk co-founder Timothy May of the EFF. "They don't represent my interests." Even those who weren't troubled by the corporate money were frustrated. They'd signed up to change the world, not attend endless meetings with Deputy Assistant Secretaries in the Commerce Department.[11]

By September 1995, the iconoclastic Speaker was getting all kinds of good high-tech press—Barlow interviewed him for the inaugural issue of John F. Kennedy Jr.'s glossy monthly *George*—and the Vice President's White House staff were fuming. Gingrich was unperturbed. "The model Gore is trying to build is a futurist vision of the welfare state. He's repainting the den; I want to build a whole new house. My project, frankly, is to replace his world."[12]

THE DECENCY PATROL

Meanwhile, Senator James Exon looked at this new online landscape, and all he saw was smut. The Nebraska Democrat had grown alarmed by the tales he'd heard of the pornography and criminality that filled the Internet's dark corners, from X-rated websites to the online copy of the "mayhem manual" that domestic terrorist Timothy McVeigh had used to blow up the federal building in Oklahoma City in the spring of 1995.

Indeed, Exon was right: porn was the early Internet's biggest growth industry, "one of the largest recreational applications of users of computer networks," reported one widely-read survey. BBSs and Usenet groups—the information superhighway of the 1980s that had grown like a weed as online access expanded in the early 1990s—were now saturated with pornography, as more bandwidth allowed users to download and share images and videos with ease. Worst of all, it was all unregulated, right where children could stumble across it at any time. The White House might be trumpeting the need for an information superhighway, but American parents hesitated to buy home Internet service because their kids "could get bombarded with X-rated porn, and I wouldn't have any idea," as one Chicago-area mother put it.[13]

As public anxiety rose about the real and perceived dangers of online content, others joined Exon's chorus. The Christian Coalition, Washington's most powerful Religious Right lobbying group, had made "Internet decency" a major plank of its 1995 platform. Other religious conservatives on Capitol Hill pushed hard for a clampdown on online content.

As Congress continued its debates on telecommunications reform that summer, Exon proposed a rider: a "Communications Decency Act" that would impose severe fines, even jail time, for creators of online material deemed "obscene, lewd, lascivious, filthy, or indecent." With this Victorian phrasing, the pipe-smoking Senator ignited a political firestorm. The CDA was censorship with the broadest brush, his critics howled, putting bureaucrats in charge of making subjective judgments about online content, and squelching free speech in the process. Washington already had been consumed with questions about keeping media G-rated for children; Al Gore's wife, Tipper, had led a successful push to put warning labels on pop music with explicit lyrics, and the final telecommunications bill rated television shows and placed a "V-chip" in cable boxes to allow parents to regulate what their children saw. Exon's proposal, however, was far tougher, and to the horror of tech companies and free-speech advocates, it passed the Senate with a thundering bipartisan majority of 86 to 14.

On nearly every other issue, Newt Gingrich lined up on the side of the Religious Right, but here he was firmly with the defenders of cyber-liberties. "I don't think it's a serious way to discuss a serious issue," the Speaker declared. He promoted another bill, the "Internet Freedom and Family Empowerment Act" (a made-for-1995 title if there ever was one). The measure reflected what the tech community wanted to see: responsibility would lie with industry, not government, to develop content standards and filtering software. Even these earnest promises of self-regulation weren't enough. Washington had veered right, family values were riding high, and an election year was right around the corner.

Instead of Gingrich's counter-proposal, Exon's more stringent, interventionist

CDA remained as a rider on the final version of the Telecommunications Act that passed Congress with overwhelming support and landed on Bill Clinton's desk in early 1996. Too many other things of importance were in the telecom bill for Clinton to say no. On the day it went into effect, Internet companies registered their cyber-protest by taking tens of thousands of screens dark. And almost immediately, the CDA was in court with a challenge to its constitutionality that was supported by everyone from the ACLU to Microsoft. By summer, its most restrictive provisions had been thrown out. But one thing stayed in: a provision, placed in the depths of the CDA as it was being hashed out in conference committee, that no Internet provider or platform would be considered the publisher or speaker of any information placed on its site by a third party.[14]

Because of this, the CDA ultimately had an immense impact—but not in the way that Senator Exon or the Christian Coalition ever imagined. By freeing Web platforms from liability for content on their sites uploaded by third parties, the Supreme Court ruling not only was a major victory for the tech companies of the dot-com era, but it was a massive win for the giant social media platforms yet to come.

By 1997, cyberspace had become the glorious next frontier proclaimed by the EFF and by the Magna Carta. More than fifty million Americans were online. Search engines were buzzing, sites proliferated, and surfing the Internet had become a daily ritual. With the technologies for secure credit card transactions now in place, e-commerce was taking off at last. Millions were pointing and clicking and buying, moving billions' worth of product through cyberspace. And for all Exon's efforts, nothing had stopped the rising tide of dirty content. Porn remained the Internet's biggest revenue producer of all.[15]

CHANGING OF THE GUARD

Amid this flurry came a somber milestone: the death of David Packard in 1996, at the age of 83. He had continued to live in the ranch house he and Lucile built in Los Altos Hills in 1957, an old-school contrast to the flash and pop of the Valley's fully-vested swagger. Leaders from the business and political worlds filled Stanford's Memorial Church to capacity on the last Friday in March, paying effusive tribute to a man who had done so much and influenced so many. A quietly grieving Bill Hewlett sat in the front row after arriving in a wheelchair. The printed program featured a photograph of Packard riding a tractor, simply captioned "Dave Packard, 1912–1996. Rancher, Etc."[16]

The death of the low-key titan coincided with changes in the Republican Party that Packard had supported so faithfully throughout his career. In the sharply partisan age of Clinton and Gingrich, the distinctive strain of liberal Republicanism practiced by Northern Californian politicians like Pete McCloskey and Ed Zschau was harder to find in Washington.

Silicon Valley Republicans bucked the tide a little longer than most. Ed Zschau had returned to industry after his 1986 Senate defeat, but he remained deeply engaged in spotting and promoting the next generation of political talent. Additional help came from real estate mogul Tom Ford, the man who had built the venture-capital corridor along Sand Hill Road, who founded the Lincoln Club, a fundraising organization for moderate GOP candidates. One of those promoted by Zschau and Ford was Tom Campbell, a former economics professor elected to Congress in 1988. Another was Becky Morgan, a Palo Alto school board member who served three terms in the State Senate.

Yet both Campbell and Morgan became increasingly rare breeds in a world of California politics that swung more liberal as national politics became more conservative. Campbell ran for U.S. Senate in 1992 and lost the primary, and the winner of the general election in the fall was Barbara Boxer, a Democrat from the liberal stronghold of Marin County. The next year, a term-limited Morgan returned home to lead a regional economic development group, Joint Venture Silicon Valley. By 1996, the most visible—and most closely tech-allied—local politicians were now Democrats. In a telling sign of how much things had changed, Ed Zschau joined an exploratory effort to form a third national party.[17]

Meanwhile, the roar of the high-tech Democrats got louder. Three weeks before Packard's Stanford memorial service, on March 9, Air Force One flew the President and Vice President to the East Bay town of Concord for "NetDay," a one-day push to wire California's public schools to the Internet. Thousands of Silicon Valley engineers spilled out of their offices and into classrooms to unroll coaxial cable and install routers. White House press staff ushered photographers into a high school library to capture a khaki-clad Clinton up a ladder, gamely threading wire through the ceiling tiles. The event had been the brainchild of Sun Microsystems chief scientist John Gage—he who had coined the phrase "the network is the computer"—and it was a high-octane kickoff for another program affixing the Clinton White House to the Internet-era Valley's shooting star.

The first year of Republican control on Capitol Hill had been a bruising one for the White House's agenda, particularly when it came to domestic social programs. Health care reform was dead. Another signature issue, welfare reform, had been pushed sharply to the right. Telecommunications reform had devolved into a squabble among carriers who wanted to hold on to local monopolies, chipping away at the grand hopes that deregulation would bring prices down and drive Internet access up. Clinton was looking toward reelection, and Gore was quietly laying the groundwork for a run in 2000. Both needed an issue that would appeal to core Democratic constituencies as well as have swing-voter appeal.

With the 1996 election less than a year away, the promise to wire every school in America hit just the right note. Who could object to kids and teachers getting access to the wonderful world of the Web? The government would provide subsidies and a bully pulpit; the private sector would bring the infrastructure and the computers and a discounted "e-rate" on telecommunications services. Educational technology, or "ed tech," was the perfect blend of Great Society and New Economy. It was something, the policy staff told Clinton and Gore, that could become "a defining issue for you and for the nation—now, through the State of the Union address, the 1996 campaign and the second Clinton-Gore term."[18]

"Every single child must have access to a computer, must understand it, must have access to good software and good teachers and, yes, to the Internet," Clinton told the NetDay crowd, "so that every person will have the opportunity to make the most of his or her own life." With unintentional irony, the program kicked off in a post–Proposition 13 California, a prime example of what happened when public education was starved for general revenue: since 1978, property tax caps had made per-student school spending plummet, leaving school buildings crumbling, teachers underpaid, and little money for anything beyond the basics. Bringing schools up to par required Herculean amounts of parental fundraising, creating a disconcerting contrast between affluent districts where parents had time and money, and poor districts where they didn't.

Enlisting private-sector money and sweat equity, the presidential "electronic barn-raising" of NetDay was the mother of all PTA fundraisers. The whole thing spoke volumes about how much these leaders were hanging their hopes on Silicon Valley technology as the cure for the inequities in education—and for the greater divides of an increasingly fractious nation. Computers in classrooms were the Democrats' moon shot and Great Society, updated for an age when government did less and information technology did more. "Technology," proclaimed the president that day in Concord, "is going to liberate Americans and bring them together, not hold them back."[19]

THE LITIGATOR

John Doerr applauded the sentiment, but he had other things on his mind that spring. The soaring performance of young and profitless dot-com companies was forging new economic rules of the game, and making VCs like him very, very rich. But the hot market had opened the floodgates in another direction: lawsuits from shareholders who'd taken a bath when a stock suddenly dipped. California-based companies were particularly vulnerable, as state law had a relatively low bar for bringing such suits. In the vast majority of these cases, the alleged wrongs wouldn't hold up to close scrutiny—people ran risks when they played the stock market, after all—but companies often settled to save the cost and hassle of going to court.

The king of California shareholder lawsuits was Bill Lerach. Scotch-swilling, foul-mouthed, and blisteringly effective, the San Diego litigator had filed hundreds of these actions, and they had made him a very rich man. Getting "Lerached" had become one of the Valley's great headaches even before the Internet boom, and it only got worse once the dot-com hits started coming. The suits seemed especially outrageous due to the roller-coaster nature of the 1990s market, but companies would settle 90 percent of the time just to get Lerach out of their hair. As they thought of the curly-haired trial lawyer luxuriating in his clifftop La Jolla estate, Valley executives were reminded of yet another reason to hate Southern California. *Wired* summed up Silicon Valley's sentiments in the title it affixed to a 1996 profile of Lerach: "Bloodsucking Scumbag."[20]

Bill Lerach also happened to be an extremely generous donor to Democratic politicians. Trial lawyers like him gave early and often, and along with their checks came exhortations to preserve the right of shareholders to sue. Class-action suits were a matter of the people versus the powerful, the lawyers would argue. Many other core Democratic constituencies—labor unions, consumer rights groups—agreed. This was about justice for the little guy: the small investor, the injured victim, the wronged consumer. Wasn't this what the Democrats were supposed to be about? Unsurprisingly, when legislation came up to make it far more difficult to bring personal-injury lawsuits—as it did on Capitol Hill in the spring of 1995—votes broke down along party lines. The Republicans were on the side of the public companies. The Democrats, by and large, were on the side of plaintiffs and their lawyers.

Of the many headaches that the Gingrich revolution had created for the Clinton White House, this was one of the most politically fraught. Two important allies were on different sides of the issue. One (tech) came with the keys to the future economy. The other (the trial lawyers) came with buckets of campaign cash. As the president agonized, the titans of tech checked out the view from the other side, meeting with Republican lawmakers and reconsidering their somewhat recent conversion to the Democratic cause. Despite the personal entreaties of his Valley loyalists, Clinton vetoed the lawsuit bill. Congress promptly overrode him. It was a win for tech, a loss for Clinton, and an ugly crack in the White House love affair with Silicon Valley. Larry Ellison was so fed up that he initially refused to endorse Clinton's reelection, and wrote a big check to Republican challenger Bob Dole.[21]

Lerach was furious as well. He decided to stop this government steamroller in its tracks at the state level, bankrolling a fall 1996 state ballot initiative to circumvent the new federal limitations entirely. California Proposition 211 was Silicon Valley's worst nightmare. Not only did it further lower the threshold for bringing a shareholder lawsuit; it also would make company officers and board members personally responsible for any damages owed to plaintiffs. Lerach's pitch was that this was a noble maneuver to save retiree nest eggs from the hustle of high-tech stocks. Silicon Valley saw it as a declaration of war.[22]

John Doerr had learned from the disaster of Clinton's veto that the Valley needed to move fast, and he took it on himself to lead the army. In the 1980s, Ed Zschau had been Silicon Valley's Congressman, crusading on the industry's behalf on trade and tax and everything in between. Bob Noyce and Steve Jobs had been the charismatic CEO-ambassadors trolling Washington's corridors of power. Dave

Packard had been the gray eminence in the background, his career an inspiration for a younger generation. Now, only a few years into his journey as the wizard of the Kleiner *keiretsu*, Doerr became the Valley's premier policy entrepreneur, a Zschau and Noyce and Jobs and Packard all rolled up into one.

Thanks to Doerr, the fight against 211 became Silicon Valley's obsession for most of 1996. He bent ears, raised money, and turned Kleiner Perkins into a campaign headquarters, with a giant banner reading "NO on 211" draped over its facade. Doerr's forces eventually raised $38 million to fight 211, nearly twice as much as what each of the major-party presidential campaigns spent in California that year. The price tag reflected the toughness of the fight. The idea of standing up for shareholder rights was very popular among voters, and California's labor unions and senior-citizen advocates were vocal supporters. The campaign took up "about fifteen percent of my day time," Doerr remembered, "but most of my sleepless nights. I was worried. I didn't think we were going to win."[23]

One thing Doerr had going for him, however, was the fact that the president was up for reelection. Clinton and Gore had just spent the past four years talking about how important Silicon Valley was to the national economy. Here was a ballot measure that threatened to send the Valley off the rails. We don't weigh in on state issues, Clinton's staff protested. This isn't any ordinary state issue, the Silicon Valley people replied, and we aren't any ordinary constituency.

Headlines started popping up in the *Mercury News* about the growing political rift. Clinton got defensive. "What do they mean, I'm against the tech industry?" the president sputtered to Larry Stone. "If you'll just talk about it with us," Stone pleaded with Clinton, "we'll explain why it's so important." A few days later, Stone got a phone call: the president would do a meeting. Keep it small, Stone was told. Clinton was coming to a San Jose school for another event, so they'd shoehorn it in. As soon as he got off the line with the White House, Stone called Regis McKenna. Can you help me get the right people there? McKenna's next call was to Gordon Moore. Then, John Doerr.

And so it came to pass, sitting uncomfortably around a middle school lunchroom table on a swelteringly hot late August day, the President of the United States told some of the most powerful tech leaders in America that he'd support them in their fight against Proposition 211. "We flipped him!" Doerr's staff rejoiced. "I think that position will earn him tremendous support throughout the country and certainly in this state," Doerr said afterward. When it came to Clinton's own bid for reelection, "I was undecided until I heard Clinton's position on it. Now I'm supporting him."[24]

Three weeks later, the president got his reward. Doerr organized an effusive endorsement from seventy-five of Silicon Valley's boldface names, who joined Clinton and Gore on a phone call to praise the administration's economic record—and pat themselves on the back for all they'd done to make it happen. "This administration really gets it," Doerr declared in his opening remarks. The Clinton era had been good for Silicon Valley, and the Valley had been good for Clinton and Gore. "I think it's notable," he added "that the California companies that are endorsing you today have created over 28,000 jobs in their companies over the last four years."

A resurgent Steve Jobs, on the cusp of being wooed back to Apple after his decade of banishment, sounded a similar note. "The past four years have been the best Silicon Valley has ever seen," he said. "Silicon Valley doesn't traditionally look for handouts, doesn't look for tax credits," said the man who'd lobbied so hard for those things fifteen years before. "I hope we see four more years." For all his travails over the previous decade, Steve Jobs remained the voice of the Valley, and he had given his blessing.[25]

Not everyone in the Valley was wild about Clinton and Gore, of course. Floyd Kvamme took on the organizing duties for a less splashy endorsement of Republican nominee Bob Dole a few weeks afterward. Most who joined him had been around since the early days of the microchip, trying to sustain Dave Packard's Republican legacy even as the Valley lurched toward the Democrats. There was Jerry Sanders, quotable as ever: "We can choose between competition, which is exemplified by the Dole/Kemp ticket, or confiscation, which is exemplified by Clinton." Kvamme added, "We don't want a bridge to the twenty-first century constructed in Washington, because we're afraid it just might be a toll bridge."[26]

And many Valley people opted for "none of the above." Ever more disillusioned with the GOP, Ed Zschau endorsed his friend Dick Lamm, a former Democratic governor of Colorado, who was running an underdog campaign against Ross Perot to become the standard-bearer of the Reform Party. By midsummer, Zschau had agreed to join Lamm on the ticket as vice presidential nominee. Cypress Semiconductor's T. J. Rodgers remained adamant that the Valley should have as little to do with D.C. politics as possible. "What does Washington really offer Silicon Valley?" he asked. "We cannot and do not want to win at their game."[27]

While not everyone agreed on who should be president, everyone agreed that Proposition 211 would be terrible for the tech industry. By October, the money, endorsements, and propulsive force of Doerr's commitment had created a formidable coalition. "Proposition 211," said Doerr's friend and mentor Andy Grove, "has mobilized this industry to an extent that nothing else has since the Japanese threat of the mid-1980s." Doerr was still nervous, but delighted. "Everyone is aligned on this issue. The only one missing is Mother Teresa."[28]

On Election Day, Lerach's proposition lost by a 3-to-1 margin throughout California—and a 4-to-1 margin in Santa Clara County. Clinton won California and reelection. The political operatives who had gotten to know John Doerr on the campaign were thoroughly impressed. "I've learned in politics that you cling to rich donors, great spokespeople, and people who get shit done," said one veteran Democratic operative. "Doerr is all three."[29]

And John Doerr already had started looking five years ahead. Silicon Valley had gotten so big, so wealthy, and so integral to the economy that it no longer could sit on the political sidelines. The battle against 211 had been a costly, uphill fight because they had started from square one. If the industry already had an organization in place, then it would be ready for the next Bill Lerach, or the next Clipper Chip, or whatever Washington or Sacramento might throw its way. Not to mention mighty Microsoft, the Death Star up in Seattle. To work, the organization should be a nonpartisan platform for a diverse set of companies and issues—a political variation of the Kleiner *keiretsu*. It was more than just a lobbying group. It was going to communicate a vision. "We need a new framework of law and thinking to help us govern in the new economy," Doerr explained. "I, and many others, will help form that new network."[30]

By early 1997, the organization had a name: TechNet. It had a staff and a multimillion-dollar budget. Longtime kingmakers like Regis McKenna and Floyd Kvamme agreed to join in, as did many other big-name Valley CEOs. The goal of staying out of partisan fundraising didn't last; within the year TechNet had established Republican and Democratic political action committees, which turned into gushing cash machines for visiting politicians. But TechNet also staged regular "graduate seminars on the new economy" so that lawmakers could better understand what was happening—and how they could best help Silicon Valley grow. It was a persuasive bipartisan pitch. "We've been able to get stuff through Congress at a time that it is hard to do that," observed Marc Andreessen; with the Valley humming, it was hard for policymakers "to be anti–high tech."[31]

John Doerr's TechNet became one conduit for younger new-economy companies—and their Gen X founders—to engage with Washington. A second was Al Gore. Few things delighted the Veep more than an opportunity to sit in a room with smart technologists, talking policy. With Doerr's help, Gore's policy aides began holding regular "Gore-Tech" meetings in California and in Washington, where awed thirtysomething Internet moguls would be seated amid the polished mahogany and gilt trim of the Vice President's ceremonial office at the White House. It was a long way from drab tilt-ups and Homebrew swap meets and all-nighters in the PARC beanbag chairs. With the Internet boom, the men and women of Silicon Valley had become establishment power players like never before.

CHAPTER 22

Don't Be Evil

As TechNet mobilized and Gore-Tech meetings proliferated, Microsoft was conspicuously absent. Awash in revenue from its total saturation of the PC platform, Bill Gates's company didn't have the same regulatory worries as the Silicon Valley crowd, and it was so large that it was a political force all on its own. Gates didn't visit the vice president's office; he made the Veep come to him.

One of the splashiest of these visits came in May 1997, when Gates threw the first of what became an annual CEO summit, bringing partners and rivals alike to Seattle for a chummy and luxe experience that underscored his position as the king of them all. Gore stayed up half the night working on his keynote speech, then joined the tech titan and other boldface names on a sunset yacht ride across Lake Washington to Gates's new, $60 million home on the eastern shore. Clocking in at 20,000 very high-tech square feet, the dazzling pile brimmed with high-definition video screens and featured a reception hall big enough to seat hundreds.

As Gore and the CEOs dined on fiddlehead fern bisque and wild salmon, topped off with chocolate soufflé and generous pours of local wines, talk turned to the extraordinary economic moment that America was experiencing, thanks in great part to tech: productivity had spiked, Wall Street soared, Internet retail seemed like it was here to stay. In his speech earlier that day, Gore had urged the tech leaders to tap into their social conscience, to think more about ways to give back and apply their know-how to the nation's broader challenges. Gates had started to think about those things, too, and he and his wife, Melinda, were in the early stages of planning how they might give away their fortune. The well-fed and self-satisfied guests nodded at Gore's message, but it was difficult to shift away from business as usual. The Internet might be surging in 1997, but Microsoft was still the richest and most powerful tech company on earth, and everyone else was hustling not to get squashed.[1]

BY 1992, NINE OF TEN PCS IN THE WORLD RAN MICROSOFT. THE COMPANY SURpassed IBM in market value the following year. "It's the Standard Oil of our era," said one analyst. Even at the white-hot peak of the NASDAQ, Microsoft towered over them all, with a market capitalization ten times that of Sun Microsystems and more than 100 times that of Netscape. When Microsoft gave $10 million to schools in its home state of Washington, Clinton flew three thousand miles west and made the joint announcement with Gates a centerpiece of a presidential visit. When Gates had a one-on-one dinner with newly installed House Speaker Newt Gingrich, it made headlines.[2]

It was mostly a one-way love affair, however. Bill Gates didn't make splashy campaign endorsements, and he generally considered Washington machinations irrelevant to his business. As Microsoft had swelled to market dominance in 1990, the Federal Trade Commission began to sniff around for possible antitrust violations. Gates had responded with open disdain. "The worst that could come of this is that I could fall down on the steps of the FTC, hit my head, and kill myself," Gates scoffed to a *BusinessWeek* reporter in early 1992. The billionaire had softened a bit in the years since—age, marriage, and the gentle prodding of his civic-minded parents had made him more conscientiously philanthropic—but he remained generally uninterested in what happened inside the Beltway. Microsoft didn't have its own Washington lobbying office until 1995, when the original antitrust investigation left it operating under a consent decree.[3]

Anyway, Bill Gates was too busy sparring with Silicon Valley to pay much attention to D.C. Despite the lovable-geek image so carefully cultivated by his PR team, Bill Gates remained the most competitive person on the planet when it came to matters involving his company. Steve Ballmer came in at number two. "At Microsoft," marketing chief Jean Richardson remembered, "the whole idea was that we would put people under." Their scorched-earth dominance of the software market had left competitors whimpering and dissuaded new entrants, contributing to Silicon Valley's wholehearted embrace of the Internet in the first place. "My firm's policy is never to back a venture that competes directly with Microsoft," John Doerr quipped. "Only damned fools stand in the way of oncoming trains."[4]

Only Sun Microsystems' comparably combative Scott McNealy dared scrap with Microsoft, fighting back when Gates & Co. tried to enter the workstation market, counterpunching with the PC-slaying programming language Java, and dropping acerbic one-liners all along the way. Microsoft's leadership was "Ballmer and Butt-head"; its operating systems were "a giant hairball." McNealy's pugnaciousness seemed reasonable, given the very real risk that Microsoft might eat Sun's lunch. "You're not going to beat Bill Gates by being conciliatory or compromising," remarked one Sun veteran.[5]

BROWSER WARS

The competition became a firestorm as the dot-com boom hit with full force. It wasn't as if the team in Redmond *hadn't* been paying attention to the Internet—Microsoft started pouring money into online networks in the spring of 1994, mere months after Clark and Andreessen started Netscape—but the velocity of the Web's growth outpaced anything Bill Gates expected. "We didn't expect," he wrote in the summer of 1996, "that within two years the Internet would captivate the whole industry and the public's imagination." By the time Java and Netscape Navigator entered the market in short order in the spring of 1995, Gates and his colleagues had realized that not only would the Internet world grow large, but it might consume Microsoft's core business altogether. Marc Andreessen was another Bill Gates: he didn't just want to build a piece of software; he wanted to create an entirely new platform that would make Microsoft's OS irrelevant. And like 1980-vintage Bill Gates, 1995-vintage Marc Andreessen wasn't subtle about his sweeping ambitions to pummel Windows so thoroughly that the market behemoth would be little more than "a poorly debugged set of device drivers."[6]

Gates decided it was time to write a memo to his senior staff explaining what Microsoft must do. He titled it "The Internet Tidal Wave," and by the time it landed on their desks in late May 1995, Microsoft already had its own Netscape-challenging browser in the works. "The Internet is the most important single development to come along since the IBM PC was introduced in 1981," Gates wrote. And unlike the pricey, data-stingy computer networks of the earlier era, connecting into the Internet on-ramp opened up a world of information for a flat fee. "The marginal cost of extra usage," he explained, is "essentially zero."

Although later brandished in court as evidence of Microsoft's monopolistic practices, the memorandum showed Gates at his visionary, aggressive best. He singled out up-and-coming start-ups building out video and voice capabilities; he talked about a future Internet that would stream television shows and feature customer-service chatbots. But, he wrote, "browsing the Web, you find almost no Microsoft file formats." This had to change. "I want every product plan to try and go overboard on Internet features," he told them. And the way to spread the use of these newly jazzed products would be to bundle them into the Windows operating system. Netscape already had a 70 percent share of the browser market, and Microsoft needed to get a toehold, fast. Gates knew Andreessen was right. Netscape had been the gateway drug for millions to become hooked on the Web. Now Netscape was on the cusp of becoming more than just a browser. It could set the standards for the entire Internet computing environment just as MS-DOS had done for the PC, with similarly market-eating results. Netscape was too dangerous to remain a direct competitor; Microsoft needed it as a partner.[7]

Less than a month after the Gates memo, a delegation flew down from Redmond to Netscape's offices, offering the Valley's hottest start-up a deal. Suspicious of their motives, Andreessen took copious notes. Microsoft was going to build its own browser, Internet Explorer (or IE). But it was willing to share. Netscape could still rule the roost on Mac, Unix, and older versions of Windows. Microsoft would get the rest, i.e., most of the market. Netscape didn't bite. Instead, Jim Clark called his lawyer: Wilson Sonsini partner Gary Reback.

A Civil War history buff with the swaggery bluster of a Union general, Reback was already a veteran of several seminal Redmond versus Silicon Valley battles. He'd been on Apple's side in the 1988 lawsuit over the graphical user interface, and he spent the first half of the 1990s nagging the FTC and the Justice Department to take a closer look at Microsoft's competitive practices in the PC market. Microsoft was operating under a consent decree, but Gates and Ballmer had bargained and battled to make sure that the language wouldn't limit new Windows features—and IE was a feature. The DOJ was still watching, but Microsoft wasn't worried. "This antitrust thing will blow over," Gates told a group of Intel executives shortly after the decree went down. "We haven't changed our business practices at all." They were right. For all Reback's pleading, the DOJ didn't stop the browser bundling plan from going forward. And in the next two years Microsoft proceeded to eat Netscape alive.[8]

It wasn't obvious at first. Windows users could still use Navigator as they always had—and Netscape's products remained a favorite of many—but right there, on the opening screen of every Windows PC, was the Internet Explorer icon. An arms race ensued, with both Microsoft and Netscape releasing new and improved versions as they battled for market share. A PR war raged at the same time, as Pam Edstrom and her troops repositioned Gates as the seer of the Internet age and Netscape's team promoted Marc Andreessen as the Internet's chief poster boy. Gates wrote a bestselling book on the online future called *The Road Ahead*. Andreessen, a twenty-four-year-old multimillionaire, posed playfully on a gilded throne on the cover of *Time* in February 1996. Gates smiled out from the same cover seven months later, over a headline that asked, "Whose Web Will It Be?" *Time*'s editors seemed to know the answer. "He conquered the computer world. Now he wants the Internet," declared the subtitle. "If Microsoft overwhelms Netscape, Bill Gates could rule the Information Age."[9]

And rule he did. Microsoft had been late to the online party, but the ubiquity of Windows allowed it to catch up very, very fast. With every version, Microsoft ate further and further into Navigator's once-dominant market share. Netscape's browser business fell to 20 percent of its revenues. Its dreams of becoming the everything platform for the Internet era were abandoned. By the end of 1997, Netscape was badly missing earnings estimates. It soon had to lay off 360 of its 3,200 employees. The Internet era's shooting star had crashed to earth in less than four years. It was a remarkably short arc, even in an industry where things always moved faster than usual. Jim Clark was characteristically blunt. "If I'd known four years ago what I know now—that Microsoft would destroy us and that the government wouldn't do anything about it for three fucking years—I never would've started Netscape in the first place."[10]

TRUST AND ANTITRUST

The way Silicon Valley responded to the browser war not only revealed the Northern California tech industry's complicated and increasingly acrimonious relationship with Microsoft. It also was yet another demonstration of the industry's contradictory politics, a strange mixtape of antipathy toward central authority and a deep, familiar relationship with certain parts of the political establishment. It wasn't the first time that the computer industry had used the courts as a weapon in gaining a market edge—the Seven Dwarfs of the mainframe industry had egged on the DOJ's antitrust case against IBM, for example—and it wouldn't be the last. Antitrust law might not move at Internet speed, but it was a useful tool for slowing the growth of a market leader; when a company got big enough—IBM, Microsoft, and later Google and Facebook—a large target would appear on its back.

Netscape was Silicon Valley in miniature: funded by a venture capital enterprise started by one of the original Traitorous Eight; represented by the Valley's most iconic law firm; its executive ranks filled by Valley veterans. Never mind that—for all the years of fierce market combat—Microsoft had always been closely intertwined with Silicon Valley, drawing on its talent pool and having long-term partnerships with Valley investors and corporate partners. (Gates in fact had tried to hire Jim Barksdale as CEO only months before Barksdale joined Netscape.) Never mind that Netscape really didn't have a Plan B if their browser strategy failed. In the minds of many who made their lives and fortunes in the Valley, Microsoft was the grasping software-should-be-paid-for capitalist, maker of vaporware, shipper of crappy first-generation products. And it now was the evil empire that had brought Netscape low. It had to be stopped before it did any further damage.[11]

For all the competitive spin when it came to talking about who had the best product, however, Netscape and its allies didn't want to draw much public attention to their campaign to get the government to help. At first, they lobbied very quietly. Even though the TechNet crowd was persuading Congress to side with them on nearly every issue those days—from more tech-worker visas to R&D tax credits to e-commerce sales taxes—they initially didn't get much headway when it came to their case against Microsoft. The White House had little relish for going up against a booming tech giant whose founder was one of the richest and most admired men in America. The thirteen-year slog of the IBM suit further dampened government interest in high-tech antitrust, as did the case of the existing Microsoft consent decree, which already was outdated by the time it went into place. The tech world simply moved too fast.

But political momentum gradually increased over the course of 1997, spurred in good part by a data dump of a white paper prepared by Reback and his Wilson

Sonsini partner Susan Creighton that enumerated all that Microsoft had done to dominate the browser business. Right before Halloween, the DOJ at last made its move. Flashbulbs popped and reporters crowded in to hear Attorney General Janet Reno and antitrust chief Joel Klein announce that Microsoft was in violation of the consent decree, and that the DOJ would fine the company $1 million per day until it stopped its browser bundling practices. Bill Gates was flabbergasted, and his allies remained defiant. "These people," declared Ann Winblad, "have no idea who they are dealing with."[12]

Microsoft v. the trustbusters now became headline news. Three weeks after the DOJ lowered the boom, consumer advocate Ralph Nader hosted a Washington conference to address the Microsoft menace. It was hardly a populist uprising—organizers charged $1,000 a head to "help support Nader's future investigations into the high-tech industry"—but it drew in an eclectic crowd. Gary Reback was on the program, pointing out the flattering entries about Gates in Microsoft's online encyclopedia, *Encarta*. Netscape lawyer Roberta Katz warned that Microsoft was boxing competitors out of its new online shopping center. Ralph Nader's star had been fading, but the fight against Bill Gates gave him a way to reclaim his old mantle as America's #1 fighter for the little guy. The software giant was unsafe at any speed, Nader warned, and it should not be allowed to take over the online future.

"Not content with its enormous market share in PC software, Microsoft wants to hold our hand as we navigate the information superhighway," Nader wrote, "and to push us—not so subtly—toward its own partners or subsidiaries by strategically placing desktop or browser links to its products and services." Somewhat blunting his argument about the danger that Bill Gates's company posed to consumer choice and free expression, Nader published these words in the Microsoft-owned online magazine, *Slate*.[13]

Despite the battering, not everyone agreed that Microsoft was the problem. Ordinary Windows users liked the features and convenience of Microsoft's software. "I think their products are top-of-the-line," one told a reporter. Admiring biographies of Gates continued to flow into bookstores and libraries; one aimed at the elementary-age market, *Bill Gates: Billionaire Computer Genius*, hit the shelves right as the U.S. prepared to announce its case. Perhaps the bad guy wasn't Gates, argued libertarian-leaning voices. "The U.S. government is a far worse monopoly than Microsoft," said one Valley engineer. The Libertarian Party condemned the "bureaucratic Lilliputians" trying to bring Microsoft down; the Cato Institute released a fifty-page brief attacking the DOJ's action. Said one Cato economist, "This is a choice between big government and big business. And we know where we come down on that one."[14]

Scott McNealy had been a longtime donor to Cato, but his love for liberty was matched by his hatred for the way Microsoft had stomped, Godzilla-like, through two decades of high-tech history. It had chewed up Netscape, and now it was again going after Sun, demanding a Windows-only version of Java that wouldn't run across other platforms. Java's power as a programming language came from that ubiquity. "We want freedom of choice—not freedom from choice," McNealy declared at Nader's conference. As Netscape went into its death spiral and other Valley companies continued to hesitate to openly criticize the world's largest software company, the libertarian McNealy became the loudest Valley voice in support of the government's move.[15]

The two-year long trial that ground out in D.C. circuit court from 1998 to 2000 battered Microsoft's reputation, tore apart Gates's carefully crafted public image, and cut the company's stock price in half. The judge presiding over the trial, Thomas Penfield Jackson, became a media celebrity, his delight in the press attention as obvious as Gates's resentment at being dragged into the courtroom. In a two-part deal announced soon after the case began, AOL acquired Netscape and then teamed up with Sun to build up an Internet software juggernaut that could keep up with Microsoft's dominance of business and consumer markets. Microsoft counsel Bill Neukom cried foul—how could the DOJ call Microsoft a monopoly when its competitors were all banding together to take it on? The government was "five steps behind the industry."[16]

Y2K

While Microsoft was getting skewered, Silicon Valley was riding high—not just on a soaring Wall Street, but especially in Washington, where politicians flocked toward the combined allure of gee-whiz technology and new-economy wealth. Scandal and partisan intrigue ruled the day in Washington, as Clinton's transgression with young White House intern Monica Lewinsky turned from titillating gossip to fodder for a special prosecutor to an impeachment vote by the GOP-led House. (The Senate didn't follow suit, and the ensuing infighting among House Republicans led to Newt Gingrich's getting kicked out of the Speaker's chair.)

The spiral of scandal in Washington only reinforced tech's golden-child reputation. About the only thing the two parties could agree upon was that Silicon Valley was a place of high-tech wonder, a miracle of economic growth, and a marvelous place to raise campaign contributions. Encouraging their glowing view of the Valley and its leaders: the fundraising prowess of TechNet's newly minted political action committees, one for the Democrats and another for the Republicans, that had started to shovel considerable campaign cash eastward from Atherton's cul-de-sacs and Woodside's country lanes.

"They are stars," said Billy Tauzin, a Louisiana Republican who now ran the telecom subcommittee. "You have to work hard to make technology issues Democrat or Republican, liberal or conservative," added Ed Markey. Long past were the fractious days of the Communications Decency Act, when so many lawmakers considered the Internet a porn-filled waste of time. They could look at the NASDAQ and read the papers; high-tech was a political winner. When each party started its own high-tech working group, members rushed to sign up. "We're the prettiest girl at the dance right now," said one industry lobbyist.[17]

Ironically, the amazing growth curve of the 1990s boom wasn't entirely due to the Internet economy over which politicians were fawning. It instead resulted from a flaw in the computers themselves, one baked into the system a half century earlier. The creators of the first computer languages, eager to save every bit of RAM and little imagining that the code they wrote might still be in use in the year 2000, had programmed in a two-digit date stamp. A four-bit "1974" became a two-bit "74," for example.

As the new millennium neared, this efficiency hack now presented a disaster scenario as fearsome as any dreamed by Hollywood. What would the computers do at the stroke of midnight on December 31, 1999, when all the stamps switched to "00"? The world was now governed by computers. If the machines thought we'd returned to the year 1900, everything from electrical grids to air traffic systems would go haywire. "The new millennium heralds the greatest challenge to society that we have to face as a planetary community," warned *The Times of India* ominously. The industry estimated that eliminating the "Millennium Bug" would cost

$1.5 trillion. Perennially cash-strapped governments faced a hefty bill as well; governments ultimately would spend $6.5 billion on the fix.[18]

The thing that was a bane for big cities and big companies was a windfall for the software services business, and the rush to reprogram created a white-hot demand for coders that far exceeded supply. Faced with high bills and a shortage of talent, corporate giants and governments looked overseas—to the large, educated, and often English-speaking workforce of India, a country whose own tech sector had exploded in size as its national government deregulated its economy over the previous decade. India also had spent heftily on broadband infrastructure, meaning that cities like the high-tech hub of Bangalore had more reliable Internet connectivity than electricity, and its homegrown software services companies could easily take on the job of programming computers on the other side of the globe. The Y2K flurry also grew the already large numbers of engineers heading across the Pacific. Over 130,000 new H-1B visa holders came to the U.S. between the spring of 1998 and the summer of 1999. Forty percent of them were Indian computer specialists.[19]

PLAYING OFFENSE

The ever-tightening friendship between John Doerr and the man he affectionately called "the commander-in-geek," Al Gore, epitomized the warm relations between high-tech and political capitals in those waning days of the twentieth century. With Clinton weakened by scandal, the Dudley Do-Right Vice President became increasingly visible, taking on an enlarging portfolio of issues as he prepared for his own presidential run. Gore's plans to follow Clinton into the Oval Office had never been much of a secret, and by the last years of Clinton's term his staff considered nearly everything Gore did through the filter of the 2000 campaign. The Veep met regularly with whomever John Doerr asked him to; he remained intensely curious about technology, but he also knew the good optics of hobnobbing with young stars like Marc Andreessen and Kim Polese. Silicon Valley also meant gold-plated fundraising, and Doerr was particularly helpful on that count as well. The tech executives who endorsed Clinton and Gore in 1992 had only given about $1,000 apiece. Now, they gave very generously. Between the 1994 and 2000 cycles, the amount of money Democrats raised in Silicon Valley grew by a factor of ten.[20]

Doerr's visibility as Gore's political wingman attracted plenty of notice on both coasts. "Gore-Doerr" campaign buttons popped up by the hundreds, printed jokingly by Stewart Alsop, founder of the dot-com era's most influential conference, Agenda. Doerr often got serious questions about his political ambitions. "No way. I wouldn't be good at it," he told the *San Francisco Chronicle*. "Everyone is out to kill you. I don't have a thick skin." But he did have an appetite for going deep on policy, including an increased interest in the environmental issues about which the Vice President was so passionate. Gore gave him an "environmental awakening," Doerr remembered, and the two men talked increasingly about how technology might address the looming climate crisis—and how "green tech" might be the Valley's next wave.[21]

Yet the embrace of the Valley by D.C.'s political establishment was not always reciprocated. Silicon Valley contained a wide range of political ideologies whose only common threads were a disdain for traditional gatekeepers and an ardent belief in the world-changing power of well-designed technology. The open-source ethos that had propelled the creation of the EFF still ran strong, especially among ordinary hackers and programmers, who remained deeply suspicious of both big government and big corporations.

Meanwhile, the relations between Washington, D.C., and Redmond, Washington, got even frostier. Judge Jackson ultimately was not persuaded by Microsoft's protestations that it was not a monopoly, and that the tech industry moved too fast to follow old rules. In the summer of 2000, his verdict came down: Microsoft must break itself into two companies—one with its OS business, the other with its applications and Internet business.

In the end, however, Jackson's appetite for publicity became Bill Gates's saving grace. After discovering that the judge had talked to reporters before he delivered his verdict, the D.C. Circuit's Court of Appeals overturned his ruling one year later. "Public confidence in the integrity and impartiality of the judiciary," scolded the court, "is seriously jeopardized when judges secretly share their thoughts about the merits of pending cases with the press."[22]

The new IBM of Microsoft remained intact, but it was a different company: humbled by the courts, battered by the market, its revenues still dependent on the

PC-era money machines of Windows and Office. The case profoundly reshaped the Seattle company's attitude toward playing politics—and it reshaped the attitude of other technology companies as well.

In the 1990s, Microsoft's D.C. government affairs operation was a one-man shop carved out of a sales office in suburban Chevy Chase. The tech giant's lobbyist had to shuttle back and forth into the city so frequently that he kept most of his files permanently stacked in the back of his Jeep Cherokee. The coming of the lawsuit showed Microsoft that lobbying no longer could be a back-of-the-Jeep afterthought. As the battle with the DOJ escalated, the company boosted the size of its D.C. office, hired a phalanx of political heavy hitters from both parties, and began shoveling six-figure donations toward both sides of the political aisle.[23]

As the Clinton years came to a close, many of the White House appointees who'd spent the past eight years on the NII and wiring schools and cybersecurity now rolled into lobbying firms, working the same issues from the industry side. Lobbying hadn't been this visible since the chip wars with Japan, and this time around there was a significant difference: in addition to the continued presence of trade associations old (SIA and AEA) and new (TechNet), the largest tech companies became lobbying forces in their own right. "Microsoft was a poster child for our industry," one tech lobbyist noted. It was the beginning of a new era.[24]

And even though millions of businesses and consumers continued to buy the shrink-wrapped boxes of Microsoft start-up disks, a new era was about to dawn in software too. It was going to be one where the software was free, and where advertising brought in the revenue. Where the word processing programs and spreadsheets and everything else lived in the cloud, downloadable at a click of a button. Where desktop PCs and closet-sized server rooms gave way to enormous, energy-gobbling server farms processing terabyte upon terabyte of data. Where new companies using these models would disrupt the software money machine that Bill Gates and Steve Ballmer had presided over for two decades, derailing its business far more than any antitrust action ever could.

THE GATES COMPUTER SCIENCE BUILDING

Forty years since his fateful tour of duty at Shockley Semiconductor, Jim Gibbons was taking his victory lap. He'd remained at Stanford his entire career, rising through the faculty ranks to become Dean of Engineering. Taking on the job as the tsunami of PC-era wealth washed over the Valley, Gibbons became a champion rainmaker, raising money for his school from every generation of engineering alumni from Hewlett and Packard on down, and persuading a number of people who'd never darkened the door of a Stanford classroom to give money as well.

One of those was Harvard dropout Bill Gates, who gave $6 million toward a sleek new computer science building that was dedicated on a windy and rainy Tuesday in early 1996, just as Gibbons prepared to retire from his post. (Paul Allen had wanted to contribute, too, but the hypercompetitive Gates would have none of it. "Name your own damn building," he told his onetime business partner.) The gift was a rounding error for the mogul, but it reflected his worries that his company was lagging behind the high-tech curve, and stronger connections with Stanford computer science might be a good insurance policy. The Valley's great nemesis now had his name on a building inhabited by Unix adherents and open-source devotees who believed that closed, proprietary systems limited innovation and advanced a particular agenda. Now that the Reagan-era Pentagon no longer loomed as an enemy, Microsoft served as Exhibit Number One.[25]

The irony of the moment wasn't lost on Gibbons. "Here is my prediction," the dean declared at the dedication ceremony. "Within the next eighteen months something will happen here, and there will be some place, some office, some corner, where people will point and say, 'Yeah, that's where they worked on the (blank) in 1996 and 1997.' And you will know it was a big deal. You will read about it." Neither Gibbons nor Gates could have imagined how spot-on that prediction proved to be, and that the kids who'd dislodge Microsoft from the top of the high-tech heap would come out of a building with Gates's name etched over the door.[26]

IN A COMPUTER SCIENCE DEPARTMENT FULL OF BIG BRAINS AND BIG EGOS, Sergey Brin and Larry Page stood out for their brilliance, their confidence in their ideas, and their inseparability. Brin, who arrived at Stanford as a graduate student in 1993, was the gregarious Russia-born son of a mathematics professor and a NASA scientist, who had emigrated with his parents at the age of six. On his ninth birthday, his parents gave him a Commodore 64; by the end of middle school, he was writing programs on a friend's Macintosh. He graduated from college at age nineteen. Page was a professor's kid as well, quieter and equally intense, who showed up on campus two years later to work under Terry Winograd. "His intelligence quotient lies four standard deviations from the mean abilities of a regular person," remarked software engineer Ellen Ullman of Page, "at the far, far right end of the bell curve."[27]

Yet the story of the two co-founders of Google isn't simply about smarts, or the Valley's remarkable ability to incubate technical entrepreneurs. It is part of the continuing saga of federal research money flowing into the Valley. For federal research dollars underwrote the costs of graduate research, and the money that supported much of Brin and Page's work at Stanford came from a key element of the Clinton-era NII: the Digital Libraries Project, a joint effort by the powerful troika of DARPA, NSF, and NASA that started up in early 1994.

Despite the name, the program wasn't about libraries at all, nor about the digitization of the books within, but about what came next: organizing the cornucopia of Internet content created when a world of paper went paperless. It was a question bedeviling information science ever since Vannevar Bush had come up with the idea of the memex in the closing days of World War II, but one made far more urgent by the Internet's commercialization. Even before the advent of the Mosaic browser, so much information was cascading around the online world that it had "come to resemble an enormous used book store," reported *Science*, and more was being added by the day. Yahoo!'s surfers didn't even crack the surface of it. The TCP/IP protocols could only do so much; a content-saturated Internet needed more. The answer came in the form of $24 million sent to six powerhouse academic computer science departments during the middle years of the 1990s.[28]

Stanford was one of them, augmenting its grant with support from Xerox PARC, HP, and tech-community influencer and tech publishing impresario Tim O'Reilly. The researchers' task: spend the next four years to build the search technologies needed for this single, integrated virtual library. A dozen internet search engines existed by 1994, and more were coming online by the minute, but the Internet already was growing at a rate that outpaced what the existing algorithms could process. They indexed URLs, not entire site contents, and they weren't terribly accurate as a result. The Digital Libraries' task was Web search like never imagined: more powerful, smarter, using machine learning.

Brin and Page were two of the first to move into the William H. Gates Computer Science Building, where in their shared third-floor office space they began work on the technology that one day would shove Microsoft's Sun King off his software throne. Their contribution to the Digital Libraries endeavor was originally going to be Page's dissertation: a system for tracing Web links backward, to determine relevance and credibility of websites by the number of other sources that had linked to them.

Page's original inspiration for this kind of peer validation came from academia—the scholarly papers cited the most often tended to be the most important in the field—but his circle of links also called to mind the ecosystem of the Valley. Those most connected were the most powerful; credibility and reputation came from knowing others in the network. The project was an upside-down Yahoo!, powered by software code rather than human surfers, and it solved the Internet's biggest problem: finding reliable and accurate information, and ranking the most valuable and strongly validated data first.

The system also departed from early Yahoo! in that its human creators were not making any value judgments. Instead, they were designing an algorithm to do it for them in an unsentimental, ostensibly apolitical way. The portal also had a spare and uncluttered interface, a striking contrast to the commercial search portals plastered with banner ads. Design simplicity was an obsession Brin and Page had picked up from their Stanford academic advisors, but it also reflected the fact that they weren't trying to make money. They were interested in Web search as a uniquely gnarly mathematical challenge; their aim was to follow in the footsteps of their academic parents, to become another John McCarthy or Terry Winograd. Yet by the time the two started collaborating, the Digital Libraries Project had already borne one innovative commercial search engine, Lycos, out of Carnegie Mellon. Before very long, it would have another.[29]

THE ENGINE

While the dot-com stocks soared and antitrust battles brewed, Brin and Page stayed in graduate school, refining their algorithm and preparing for brilliant academic careers. But it was getting harder for denizens of the Gates Building to resist the siren's call of the 1990s gold rush, and less than two years after they'd gotten started, they let Stanford's Office of Technology Licensing in on their secret. The university began to field offers for the technology from other search companies. But no one was biting, or at least offering enough to make it worth dropping out of school. Yahoo! and Excite and all the other dot-com darlings were focused on adding more and more features to keep visitors sticking to their portals and looking at their banner ads. A really good search bar would take users elsewhere; why invest in one? The two students kept waiting.

As a new school year began in the fall of 1998, the pair decided to launch the company themselves. Page's dorm room became the first offices of Google, Inc. Soon, one dorm room wasn't enough, and the chaos spilled out into Brin's pad next door. They had to buy a terabyte of memory—which then cost a cool $15,000—and put it on their credit cards. Bootstrapping on a graduate school stipend would only get a company so far, but luckily the two won over a key angel investor: Sun Microsystems' Andy Bechtolsheim, who only had to hear thirty minutes of their pitch before he wrote a $100,000 check on the spot.

Bechtolsheim's investment was like a flare sent up into the Palo Alto sky, alerting Silicon Valley's tight network that the Next Big Thing was in town. Their last great hope, Netscape, was floundering. Amazon chugged away up in Seattle, but it hadn't translated a high stock price into actual profit. Yahoo! and the other search portals were choking with garish ads. Here came Google's cool expanse of white, a visual oasis to the user. Its search algorithm, honed over four years of academic testing, was a leap forward in the state of the art.

With seed money in hand, Brin and Page moved operations out of the dorm and into their friend Susan Wojcicki's nearby garage by the start of 1999. (Wojcicki was another child of academics; the professors' kids were taking over the world.) They snagged Ram Shriram, a former Netscaper now at Amazon, as an advisor; Shriram persuaded his boss Jeff Bezos to make a personal investment too. Wilson

Sonsini became Google's counsel. Very soon the search company had outgrown the garage and moved into more grown-up digs on University Avenue in Palo Alto. They now had six employees. Down the hall was Peter Thiel's equally tiny Confinity, soon to be renamed PayPal.

In June 1999 came a stunning deal: Brin and Page scored a cool $25 million in venture funding, split evenly between two of the Valley's heaviest dot-com hitters, Kleiner Perkins's John Doerr and Sequoia's Michael Moritz. Bezos had advised them where to go for the money, and with confidence that bordered on arrogance, the two graduate students had pitted the VCs against one another in a bidding war. The gamble paid off brilliantly. Now they helmed an operation valued at $100 million, and they each had retained 15 percent ownership. Brin and Page were no longer the sweet and goofy guys throwing spitballs on the third floor; they had the swagger of future masters of the universe.[30]

Still, it was hard to leave academia behind. As news of Google's funding spread, Terry Winograd got an e-mail from the administrator who assigned space in the Gates Building, wondering if the two entrepreneurs would be dropping out, thus freeing up their cubicles (few things in Silicon Valley are more precious than Stanford office space). Yes, we'll be leaving, Page reluctantly confessed to his advisor. It still took them a year to fully move out. Brin never took his student home page off Stanford's servers; twenty years later, it remained live, listing his academic papers, his recent teaching, and a simple note at the top: "Currently I am at Google."[31]

Google's growth spiked at an eye-popping pace, with 3.5 million searches per day in September 1999, and 6 million by the end of the year. Nurtured and funded by seasoned Silicon Valley hands, the young company echoed the Apple II in its clean lines and the sophisticated architecture of its product. It carried on the legacy of HP and Intel in its belief that engineering came first. It sounded a now-classic Valley message of rebelling against authority, of thinking different. Giant computer companies roamed the earth and the Valley was saturated with Wall Street money, but Brin and Page promised a return to simpler and more idealistic times.

Microsoft's overreach and comeuppance seemed further validation that Silicon Valley had it right all along: promote the small, the entrepreneurial, the agile and collaborative. Don't get big, don't close yourself in, and, as Google's widely touted corporate motto put it, "don't be evil." This mythos—the story that the Valley had told to itself again and again since the start of the 1960s, and then broadcast to the world ever since the days of Don Hoefler—overlooked the inconvenient reality that every start-up company in the region's history eventually did one of two things. Most often, it went out of business. More rarely, it found success—and success meant bigness. Start-ups either grew large on their own, or they became absorbed by other large companies. HP and Intel and Apple started tiny and became giant; so did Sun and Microsoft.

Even as they grew large, these companies stubbornly continued to think of themselves as scrappy outsiders. That was Microsoft's great vulnerability; still run and controlled by the people who had been there from nearly the beginning, the company still thought of itself as a start-up, not as a big multinational corporation. As hard it was to believe, though, bigness was inescapable, and every successful start-up had the potential to become another Big Brother. Silicon Valley's next two decades made that clearer than ever.

Arrivals

PALO ALTO, 2000

Chamath Palihapitiya came to San Francisco because he was tired of checking all the boxes. Born in Sri Lanka, he'd emigrated to Canada as a young child and spent the next two decades dutifully meeting parental expectations. Brainy and awkward, he'd graduated high school at sixteen and headed straight into the prestigious electrical engineering program at the University of Waterloo. From there came a job in investment banking and the safety of an ample salary far beyond what this son of a financially struggling immigrant family had ever dreamed he'd earn.

Yet he was restless. The work he was doing—derivatives trading—appealed to his appetite for risk, but he itched for freedom from the corporate grind. "I was basically living my parents' life," he lamented. It was 1999, and the dot-com boom was white-hot. And there he was, marooned in the snows of Toronto, gazing from afar as college friends moved to California and joined the party. He printed out a slew of résumés and mailed them west.

Most of the places he applied took a pass, but he ended up with two offers—one at eBay, and another at Winamp, maker of an audio player software that allowed users to download and play music files on their computers. Stock-rich eBay was tempting, but Winamp had buzz. Plus, the company's offices were in a hip neighborhood of San Francisco, thus avoiding a grinding commute to eBay's bland precincts forty miles down the freeway in San Jose. So, to the further consternation of his parents, the Sri Lankan Canadian plunged into the world of online music.[1]

It turned out that he had joined the leading edge of a new, even more disruptive, phase of the software revolution. "The music industry should be afraid—very afraid," *Billboard* magazine had warned soon after Winamp's software came on the market in 1997. Beefed-up Internet infrastructure and the replacement of slow dial-up modems with cable broadband—one big consequence of the mid-1990s telecom reforms—now made it feasible for fans to swap and share megabytes of music and video with ease.

By the time Palihapitiya came on board a couple of years later, every music geek knew about Winamp, jokingly familiar with the nonsensical sound clip that would play upon launch—a rich baritone declaring, "Winamp, it really whips the llama's ass." And online music's niche market was blowing up to massive proportions thanks to Napster, a peer-to-peer music-sharing network created by a teenager in his Los Angeles bedroom. Ripping and sharing music files wasn't exactly legal, but the technology had far outpaced the ability of copyright law to keep up, and tech-savvy young people flocked to the service. "Napster is like a candy store for a music fan," wrote *The Wall Street Journal*'s tastemaking tech columnist Walt Mossberg.[2]

As users happily swapped free music files by the thousands, leaving record stores desolate and music-industry executives steaming, the larger tech companies began to sniff around for a piece of the online music action. A few months after Palihapitiya joined the company, Winamp hit the jackpot, getting acquired by America Online for a deal worth close to $100 million. Six months after that, in January 2000, AOL bought media giant Time Warner in a jaw-dropping $165 billion deal that was a harbinger of the information society to come—one where tech companies became powerful platforms for news delivery and social interaction.[3]

The AOL-Time Warner octopus now encompassed everything from the instant-messaging services used by teenagers to the newsmagazines read by their parents to the granddaddy of all cable-news networks, CNN. Now, instead of working for a scrappy start-up, Chamath Palihapitiya, the risk-seeking migrant with start-up fever, was working for one of the world's largest media companies.

But perhaps it was a lucky break. For within months of the AOL-Time Warner deal, the dot-com fever broke and Wall Street and venture investment abruptly stopped flowing into San Francisco's feverish start-up scene. Without cash, all kinds of young companies—online delivery services, search engines, enterprise software companies—bit the dust. AOL's stock price tumbled along with the rest, and the merger with Time Warner rapidly turned toxic. People were getting fired left and right, and Palihapitiya knew he couldn't be one of them: leaving AOL would mean losing his visa. Things might have been bleak in San Francisco, but he was going to stay.

CHAPTER 23

The Internet Is You

At the dawn of Silicon Valley's second century, the glory years shuddered to an abrupt halt. After peaking at over 5,000 for a few glorious days in March 2000, the NASDAQ composite plummeted below 2,000 one year later. The Dow Jones Internet index was one-fifth of what it had been the year before. "We never built models to anticipate something of this magnitude," confessed Cisco's John Chambers. Companies that had raised tens of millions in venture capital had burned through it all and couldn't raise another dollar. The market had gone cold, investors were holding their money close, and the last thing you wanted to be was a young company with dot-com at the end of your name.[1]

Wall Street's tech analysts had been brought low as well, accused of pumping up the market to an overheated state by staying so bullish for so long. "The Age of the Analyst is dead," announced *The New Yorker* in 2001, less than two years after it had published a long and adulatory profile of Mary Meeker. Although the Queen of the Net had started modulating her excitement as the millennium approached, many of her big picks plunged by more than 90 percent. As evidence of how irrational Meeker's exuberance seemed by September 2001, her favorite "buy," Amazon, had fallen by 94 percent and now hovered at under $6 per share. Dot-bomb blame heaped on Amazon's other great champion, Henry Blodget, a preppy Merrill Lynch analyst launched from near-obscurity after he predicted that the online bookseller's stock would one day hit $400.[2]

The pain spread all the way from the sunny flatlands of San Jose, where big chipmakers and hardware companies sheared off employees by the thousands, to the "Multimedia Gulch" of San Francisco, where smug e-commerce start-ups with cheeky names collapsed like dominoes. The confident young people who'd so recently swept into town for dot-com jobs now spent unemployed days in packed cafes and tipsy evenings at "pink-slip parties," or cashed in their severance checks to take long surfing vacations in Bali. (Life wasn't *all* that dire for the young and college-educated.)

Not even the most seasoned had escaped: David Morgenthaler saw a $500 million return on one investment evaporate into pennies on the dollar as the company's stock price sank 96 percent in the first quarter of 2001 alone. Burt McMurtry already had stepped back from day-to-day investing in the second half of the 1990s, and the downturn seemed like a fine time to embrace retirement for good. Ann Hardy's software contracts shrank to nothing as the bust spread across the financial technology field. Lacking the fortune of Morgenthaler or McMurtry, she moved down to Mexico, living like the hippie she never was: hopping the bus to travel at whim, wandering through the artists' colony of Oaxaca, relaxing after decades of work.[3]

To the national media, the rapid descent of highly visible dot-com names made for particularly juicy cautionary tales of Californian tech excess. The splashier the brand, the greater the public relish in its fall. Exhibit A in this regard was Pets.com, online purveyor of cat food and doggie toys, which had gone public at the tail end of the boom with a valuation of close to $300 million. It threw most of its money into marketing, saturating television airwaves with ads featuring a canine sock-puppet mascot who bantered with pets and crooned Blood, Sweat & Tears to a mailman ("What goes up, must come down . . ."). And come down it did—less than a year after the Pets.com puppet appeared as a float in the 1999 Macy's Thanksgiving Day Parade, and 268 days after its IPO, the dot-com retailer shuttered for good. The sock puppet later resurfaced from the dot-bomb murk as a spokesman for a loan company.[4]

The national political scene shifted dramatically too. Dogged by Bill Clinton's second-term scandals and unable to settle on a compelling campaign message, Al Gore eked out a popular-vote victory in the 2000 presidential election, but no clear Electoral College win. The outcome hung in the balance for several agonizing weeks until the U.S. Supreme Court ruled in favor of George W. Bush. The commander-in-geek was out of a job, and the Valley's brief love affair with Washington sometimes seemed like it never happened.

FOR ENTREPRENEURS AND INVESTORS WHO HAD RIDDEN TECH'S BOOM-AND-bust cycle for decades, however, times were bleak, but endurable. The boom this go-round had been so long-lived, and so enormous, that the Valley remained larger and richer even after the market crashed and burned. Pets.com and its peers might have bitten the dust, but Internet retail sales continued their steady upward climb. People had gotten comfortable with buying things online, and they weren't turning back.

Four years of fevered deal-making had made rich investors even richer, even if billions of dollars in paper wealth had evaporated overnight. Santa Clara County had 200,000 more jobs than it had at the start of the Internet era—and software and semiconductor firms were actually *adding* employees by mid-2001. And new talent kept coming to town—hungry young people like Chamath Palihapitiya, who sensed that the worlds of tech and finance and media were now intertwined so tightly that no bear market could untangle them.[5]

If you looked a little closer in those twilight days of 2001, past the armies of résumé-wielding MBAs and the acres of empty cubicles, you could see a next generation of Valley companies confidently gaining their footing—and the pop of the market bubble was the best thing that could have happened to them. Silicon Valley not only didn't die, it became wealthier and more influential than ever in the first two decades of the twenty-first century, propelled on overlapping waves of software-powered businesses: search, social, mobile, and cloud computing.

The mercenary bankers and MBAs-who-would-be-millionaires left town; the missionaries stayed. The post-dot-com era was the revenge of the nerds, as smart software engineers built the tools that finally turned the Internet into a money machine. The Web enterprises of the 1990s already had been edging away from the business model guiding capitalism since the Industrial Revolution (make something; sell it for a certain price; pocket the profit). Those free Netscape browsers and Yahoo directories generated revenue from advertising, but the interface was clumsy, annoying, and often missed its target.

The companies of what became known as Web 2.0 engineered a more elegant, less intrusive, and far more lucrative approach. Building on sixty years of discoveries in AI and human-computer interaction, they built giant user bases, then drew on data about these users to precisely deliver the information they wanted to see—and to send carefully targeted ads alongside that information. It was what one reporter called "the Holy Grail of web commerce": you could reach potential customers at the exact moment they wanted to buy what you were selling. "You're not the customer," quipped one coder, using a turn of phrase bandied about with increasing frequency in the early '00s. "You're the commodity." And the company that did this first and best of all was Google.[6]

THE GOOGLEPLEX

By the time the NASDAQ crested in March of 2000, Google's employee base had grown from six to sixty. It performed more than seven million searches a day, a figure that doubled in June, when Yahoo! stepped away from trying to run its own algorithmic search and made Google the featured engine on its portal.

AltaVista, a comparably sophisticated engine nurtured within the research operation of the now-enfeebled Digital, remained Google's most serious competition in the search game, but Brin and Page's company was catching up fast. By September, the founders announced that Google now indexed 560 million Web pages and would run a version of its portal in ten languages. By the following January, encouraged by John Doerr, a curious Al Gore made a visit to Google during his first trip out to Silicon Valley after the election debacle. Gore declined an offer to join Doerr on Google's board—he hadn't yet decided whether another presidential run was in his future—but he agreed to sign on as an advisor, taking a hefty chunk of stock options in exchange.[7]

As the market plummeted and venture investors closed up their wallets in the early days of 2001, Google kept growing. It was still a private company, mostly unruffled by the walloping of the NASDAQ. Instead, the market crash worked to its advantage. Prices dropped for the considerable amount of computer hardware that Google required to power its searches. Silicon Valley office space became easier to find and relatively cheaper to rent; Google soon moved down to Mountain View into a luxe campus vacated by the now-faded Valley superstar Silicon Graphics. Most important of all, the mass layoffs of the dot-com bust gave Google access to top-notch engineering talent that it otherwise might not have been able to afford. Now the small company had its pick of engineers, who were so eager for good work that they'd take stock options over a hefty salary.

In-the-know Valley people clamored to work for Google for other reasons too. Amid scornful chatter about silly dot-coms and corporate Internet profiteers, the clean and uncluttered Google interface became the cool kids' search engine, a soothing escape from the pop-ups and strobe-flashing ads that swamped the late-'90s Web. The portal's disdain for raw commerce extended to the corporate culture: Page and Brin still exhibited the earnest idealism of the Stanford graduate students they so recently were, determined to keep information free, the Internet transparent, and avoid suits and ties—and all they stood for—as long as possible. They hired by referral to keep that ethos intact. So distinctive was Google's sense of itself that employee number 50 had the title "Chief Cultural Officer."[8]

The two co-founders turned the Googleplex into a Stanford graduate student's dreamland, filled with ping-pong tables and comfy office furniture, where the sun was always shining and you never needed to wear a bike helmet. To entice PhDs and faculty wannabes, they mirrored the faculty consulting model Fred Terman had introduced at Stanford: you did your day job 80 percent of the time, and the other 20 percent was yours to play around with new ideas and innovations. The floor plates were large, the office spaces were shared, the people were packed in; it was just like the Gates Building, except the restrooms had $3,000 Japanese toilets with heated seats. The chef cooking up free food in Google's employee cafeteria had previously worked for the Grateful Dead. The quirks and perks were typical Silicon Valley style, an update of HP's back-patio horseshoes and Tandem's swimming pool, but the primary-colored playfulness of the Google campus topped anything the tech world had seen before.[9]

Lest Googlers become distracted by all the increasingly luxe bells and whistles around the Plex, the founders continually emphasized the boundary-defying significance of what they were trying to do. Earlier self-actualizing Valley generations had Esalen; the post-2000 crowd had Burning Man. The annual festival of art and drugs and free expression in Nevada's Black Rock Desert—a self-described "catalyst for creative culture in the world" that the co-founders attended faithfully each year—became metaphor and motif for all things Googley. An homage to the Man adorned the foyer of one of the buildings on Google's campus. The company sponsored shuttle buses to take Googlers to Black Rock each year. In 2001, a key factor in Brin and Page agreeing to bring in Valley veteran Eric Schmidt as CEO was the fact that Schmidt was already a Burner.[10]

Schmidt's hire was a long-fought victory for John Doerr and Michael Moritz, who had insisted the two founders bring on an experienced chief executive as a condition of their first investment and watched in frustration as Page and Brin rejected close to fifty candidates before they settled on Schmidt. The new CEO brought more Valley DNA into the Googleplex, forging connections between the new-era company and the people and firms that had come before. In his mid-forties at the time of his arrival, Schmidt sported a gold-plated résumé featuring a PhD from Berkeley and stints at Bell Labs and Xerox PARC. Then came fourteen years at Sun Microsystems, where he was one of the first employees, rising high enough on the org chart to be a target of the staff's famous April Fools' Day pranks. He moved from there to lead networking software company Novell. Schmidt hadn't been the VCs' first choice, but his technical smarts and management credentials made him the perfect fit for a company that venerated engineering above all else.

With Schmidt came others from beyond the intimate, Stanford-centric world of early Google. One notable hire came from Washington: Sheryl Sandberg, former chief of staff to the U.S. Treasury Secretary, whom Schmidt brought in to grow Google's advertising operations. Then there was Bill Campbell, the Kodak executive who had come to Apple at the start of the Sculley years and had gone on to helm the business software powerhouse Intuit. Campbell was now Silicon Valley's beloved "Coach," often brought in by VCs to give encouragement and enlarge the worldview of boy-wonder founders.

Coming on as an advisor to Brin and Page, Campbell became a familiar fatherly presence in Google's conference rooms as 2001 turned into 2002. The Coach had formed tight bonds with many Valley legends. Like his fellow Pittsburgh native and dear friend Regis McKenna, Campbell was very close to Steve Jobs. But his feelings for the Google guys were particularly heartfelt. "This is family for me," he told author Ken Auletta, with visible emotion. "There's innovation daily. They think about changing the world."[11]

THE AD ENGINE

Despite the hockey-stick growth and feverish media coverage, Google entered its fourth year in business without having turned a measurable profit. Engineering had been prioritized over all else. The company was burning cash on those ping-pong tables and toilet seats, and new VC money wasn't forthcoming while the dot-com crash smoldered. The founders needed a way to monetize their search engine without turning its Zen-like simplicity into an ad-choked mess.

That was already happening to AltaVista, the only Web crawler whose sophistication had rivaled that of Brin and Page's creation, and which quickly had descended into banner-ad hell after a series of acquisitions that sent it from the research labs of Digital to the online advertising giant Overture (first known as GoTo.com). But all that Overture lacked in design purity it made up for in its new model for monetizing search, which was to integrate advertising into the search itself. Instead of traditional ad buys, which plastered a banner or pop-up and hoped that the rare someone might click through to what you were selling, companies would buy rights to a *keyword*, bidding to have their product show up at the top. Advertisers only paid when the searcher clicked on their link.[12]

The eagle-eyed Googlers saw what Overture was doing and knew that this was where the future lay. But they didn't like the other company's practice of selling search results out to the highest bidder, making it difficult for the user to distinguish a truly relevant site from one that paid for its spot on the list. That definitely wasn't Googley. Instead, they produced a system (similar enough that Overture sued for patent infringement) that adopted the keyword-auction concept to generate paid results, subtly but clearly marked "ad," that appeared atop or to the side of a regular search. The site would stay as clean as ever, with its core principles intact. "You can make money without doing evil," proclaimed Sergey and Larry in the "Ten things we know to be true" that they posted on their corporate website around this time. The keyword technology they called AdWords—which they then sold to other websites under the brand-name AdSense—made Google a mint, and it changed the software business model for good.[13]

With every letter entered into a search box, Google or its customers learned a new scrap of information about their users, information that they could use to deliver ads that were targeted and relevant—and thus far more likely to result in a purchase. As visitors waited in the lobby of the Googleplex, they could see a scrolling report of the searches people were typing in at that very moment, projected on the wall like a particularly intriguing piece of panoptic video art. When Google rolled out new products—like Google News in 2002, Gmail in 2004, and YouTube, which it purchased for $1.65 billion from three PayPal mafiosi in 2006—the ad engine became even smarter.

By expanding its reach and monetizing its huge user base, Google turned the existing software ecosystem upside down. Microsoft might have come late to the Internet revolution, and it was sputtering to catch up with Google in search, but it still earned billions from selling packaged software. When Google unveiled online word processing and spreadsheet applications, however, consumers flocked to them. They were getting something a lot like Microsoft Office, but all for free! Microsoft continued to rule the business market—corporate IT managers had little interest in switching away from something so reliable and familiar—but Google's incursion put Gates and Ballmer on notice: the PC platform wasn't going to last forever.

"Free" software and content did have a price, of course, and some early software releases made this a little too clear to the user. Gmail initially featured small ads generated by the keywords being typed into an e-mail message. After outcry and further refinement, the more obvious signs of surveillance disappeared. Internet users knew that they were leaving a trail of information online, but the downsides of that openness were hazy, and the upsides of the services were tremendous. Google's founders were missionaries, not mercenaries, John Doerr would remind anyone who'd listen; all they wanted to do was make information free.[14]

By 2004, Google's revenues soared, its operating income topping $320 million, comparable to online auction giant eBay and far surpassing Yahoo! and Amazon. Google went public in August of that year, "one of Wall Street's most eagerly awaited births ever," declared *Fortune*, apparently forgetting how Intel, Genentech, and Apple once had set the Street atwitter. By December, the stock had doubled in price, to $165; two years later, it neared $300. Google wasn't going to be another dot-bomb after all, and its runaway success increased investors' appetite for finding the next big thing.[15]

THE HACKER WAY

"Classes are being skipped," marveled the *Stanford Daily*. "Work is being ignored. Students are spending hours in front of their computers in utter fascination. The facebook.com craze has swept through campus." In the early spring of 2004, the Silicon Valley business scene had begun to froth once more, but the undergraduates living right at the center of it suddenly were glued to their screens, checking out prospective dates and sharing likes and dislikes with their circle of online buddies—a circle that might include people you really didn't know in real life, but with whom you suddenly had a new, strange sort of familiarity and intimacy. "It provides a way to find out about someone without even approaching them," marveled one Stanford sophomore, who boasted an impressively long list of 115 Facebook friends.[16]

This was a new kind of online connection, one with real names and pictures rather than usernames and avatars. And it felt okay to put your real information up there—and your hobbies, favorite films, and relationship status—because Thefacebook (renamed Facebook the next year) was limited only to college students, and only at a few elite campuses. It had originated at the most elite one of all, Harvard, where a preternaturally focused nineteen-year-old named Mark Zuckerberg had launched the site out of his dorm room one month earlier. (The dorm room was rapidly supplanting the garage as the mythic birthplace of iconic Valley brands.) One month in, Facebook had nearly ten thousand users at Harvard and Stanford alone.[17]

Facebook was the latest entry in a great wave that rose up after the tech bust: social networking. These companies popped up in dizzyingly rapid succession in 2003 and early 2004, a speed and frenzy unmatched since the blossoming of the first microcomputer companies out of the homebrew scene a quarter century earlier.

Longstanding habits of networked connection now had a technological accelerant, as bandwidth grew and users could create jazzy custom pages to accompany their online personae. Los Angeles–based MySpace had a million users at the time of Facebook's launch; online dating had become the decade's new singles bar. The Valley already had two popular and growing social networks, Friendster and LinkedIn, which had attracted blue-chip VC backing and distinguished themselves from their rivals by requiring people to use their real names. Despite sky-high valuations, however, no one had figured out how to make real money on the social networking phenomenon. And it was hard to shake the icky feeling that many people had about sharing their lives with online strangers.[18]

The college students of the early 2000s had fewer qualms. They had grown up doing homework on a computer and sneaking late-night hours in the chaotic social world of Internet chat rooms. They file swapped on Napster until it got shut down; they added HTML flourishes to their MySpace pages. Still, Facebook started as a college kid's side hustle, a vehicle for the silly, ephemeral musings and gossip of his fellow students, who presumably would move on to more serious things after graduation. In those early months the site's server space was paid for by a wealthy roommate and, at one particularly cash-strapped point, by Zuckerberg's parents. It seemed unlikely to become the next world-changing tech company.[19]

That was, of course, before Mark Zuckerberg and his roommates moved to Palo Alto, secured money and mentorship, and became the runaway start-up success story of the decade, fodder for countless magazine cover stories, books, and one big-studio Hollywood film. Tech investors (and tech journalists) were perpetually seeking the next Steve Jobs or Bill Gates, and Mark Zuckerberg fit the bill—extraordinarily driven, far-seeing, his technologist's eyes on the prize even before he reached the legal drinking age.

Facebook soon expanded from college campuses to high schools, then opened its gates to the world. It turned out that people over the age of twenty-one also liked to share party pics and quotes from their favorite movies, and the stunning rate of growth never slackened. By the end of 2006, the elder generation of tech and media giants—Microsoft, Yahoo!, MTV, AOL—were descending on Facebook's University Avenue offices in an eager horde, desperate to get a piece of the company and its young, educated, affluent market. Zuckerberg unnerved and impressed his advisors by turning most of the new suitors down, including Yahoo!'s offer to buy the company for $1 billion. No thanks. Silicon Valley's latest star entrepreneur had decided he'd make history on his own.

THE SOCIAL NETWORKS

Google and Facebook were the biggest, but not the only, tech success stories in the first years of the new millennium. Joining them were other online phenomena that leveraged the Internet's speed and market penetration to build new online communities. They included a raft of social news aggregators like Reddit, whose blend of customized news feeds and passionately opinionated comment threads anticipated what Facebook would later become. Volunteer moderators powered these forums, just like the BBSs that came before them. An army of passionate volunteers also ran Wikipedia, a nonprofit online encyclopedia that by 2018 became the fifth most-visited destination on the Internet, behind only Google, YouTube, Facebook, and the Chinese search giant Baidu.[20]

The common thread binding all these enterprises, large and small, was that the content on their sites came from users, not journalists or scholars or "experts." Internet media had already derailed the music industry, and now it took down traditional print journalism as well. Newspapers had gamely put their content on the Web, for free, and now they competed with thousands of blogs and online outlets in a brawling battle for users' attention. Even new ventures by more established tech companies couldn't get traction. Wikipedia's surging growth helped kill off Microsoft's exquisite and expensive encyclopedia project, *Encarta*; the algorithmically curated Google News outpaced the all-purpose headlines provided by Yahoo! and AOL.

There was now so much content surging around the Web that it made Alvin Toffler's predictions of "information overload" seem quaint, and large quantity did not necessarily mean higher quality. The changes also encouraged a growing tribalism in a nation already fractured by war and economic insecurity, by race and gender, by faith and politics. People rallied online around a common interest or cause. They also came together because of their opposition to or outright hatred of something or someone else.

In the early days of social media, there was great hope that the new networks would cure divisions rather than increase them. In 2006, *Time* magazine, reliable bellwether of the zeitgeist, had a surprising choice for person of the year: "You." The year's story, wrote *Time*'s correspondent, isn't just "about conflict or great men. It's a story about community and collaboration on a scale never seen before. . . . It's about the many wresting power from the few and helping one another for nothing and how that will change the world, but also change the way the world changes." This was the great hope that so many in Silicon Valley had held for so long, the thread winding through Community Memory and Homebrew and the WELL, the thing that propelled the Liza Loops and the Terry Winograds and the thousands who flocked to watch the Man burn in the Nevada desert every September.[21]

All of that wide-open empowerment floated upon the loose, fragile, unpredictable framework of the commercial Internet. This was a system that the political debates of the early 1990s had ruled would be "liberat[ed] from Second Wave rules," as Esther Dyson and her collaborators put it in 1994—meaning that it was as lightly regulated as possible. The extraordinary new generation of thinking machines channeled the spirit of Mitch Kapor's Jeffersonian Internet: an independent, decentralized forum of many voices. Their designers remained resolute in their commitment to not take sides. As these powerful tools reached into the worlds of media and politics, however, sides would have to be taken.[22]

HEAD OF GROWTH

Facebook was a little more than five years old when it moved into a building on the fringe of the Stanford Research Park that once had housed part of Hewlett-Packard. The platform's growth had left all its competitors and predecessors in the dust. An expansionist, earnest, set-the-defaults-to-public spirit reverberated through the campus. By connecting the world through software, and doing so at massive scale, the company was accomplishing something the Valley had been trying to do for generations. Posters emblazoned with the company's de facto motto adorned the walls surrounding Facebook's expansive open-plan bullpen: "Move fast and break things."

Mark Zuckerberg remained in charge, owning over 24 percent of the company and controlling three of its five board seats. The Valley's Internet-era inner circle had become funders and close advisors. Peter Thiel had given Facebook its first big investment back in 2004 and was a board member. Marc Andreessen was a mentor as well, meeting Zuckerberg regularly for hash-and-egg breakfasts at a local diner. Star executives had joined from Yahoo! and Google, including Sheryl Sandberg, who became the company's chief operating officer in 2007. Gone was the hypermacho culture of earlier Valley giants like Intel and Sun; Facebook's top people were a tight and friendly team, passionate about the value of their product. "Technology does not need to estrange us from one another," declared senior executive Chris Cox, well-known around the company for the upbeat speeches he delivered to new hires. "In the grand scheme of things, communicating with each other changes everything."[23]

Chamath Palihapitiya was one of the techno-optimists in the C-suite by then as well. He'd known Mark Zuckerberg since Facebook's earliest days, when he was still at AOL, having become the youngest vice president in the company's history. Although failing to persuade his bosses to buy or invest in a little company helmed by a twenty-year-old who liked to wear shorts and flip-flops to business meetings, Palihapitiya brokered a deal that allowed Facebook to feature AOL's wildly popular instant-messaging service (which Zuckerberg and his team already used religiously). Soon after, Palihapitiya moved to Palo Alto for a job at the Mayfield Fund, the VC firm founded back in the late 1960s by Tommy Davis and Stanford's Bill Miller. He and Zuckerberg became regular dinner companions.

Zuck's cool intensity smoothed out Chamath's hyperkinetic style. "I've never met anyone at such a young age who would truly listen," observed Palihapitiya. "He doesn't need to talk a lot." When the young CEO suggested that the former media executive come to Facebook, it was an easy decision. Palihapitiya had been wanting to return to start-ups for the better part of the decade. He felt out of place in the country club of old-school venture capital. Here was a great opportunity to switch gears, and he might make a couple of million dollars on the way. His job title—Vice President for User Growth—spoke volumes about where Facebook's priorities lay. "There is so much accidental tourism in great things in life," Palihapitiya later reflected, and he had hopped on the tour bus at exactly the right time.[24]

In 2007, Facebook opened up its network to third-party apps, bringing in games and quizzes and other content to its newsfeed, and allowing developers to tap into the treasure trove of knowledge about users' connections and likes that Facebook called the "social graph." In 2010, Facebook announced "Open Graph," which connected a user's profile and network to the other places she traveled online. It wasn't just a social network atop the Web anymore. Facebook had remade the Web itself into something, as Zuckerberg put it, "more social, more personalized, and more semantically aware." The company allowed academic researchers to tap into its troves of information as well, underscoring its made-in-Silicon-Valley belief that freer and more transparent flows of information served the greater good.[25]

Facebook and its founder were remarkably young and relentlessly future tense, but Zuckerberg had a deepening sense of his place in Valley history as the company's wealth and influence grew. In the new digs, he adopted Steve Jobs's famous habit of holding "walk and talk" meetings, taking a prospective employee or business partner on a short ramble behind Facebook's building, up a steep and winding path through the eucalyptus trees to a hill that loomed above. Although it wasn't that high, the view from the top was sweeping, from Stanford's sandstone and tile just to the north, across the haze of the Bay to the eastern mountains, and down south where amid the sun-dappled sprawl lay the birthplace of so many of the Valley's iconic names: Shockley and Fairchild, Intel and Apple, Netscape and Google.

Zuckerberg would point out these sites, gesture to the building below, then turn to his companion to make his pitch. Facebook "would eventually be bigger than all of the companies" he had just mentioned, one recruit later recounted him saying. "If I joined the company, I could be part of it all." *Time* agreed that the young CEO was making history, making him Person of the Year for 2010. "We have entered the Facebook age," its reporter wrote, "and Mark Zuckerberg is the man who brought us here."[26]

THE SOCIAL MEDIA PRESIDENT

Like generations of tech companies before it, Facebook owed its success not only to the talents of its creators but also to the historical moment in which it grew. The long-brewing distrust of government, dislike of traditional gatekeepers, and decentralization of American mass media accelerated rapidly in the post-9/11 era, aided by (but not solely because of) the Internet. Added into the already frantic spin of cable TV came the cacophony of online outlets and the you-may-also-like curation of RSS feeds and Google News. From Capitol Hill to town council meetings, political discourse divided into sharply partisan echo chambers; rural-to-urban migration and political redistricting sharply separated Americans by class, race, geography, and party. The age of terror and grinding war in the Middle East caused a longing for familiar realms of family and community, and it increased suspicion of foreigners and religious minorities, the "them" versus "us." When real life felt terrifying, social media was a welcome retreat.

But Facebook and other social networks also filled a cultural void created by a half century of political liberation and economic dislocation, the vanishing of the bowling leagues and church picnics and union meetings that had glued together midcentury America in conformity and community. Social media became a more cosmopolitan town square, one that crossed national borders, launched new voices, and created moments of joyous connection that could morph into real-life friendships. It turned everyone into a diarist, a philosopher, an activist—even if that activism was merely clicking a "like" button.

Both Facebook and Twitter, a social platform originally designed for 140-word "microblog" status updates, became powerful mechanisms for political organizing and communication during the Arab Spring and Occupy Wall Street movements of 2011. Twitter swiftly gained a disproportionate number of African American users and "Black Twitter" became a powerful platform for both civic activism and cultural exchange; the most powerful racial justice movement of the century's second decade, Black Lives Matter, began as a Twitter hashtag. And in the 2008 and 2012 presidential races, candidates used social networking as a powerful tool to reach sharply targeted groups of likely voters, as well as providing the ultimate free-media platform for unfiltered campaign messaging.[27]

Few did this earlier and better than Barack Obama. Like Mark Zuckerberg, the onetime state senator from Illinois had been a virtual unknown in 2004, shooting into the international spotlight because of his remarkable charisma, singular vision, and lucky timing. Silicon Valley power players had been searching for a new boy wonder in the wake of Brin and Page's success, and they found it in Zuckerberg. Similarly, Clinton-weary Democrats who opposed the Bush Administration's decision to go to war in Iraq (and 2008 frontrunner Hillary Clinton's vote in favor of it) found in Obama a fresh face and compelling voice.

Just as Franklin Roosevelt had done with radio and John F. Kennedy with television, Barack Obama leveraged social media more thoroughly and creatively than his political rivals, and he formed a close and convivial relationship with the Valley in the process. Google's Eric Schmidt became an early donor and advisor. Chris Hughes, a member of Zuckerberg's original Harvard team, took a sabbatical from Facebook to serve as Obama's new-media guru, helping the campaign deliver targeted messages as cool and crisply designed as Web 2.0 itself.

Traditional direct-mail operations couldn't hold a candle to inexpensive and viral Facebook pages; a well-turned tweet by the candidate reached more voters than any stump speech. Bill Clinton might have won the tech community's votes in the early 1990s, but the new generation's hearts and wallets were with Obama, the Unix to Hillary Clinton's MS-DOS. As eager Stanford student volunteers swarmed the Palo Alto field office and tech executives lined up to give high-dollar donations, one reporter quipped that the Obama campaign had become "the hottest start-up in the Valley."[28]

After entering office in 2009, the commander in chief became a familiar presence in town, holding town hall meetings at Facebook and LinkedIn, convening big-ticket fundraisers, and enjoying private dinners with tech titans. One CEO gathering at John and Ann Doerr's Woodside home featured one of the most staggering assemblies of net worth in human history, with Zuckerberg, Eric Schmidt, and Steve Jobs all joining Doerr and Obama around the table.[29]

Back in Washington, the president pushed for wiring schools and reinventing bureaucracy with new software. He called on his tech allies and donors after the disastrous rollout of the enrollment website for his health care plan. Obama hired the nation's first chief technology officer, beefed up the White House Office of Science and Technology Policy, and staged science-fair photo ops to encourage kids to pursue engineering. He hosted a Reddit Ask Me Anything ("Hey everybody—this is barack," the president began), had millions of followers on Twitter, and hired a mind-boggling number of people who had once worked at Google. Obama aides, in turn, often made their way to Silicon Valley after their stint in public service was up.[30]

Toward the end of his time in office, in one final and important victory for the information-should-be-free crowd, Obama's FCC sided with the Valley (and against telecom companies) on the hot-button issue of "net neutrality," which prevented Internet service providers from blocking or charging higher prices for certain content. But it was tech's great capitalists whom Obama seemed to admire and rely upon the most. He quietly conferred with Doerr, Schmidt, and others as he began to mull his post-presidency life, and at one point floated the notion of becoming a venture capitalist himself.[31]

America had become even more fractured and fractious over the course of Obama's presidency, yet he remained optimistic about social media's potential to bridge the divide. Even a rising swell of foreign hacks and online security breaches did not dim the president's hope that much could be overcome if tech and government were both at the table. "I'm absolutely confident that if we keep at this, if we keep working together in a spirit of collaboration, like all those innovators before us, our work will endure, like a great cathedral, for centuries to come," he exhorted an admiring Stanford crowd during a cybersecurity summit the White House held on campus in early 2015. "And that cathedral will not just be about technology, it will be about the values that we've embedded in the architecture of this system. It will be about privacy, and it will be about community. And it will be about connection." Mark Zuckerberg couldn't have said it better himself.[32]

CHAPTER 24

Software Eats the World

For all the chatter about Brin and Page and Zuckerberg, Steve Jobs remained uncontestably the most important person in Silicon Valley in the century's first decade. A legend in his own time, Jobs had returned to Apple in the summer of 1997 as it clawed for its share of the dwindling fragments of a desktop market utterly dominated by Microsoft and the PC platform. Then he brought the company back from the dead. Adding a theatrical flourish to the whole resurrection, Jobs reached détente with his fiercest business rival, Bill Gates, who agreed to make a $150 million investment in Apple that saved the company from going under.[1]

In the decade that followed, Apple roared back into the center of the Silicon Valley story, with Jobs headlining one momentous product reveal after another—the bulbous and playful iMac, the sleek and intuitive iPod, and the market-upending iTunes, which harnessed the anarchic file-swapping energy of Napster to create a legitimate and immensely lucrative music platform. By the mid-2000s, the Apple team had shifted its focus to the biggest hardware challenge—and potentially biggest moneymaker—of them all. They were going to make a mobile phone. Cell phones were already a massive market, but Jobs was less interested in imitating what was already out there than he was in creating something quite different: an intuitive, elegantly designed handheld computer.

THE SUPERCOMPUTER IN YOUR POCKET

Silicon Valley technologists had been trying to build such a device since before the Apple II. It had been an arduous quest. In 1972, Xerox PARC's Alan Kay had mocked up a prototype of a mobile companion for young children that he called the "Dynabook." In 1991, an all-star roster of Silicon Valley insiders came together to launch Go Corp., developing software for a notebook-sized computer that used a stylus instead of a keyboard. Despite having Bill Campbell as CEO and John Doerr as a major investor, Go was too far ahead of its time. Apple made its own foray into stylus-and-notebook computing with the Newton MessagePad. But that device had an early death as well, felled by glitchy software and by the fact that it was John Sculley's pet project. As soon as Jobs got back into the CEO suite, he axed it. "God gave us ten styluses," Jobs's biographer Walter Isaacson recounted him saying as he waved his fingers in the air. "Let's not invent another." The closest the Valley came to realizing the dream was the ill-fated adventure of General Magic.[2]

By the early 2000s, other companies had achieved tremendous success with cell phones that featured e-mail and some very rudimentary Web browsing. The BlackBerry, a mobile phone featuring a tiny keyboard, became an indispensable device for legions of businesspeople in the first years of the decade, turning swift thumb typing into a badge of workaholic honor. The Palm Treo featured e-mail, a calendar, and a color screen.

Then there were the mobile phone giants—Motorola, Nokia, Samsung—who with every year made their phones "smarter" by loading on features and Internet access and ever-smaller keyboards. Progress in microchip technology fueled the market, as the advanced reduced-instruction-set microprocessors (or ARMs) that had been helping make computers faster and cheaper for a decade were now capable of powering a device small enough to hold, capable of surfing the Web, and possessing enough battery life to be useful.[3]

However, the way these got made infuriated Jobs and fellow design purists. The telecom companies held great leverage over phone design and packed them with applications users didn't want or need. Carriers fiercely resisted mobile devices that tried to deliver a richer Web browsing experience, protesting that phones that smart would hog too much network bandwidth. Unsurprisingly, Jobs had a very clear idea of what he wanted in a phone, and because he was *Steve Jobs*, he and his team were able to wrangle control from the wireless providers to make it happen.

The Apple iPhone that Jobs unveiled to the world in January 2007 was a mobile phone unlike any before it: a sleek bar of metal and glass, no keyboard, no buttons, no antenna. It had a touch screen, a phone, and GPS. Before too long, it would have voice recognition software. The iPhone looked like a palm-sized version of the mysterious black monolith that so fascinated the apes in *2001: A Space Odyssey*, and it garnered nearly as much chattering excitement.[4]

The order of show that day was designed to impress, as other titans of the Valley gathered around Jobs on stage to endorse the effort. Eric Schmidt joked about an Apple-Google merger—"we can call it Applegoo"—and Jerry Yang enthused like a star-struck teenager. "I would love to have one of these too, what a great device!" The final touch came via an iPhone voicemail from Apple board member Al Gore, relaying his congratulations on the achievement.[5]

Those outside the convention hall weren't as easily convinced. Microsoft CEO Steve Ballmer dismissed the phone out of hand. "There's no chance that the iPhone is going to get any significant market share. No chance," he told *USA Today*. The $500 retail price seemed ridiculous to Ballmer, as it did to others at the time. Plus, the first version of the phone only featured applications made by Apple. Steve Jobs's attitude about third-party software was the same as it had always been: he didn't want any of it gumming up the device's beautiful simplicity.[6]

Fortunately for iPhone users and for Apple's revenue stream, Jobs was eventually overruled. The App Store launched a year later. Apple remained firmly in control, approving any app before it could appear in the store, and taking a whopping 30 percent cut of the profits. The tactic was wildly successful and wildly profitable. Developers flocked to build for the iPhone, choosing it over other competing platforms. With a plethora of interesting applications, consumers got over their trepidation about the iPhone's high price; this wasn't just a beautiful piece of hardware, it was *useful*. Apple embraced its new role as Pied Piper for an entirely new mobile ecosystem. Its advertising tagline, "There's an app for that," became so popular that the company had it trademarked.[7]

The spread of the iPhone and its App Store sent ripples through the entire Internet world. Websites had to be rebuilt to look as good on mobile as they did on a desktop; social and search giants had to scramble to build mobile apps. Over at Google, leaders like Sun veteran Eric Schmidt and Netscape godfather John Doerr heard alarming echoes of platform and browser wars, when Bill Gates made a killing on proprietary software and boxed nearly everyone else out of the business. Microsoft's Ballmer was already mocking Google as a "one-trick pony" for its continued reliance on search for most of its revenue, and the executives over at the Plex knew they couldn't miss out on the mobile moment. Google's don't-be-evil answer to the dilemma was to release an open-source OS for smartphones, called Android, giving it away for free to any handset maker who wanted to use it. The move was also a boon for Google's business, providing a well-matched platform for mobile versions of its products. The Android platform spread like wildfire, becoming the standard OS in nearly any mobile that wasn't an iPhone. By the end of 2016, Android phones made up over 80 percent of the global market, and over half of Google's revenue came from mobile.[8]

The entry into the phone market was even more profitable for Apple. Ten years after its introduction, over one billion iPhones had been sold worldwide. It was the bestselling consumer product in human history. Having a geolocated, camera-equipped supercomputer in millions of pockets jump-started whole new business categories, such as ride-sharing (Uber and Lyft), local search (Yelp), and short-term rentals (Airbnb). It further spiked the growth of social media, launching born-mobile apps (Instagram, Snapchat) and turning existing networks into even more potent vehicles for advertising and sales. The switch to mobile made Facebook's user base grow even faster. By 2018, three out of four Americans owned a smartphone.[9]

With so many addictive morsels right at people's fingertips, the daily hours spent staring at tiny screens rose so sharply that a new and popular category of apps appeared, reminding users to put their phones down. By 2017, the mobile app business was larger than the film industry, and payments to app developers alone totaled $57 billion. Apple became the world's most valuable company, raking in nearly $230 billion in sales. Although the secret to Apple's stratospheric earnings was that it remained a hardware company—and very expensive hardware, at that—the iPhone's greatest contribution was in unleashing software from its desktop anchor and placing it on a candy-bar sized supercomputer. The iPhone was always on, always accessible, and, so very quickly, impossible to live without.[10]

THINK DIFFERENT

Steve Jobs had been diagnosed with pancreatic cancer four years before the iPhone's release. Although he pronounced himself healthy after surgery in 2004, his increasingly gaunt appearance caused rumors to swirl in the years that followed.

"Reports of my death are greatly exaggerated," he'd quip, channeling Mark Twain, but by 2009 it was impossible to keep up the front. He took a medical leave from Apple to have a liver transplant, returning shortly after, only to take another leave by early 2011. This time it was for good. On October 5, he died. Jobs was fifty-six. "Ahead of his time to the very end," eulogized the San Jose *Mercury News*.[11]

No other tech leader had been so iconic, so enduring, bridging the generations and becoming the face and the personality behind so many legendary Valley moments and high-tech products. Even the tales of Jobs being an arrogant jerk—moments balanced by the warmer and humbler man remembered by close confidantes like Regis McKenna and Bill Campbell—were an important part of the Valley legend. His death prompted an extraordinary outpouring of grief, not only from those who knew him personally but from the millions of Apple users who felt that they knew him nearly as well. "Steve was a dreamer and a doer," one wrote in a tribute wall on the company's website. "I am grateful for the gift he was in his creative genius," wrote another. At Apple retail stores throughout the world, people brought flowers and personal notes in tribute.[12]

At the private memorial service held on Apple's Cupertino campus a few weeks after his death, new CEO Tim Cook played a recording for the assembled crowd of company employees, celebrities, and Valley power players. It was Jobs's voice that boomed out through the speakers, reading the copy for the 1997 ad campaign—titled "Think Different"—that had started airing soon after his return to the company he'd founded. "Here's to the crazy ones," Jobs said. "The misfits. The rebels. The troublemakers . . . Because the people who are crazy enough to think they can change the world, are the ones who do."[13]

NOT EVERYONE AGREED ABOUT THE SAINTLINESS OF STEVE JOBS. IN A SOCIAL-media-driven moment, the critiques began even before the mourners had filed out of the memorial service. Jobs was a jerk, a greedy capitalist, a terrible boss, cried tweets and blog posts. The back-and-forth about "Good Steve" versus "Bad Steve" was only partly about Jobs. It also was about the place and the industry that he had come to symbolize. By 2011, the largest tech companies had transformed the way people across the globe worked, played, and communicated. They had opened up access to information like never before. The answer to nearly any question was just a Google search away. Long-lost friends and family reunited thanks to Facebook. Smartphones made the dream of a "computer utility" a reality at last.

Yet the greatest beneficiaries of the new tech companies seemed to be the very rich people who led and invested in them. Silicon Valley's titans had more money than God and an unimaginable amount of data on ordinary people. America was still climbing out of the market-shattering Great Recession, and sharp levels of income inequality had spurred populist movements on both Left and Right. While Jobs was being eulogized in Cupertino, the protesters of Occupy Wall Street had taken over New York City's Zuccotti Park, railing against "the 1 percent." Tech moguls were the 0.001 percent, and all their change-the-world promises seemed to have done nothing except encourage smartphone addiction.

Chasing fast money on apps and games that appealed to a narrow demographic of young, educated urbanites, the Valley seemed to be out of ideas. Even leaders within the industry saw a place that was falling short of its promise. Peter Thiel became one of the more outspoken critics. "What Happened to the Future?" asked a 2011 manifesto issued by Thiel's VC firm, Founders Fund. "We wanted flying cars; instead we got 140 characters."[14]

DAY ONE

Jeff Bezos also believed the Internet economy could do more. Visionary and relentless, Bezos already drew comparisons to Jobs for his intense management style and insistence on high standards. As Amazon grew large, his mantra remained largely unaltered since his early bookselling days: think long-term, put customer satisfaction first, and be willing to invent. To underscore Amazon's continued fidelity to its founding mission, Bezos attached his original 1997 letter to shareholders to every one of the company's Annual Reports. Its signoff was one of the CEO's favorite catchphrases: "It's still Day 1!"[15]

In contrast to the technicolored playgrounds of the Valley, Amazon remained a realm of spec buildings and door desks, and leanness of operations was at the core of its business model. Like his friend and board member John Doerr, Bezos was a faithful follower of Japanese manufacturing principles, dogged in his pursuit of cutting *muda*, or waste, at all points in the production chain. Bezos didn't issue

Zen koans like Jobs; while giddily passionate about innovation's promise, he remained a quant. Numbers, not emotion, guided him. "There is a right answer or a wrong answer," he once wrote, "a better answer or a worse answer, and math tells us which is which." He was careful about where he gave interviews, wasn't much interested in corporate PR, and Amazon still didn't advertise on television. The product, Bezos believed, spoke for itself.[16]

Amazon's dot-bomb days had faded into distant memory, displaced by its new identity as an unstoppable retail behemoth. The company had upended the publishing industry and was pushing into new realms, delivering value and convenience for its customers while it chased brick-and-mortar stores out of business. A big part of its growth came from turning itself into a platform for third-party buying and selling, giving businesses small and large an opportunity to reach Amazon's enormous audience. Now Amazon was branching out further into large-scale software platforms. The biggest of them all was Amazon Web Services, or AWS.

When he talked about AWS, Bezos sounded a lot like Steve Jobs talking about the Apple II. "The most radical and transformative of inventions are often those that empower *others* to unleash *their* creativity—to pursue *their* dreams," he told shareholders in 2011. Amazon had launched the service without much hoopla in 2006, targeting a new set of customers: software developers in search of storage and sophisticated computer power. But AWS's origins went back to the dark days of the dot-com bust, when analysts were getting sacked for having rated Amazon a "buy." Part of the company's path out of that mess was to turn itself into an e-commerce platform for other retailers to sell their wares, and it had to rebuild its technology infrastructure in order to do it.[17]

The result was a cleanly designed and resilient suite of software services, linked into a national network of humming data centers with enormous computing capacity. Bezos had moved to Seattle partly to be close to the book distribution centers of Washington and Oregon. These states' vast rural hinterlands now became fertile fields of server farms powering data-intensive operations. The region had an abundance of cheap hydropower, courtesy of the New Deal dams that straddled its great river systems, making it one of the best places on the continent to consume the vast amounts of electricity that were necessary for something like AWS. On the East Coast, Amazon repurposed older data centers in the national-security corridor of Northern Virginia, close to the original backbone of the Internet. From coast to coast, plenty of computing power, and very little *muda*.[18]

The name given to this platform—"cloud computing"—was new, but the underlying concept was as old as the UNIVAC. The cloud was time-sharing, twenty-first-century style, in which a data center substituted for a mainframe, laptops subbed in for "dumb" teletype machines, and the virtual machines ran Linux as well as everything else. Instead of the telephone wires of time-sharing and its inheritors, the network on which it ran was broadband-capacity Internet. The value proposition wasn't all that different from what Ann Hardy had built on that SDS mini at Tymshare four decades earlier: an OS that allowed clients to access computing power on an as-needed basis, lowering cost and increasing efficiency.

Infrastructure technology had deep roots, but the market opportunities were new. Open-source software made it possible for tiny teams to develop and run new systems and apps. A raft of companies had sprung up to build mobile apps, enterprise tools, and video and music streaming services. These entrepreneurs and developers had skills, laptops, and fast broadband connections. They needed server space and computing power, and AWS provided it.

AWS might have looked like a happy accident, a step outside Jeff Bezos's carefully cultivated long-range plan. Yet it made complete sense. Amazon had always been a big-data company, even when Bezos was the guy with the goofy laugh in the aging Honda, selling books on the Internet. He had stocked Amazon with crack engineering talent, people capable of creating a sophisticated and knowing platform that encouraged people to buy again and again, and a shipping-and-fulfillment system of unmatched logistical sophistication. Decades of dealing with seasonal fluctuations of American retail—pandemonium at the Christmas season, gentle undulations the rest of the year—gave Amazon smarter infrastructure, capable of accommodating sharp spikes in usage. The online giant's brick-and-mortar retail competitors had seriously underestimated what a software-powered platform could do to upend their business model. AWS scaled this sensibility upward, becoming a service that helped a whole new crop of software-driven companies take on traditional market incumbents, from hotels to taxis to broadcast TV. It wasn't quite flying cars (yet), but it certainly was more than 140 characters.

AWS provided the back-end support for some of the biggest new consumer companies of the century's second decade, and it also brought Amazon into the realm of big-ticket defense contracting, providing storage and analytics services for a data-gobbling intelligence community. "We decided we needed to buy innovation," explained one Pentagon official, and in 2014 they bought it from Amazon: a $600 million contract to build a cloud platform for seventeen national security agencies. The CIA's chief information officer called it "one of the most important technology procurements in recent history."[19]

Recognizing an enormously lucrative opportunity, other tech giants crowded in, further eroding the hardware-driven business that had ruled enterprise computing for decades. Large corporate customers no longer needed to buy standalone computers to fill their own data centers or server rooms, nor did they need to buy packages of enterprise software. As a signal of the changing times, IBM—the company once synonymous with mainframe computing, the fearsome hardware juggernaut that once considered software to be an add-on rather than a standalone product—launched its own cloud computing division.

While IBM was able to become a dominant provider for the Fortune 500, the very biggest players in the cloud market had been software companies from the start: Amazon, Google, and Microsoft. After years of struggling to find a new hit product, Microsoft hit the jackpot with its cloud service, Azure, whose growth spiked over 70 percent between 2016 and 2017 alone. In a remarkable turn of events for a company that had once been the cathedral of proprietary software, 40 percent of the virtual machines in Azure ran on open-source Linux. It felt like Day One in Redmond, too.[20]

THE SEEING STONE

Search and social, mobile and cloud: the tech story of the 2000s seemed to be more of a free-market success story than ever before. Yet the nine-figure defense contracts secured by AWS and other cloud computing giants were one sign of renewed collaboration between the tech industry and the Pentagon after two decades of relative chill. This time, it was all about software.

The Valley's defense economy had never disappeared, of course. The Cold Warriors still held court at the Hoover Institution; the quiet hulk of Lockheed still brooded alongside Highway 101. But the investors and entrepreneurs of the Internet era had seen little upside or need for defense work, and the Pentagon's investments in academic research had dwindled. Lockheed moved much of its missile and space division to Colorado. Cold War–era names—Raytheon, Boeing—continued to dominate the list of the Pentagon's biggest contractors. While Internet giants pushed into new frontiers, military software lagged behind. "The cutting edge in information technology," Harvard physicist and future Secretary of Defense Ash Carter observed in 2001, "has passed from defense to commercial companies."[21]

The U.S. thus entered the post-9/11 era with a technological dilemma. The emergence of stateless, widely dispersed terrorist networks meant that the American way of war demanded higher-tech tools than ever. Added into that were large-scale data breaches and surveillance by foreign agents that had started to plague American corporations and government agencies. Hacking wasn't about teenage cyberpunks in suburban bedrooms anymore; it was information warfare waged by the West's most dangerous foes. With the outlays for conventional warfare surging, military and intelligence leaders needed a fast, relatively cheap way to ramp up the military's technological capacity.

To do it, they turned once again to Silicon Valley, but flipped the Cold War supply chain on its head. Instead of government-funded academic labs and contracts producing military tech that later could be commercialized, now the defense establishment created VC firms to seed private software companies that could one day become contractors. Instead of the traditional research and procurement process, the Pentagon sponsored hackathons and design charrettes to get government bureaucracies to behave more like start-ups.

The swelling number of defense contracts were hard to see at first—just as in the Valley's early years, the top secret nature of so much of this activity kept observers from fully understanding its size and scope—but it soon became impossible to ignore the growing amount of work Big Tech had started to do for the military. What also became clear was a continuing irony: that some of those most enriched by the new-style military-industrial complex were also some of the tech industry's most outspoken critics of big government, and champions of the free market. In the space-age Valley, the person embodying this contradiction was Dave Packard. In the cyber age, it was Peter Thiel.

In contrast to his tech brethren who rallied around Barack Obama in 2008, Thiel remained unwavering in his belief that modern politics was a dead-end pursuit. "Politics is about interfering with other people's lives without their consent,"

Thiel wrote not too long after Obama entered office. "I advocate focusing energy elsewhere, onto peaceful projects that some consider utopian."[22]

Thiel wasn't alone. He was one of several tech titans making bets on private space travel, "a limitless possibility for escape from world politics." He was one of a number who had bought compounds in New Zealand as extra insurance in case of social collapse. (A few years later Thiel went one step further and became a New Zealand citizen, just in case things really went sideways.) Thiel also spent millions on more-personal pursuits: fellowships to encourage smart young people to drop out of college and try entrepreneurship instead; a foundation dedicated to reversing human aging; a think tank devoted to preparing for the "singularity"—the moment when thinking machines would be able to self-replicate and, possibly, displace humans altogether. He became a major backer of a "seasteading" effort to build a floating city, free of government control, in international waters. The libertarian utopia was the brainchild of a former Googler who happened to also be a grandson of economist Milton Friedman, and perhaps the ultimate expression of the techno-libertarian project to escape the tentacles of bureaucratic control.[23]

That had been the original idea fueling PayPal, of course: an alternative system of Internet-based currency, unharnessed from government-controlled money. The fact that the company had devolved into a mere online payment-processing system had been extremely lucrative for Thiel and his colleagues, but he never had let go of the notion that a bigger and more disruptive system was possible. Along with the millions he pocketed from the PayPal sale, Thiel also took away an idea for fraud-detection software that he believed could be repurposed to root out potential terrorist attacks "while protecting civil liberties." Mining his networks of old friends and Stanford connections to build a sharp, young leadership team, Thiel in 2003 bankrolled a new data-mining company he called Palantir, after the "seeing stones" in Tolkien's *Lord of the Rings*. (And a hat tip to Silicon Valley's legend, referring back to those Tolkien-themed offices at early Xerox PARC.) Established VCs were lukewarm on the start-up—Sequoia passed, as did Kleiner—despite the fact that the team believed it was, in the words of CEO Alex Karp, "building the most important company in the world."[24]

Then in swooped an unlikely angel: the U.S. Central Intelligence Agency. Keen for access to topflight software engineering, the CIA had gotten into the venture business at the tail end of the dot-com boom, creating an entity it called In-Q-Tel, and hiring the former CEO of Lockheed to run it. The CIA became Palantir's first and sole customer from 2005 to 2008; after seeing the impressive results delivered by the company's tracking software, other parts of the intelligence community joined as Palantir clients. Former FBI Director George Tenet became an advisor, lamenting that the intelligence community hadn't had "a tool of its power" before 9/11. Condoleezza Rice signed on as an advisor as well. While the exact nature of Palantir's intelligence work remained top secret, rumors swirled that its software was the literal "killer app" that helped track down Osama bin Laden. Palantir executives did little to quash the speculation.[25]

Soon other government clients signed on. Large police departments sought out Palantir's visualizations and graphs and data-mining techniques to track criminals. The U.S. Immigration and Customs Enforcement agency, or ICE, bought software to profile its targets. To help break into the tight circle of favored government contractors, Palantir spent heftily on Washington lobbyists and cultivated House and Senate lawmakers responsible for defense appropriations, ultimately landing over $1 billion in federal contracts. Data-hungry and privacy-obsessed corporations swelled Palantir's billings even further; by 2013, the company was drawing in 60 percent of its revenue from the private sector.[26]

By 2016, Palantir had a $20 billion valuation, the third largest of any private company in the Valley, and thousands of employees. Its corporate culture was as distinctive and quirky as early Google. *Lord of the Rings* references abounded; Palantir's downtown Palo Alto offices were "the Shire" and its two thousand employees "Palantirians." The hiring process was notoriously demanding. "I interviewed at Facebook, Google, D. E. Shaw, and a bunch of other places," reported one engineer. "Without a doubt, Palantir's questions were the hardest and they asked more of them than anyone else." If you make it through our gauntlet, Palantir's leaders seemed to be saying, you are the best of the best.[27]

Underneath the geeky normalcy, the firm unnerved the industry's privacy watchdogs. "They're in a scary business," remarked one EFF attorney. And it was a business made scarier by how connected its founders and funders were to the Valley elite. Chamath Palihapitiya was an investor; Thiel remained on the board of Facebook. One year, to some members' horror, Palantir sponsored the EFF's annual awards ceremony. Questions swirled about whether the company's surveillance software used too broad a brush; were the innocent also getting ensnared in a net designed to catch terrorists and thieves? But the contracts continued to roll in.[28]

Peter Thiel always had been a figure of contradictions: a gay man who rejected special treatment for minorities; a defender of free speech who funded a lawsuit that drove a prominent online publication out of existence; a steadfast libertarian who was close to some of the Valley's biggest liberals. His halting manner and sparing public pronouncements added to the mystery. "He really is like a chess master," said one young admirer, "planning his moves several steps ahead." Now, as Palantir soared, Thiel became a latter-day H. Ross Perot: a champion of free enterprise who was simultaneously reaping a great fortune from the government he disdained.[29]

CHAPTER 25

Masters of the Universe

"I wanted to see with my own eyes the origin of success," Russian President Dmitry Medvedev declared from the Stanford stage one bright summer's day in 2010. American and Russian flags stood at attention behind the jeans-and-blazer-clad leader, looking as sleekly casual as any Valley venture capitalist. A member of the *glasnost* and *perestroika* generation who became president at forty-two, Medvedev was on a quest to end the brain drain that had plagued Russia since the end of the Cold War. His country was one of the fastest-growing Internet markets in the world, with sixty million citizens online and counting. Its financiers and oligarchs were investing millions in American tech giants. Now it was time for this onetime technical superpower to build some world-changing companies of its own.

Only weeks before, the Russian leader had rolled out splashy plans for a high-tech "Innograd" on Moscow's suburban outskirts, little more than a twenty-minute drive away from the auditorium where Ronald Reagan had praised the high-tech revolution two decades earlier. He then hopped on a plane to San Francisco, following the trail of so many world leaders before him. He visited Twitter and sent his first presidential tweet (handle: @KremlinRussia). He met Steve Jobs, then back at Apple after his liver transplant, although visibly ill. He sat down with Stanford's leadership. "Unfortunately for us," he confessed to Provost John Etchemendy, "venture capitalism is not going well so far." There simply wasn't enough appetite for risk. "It's a problem of culture, as Steve Jobs told me today. We need to change the mentality."[1]

The president's road show generated plenty of skepticism. No sooner had he sent out his first tweets than a parody account called @KermlinRussia began mercilessly poking fun at the earnest high-tech experiment. "One needs to understand that money given to modernization and innovation will be spent on corruption and swindling," read one post. "We are aware that you are aware that we are thieves," read another. In the raucous world of social media, nothing—and no one—was sacred.[2]

Medvedev was discovering that the quest to build another Silicon Valley rarely went as planned. Despite the billions spent around the world by national governments on high-tech ventures—research parks, venture funds, broadband networks, even entire cities—the United States continued to out-innovate them all, pumping out one market-altering tech company after another. China was the only place that had managed to produce companies of comparable size and reach to Google or Facebook, and that was largely because the government placed stringent and often censorious barriers that kept American tech giants from entering.

The mostly fruitless global chase showed exactly how much politics still mattered, even for an industry so long understood as a free-enterprise success story. A light government touch when it came to the online world had enabled Google and Facebook and Amazon to grow immense and ubiquitous without much worry about regulation or antitrust action. (America's laissez-faire status quo was a striking contrast to the European Union, where tech giants faced continual pushback from the courts for privacy violations and anticompetitive practices.) Open doors to the world allowed Silicon Valley to draw from a global talent pool, even as politicians had begun to hotly debate the nation's immigration policy. Over half of the companies founded in the Valley between 1995 and 2005 had a foreign-born founder. About 40 percent of the engineering degrees at Stanford were earned by international students. Drawing in this talent helped keep U.S. universities the best in the world.[3]

This was a distinctly American tale of state-building by stealth, over many generations—through defense contracts to private industry, grants to academic labs, tax breaks to venture capitalists, sustained boosterism by politicians of both parties, and more. No high-tech city built by presidential decree could match the exuberantly capitalist, slightly anarchic tech ecosystem that had evolved over several generations.

The thing that made Silicon Valley so alluring to foreign leaders like Medvedev was the same thing that allowed Twitter trolls to gleefully skewer his technotopian ambitions. Valley culture was *American* culture, allowing free flows of people, capital, and information like no other country in the world. And in the years immediately before and after Medvedev's visit, it made a small group of people in Silicon Valley and Seattle very, very rich.

THE POWER OF PLACE

By the time Marc Andreessen took to the opinion pages of *The Wall Street Journal* in the summer of 2011 to pronounce that "software is eating the world," the new tech platforms were not only altering entire industries. They were transforming the geography of tech as well.[4]

Across North America and beyond, tech had inhabited a sprawling, suburbanized landscape of research parks and corporate campuses since the age of Eisenhower. It had continued to do so even as other white-collar industries and middle-class residents began returning to denser urban environments at the century's end. Part of this suburban persistence had to do with the outsized influence of Silicon Valley, leading its imitators to assume that high-tech magic required extensive landscaping and low-rise buildings sheathed in mirrored glass. It also, however, reflected tech's need for plenty of square footage to accommodate scores of coders in cubicles, roomfuls of server blades, and closets full of coaxial cable and routers. The up-front costs for real estate, personnel, and equipment were considerable, demanding early venture rounds of $10 million, $20 million, or more.

This changed in the new era. After two decades of concerted government-directed effort, broadband penetration was extensive not only in the U.S. but across the world. After a decade-plus of software and hardware innovation by tech's largest operators, supercomputing power became ever smaller: the long-dominant desktop PC market shrank as lightweight laptops, tablets, and mobile devices delivered comparable amounts of computing power. The iPhone and other mobile platforms allowed a popular app to build a massive user base very quickly; the dreaded "valley of death" that tech start-ups had to travel between idea and go-to-market became a dip in the road. On top of it all, cloud services freed new companies from having to fork over precious venture capital for computing power and server rooms.

The cost of starting up plummeted and the variety of entrepreneurial enterprises exploded. Want to build an app for the iPhone? All you needed were some sharp coding skills, a good laptop, and a little cash each month for some AWS server space. Support services of all kinds could be outsourced to hourly contractors who went from gig to gig (and a whole new wave of software-powered start-ups had emerged to facilitate this matchmaking). Founders didn't have to search for a suite in a fully-wired tech park anymore. They could rent a desk or two in a tech incubator or co-working space. Unleashed from real estate constraints, tech start-ups left the suburbs for the cities, popping up in high-density, high-rent districts favored by the young and hip from Brooklyn to Boulder, Munich to Melbourne to Mumbai. A small slice of entrepreneurs chose not to settle down anywhere, becoming high-tech nomads who could do their work from anywhere in the world with a decent Internet connection.

Cities that had been pining away for some of Silicon Valley's magic thrilled at the opportunities the new era presented, sponsoring makerspaces and demo days and holding seminars on how to lure in more venture investment. While the list of places with viable high-tech clusters expanded, however, high-tech investors remained firmly concentrated in the same places they had been in the 1980s. Urban theorist Richard Florida, whose widely read work on the "creative class" fueled cities' high-tech hopes, found that San Francisco and Silicon Valley firms together accounted for over 40 percent of the VC investments and over 30 percent of the deals made nationally in 2013. Seattle came in at a feeble seventh place. Start-ups were blossoming in the home of Gates and Bezos, but it was too easy to fly down to Sand Hill Road to raise money. "There are vanishingly few growth capital sources for tech innovation that don't flow through San Francisco and New York," one Seattle investor observed in frustration.[5]

Despite the homegrown hackathons and co-working spaces and incubators sprouting up in cities large and small, the wealth and personnel of the innovation economy were increasingly monopolized by five companies: Amazon, Apple, Facebook, Google, and Microsoft. Soon, it became clear that the surest way for a company to make money in the world of the new titans was to be acquired by one of them. By mid-2018, Facebook had made 67 acquisitions, Amazon had made 91, and Google had made 214. Valley veterans were baffled by the rush to the exits. "Doesn't anyone want to build a company anymore?" Regis McKenna wondered. The tech business long had been about elbowing out your competition to grab market share. Now, the game became one of building a platform so unique and so dominant that it *was* the market.[6]

The concentration of wealth rippled out into real estate too. Tech's biggest companies abandoned spec buildings and dun-colored tilt-ups for dazzling, custom-built urban and suburban campuses. Facebook converted the old Sun Microsystems campus in Menlo Park into a spectacular complex rivaling only the Googleplex in its playfulness and perks, with an open interior courtyard that was like Palo Alto's University Avenue in miniature, except that you never had to find a place to park your car, and all the food and beverages were free. In 2015, the company opened an enormous, Frank Gehry–designed building across the street, designed to be what Mark Zuckerberg called "the perfect engineering space" and to tell a story. "We want our space to feel like a work in progress," Zuckerberg wrote. "When you enter our buildings, we want you to feel how much left there is to be done in our mission to connect the world."[7]

Even stripped-down Amazon couldn't resist adding a grand architectural flourish to the generally undistinguished set of buildings that made up its headquarters in central Seattle, building a striking pair of "biospheres" housing indoor gardens for Amazonians to enjoy. On the other side of Lake Washington, Microsoft tried to keep up with its crosstown rival's riff on Buckminster Fuller by building treehouses for employees' midday retreats. But the most stunning monument of them all was Apple's massive new Cupertino headquarters, a sleek ring of glass and steel housing twelve thousand employees. "Apple Park" had been one of Steve Jobs's last ideas before he died. In homage to their founder and the Valley that once was, Apple planted an apricot orchard in the building's shadow.[8]

THE NEW MONEY MEN

The Silicon Valley money machine seemed unstoppable. Within a few years of Google's IPO, a thousand of its current or former employees had a net worth of $5 million or more, including the in-house massage therapist the founders hired back in 1999. Page and Brin were worth about $20 billion apiece. The dot-com kings of the 1990s who reinvented themselves as angel investors and venture capitalists saw their net worth climb. At the head of the pack was Marc Andreessen, who in 2009 founded a new-style VC firm with partner Ben Horowitz designed to nurture young technical founders into savvy company leaders, rather than shoving them aside to bring in adult supervision. Mark Zuckerberg was a perfect example of how it could be done. "The Valley's fearlessness is coming back," Andreessen told a reporter.[9]

Also coming back: the star analysts of 1990s Wall Street. Mary Meeker had never flagged in her faith in the Internet, even amid the plunging prices and shareholder lawsuits of the dot-com bust. Nor had Ruth Porat, who surged to the top of the technology-banking ranks after taking Google public in 2004. Nine years later, Porat made even bigger headlines by moving to Google to become its CFO. Meeker already had moved to California by then, too, departing Morgan Stanley in late 2010 to become a partner at Kleiner Perkins, bringing along her now-legendary annual slideshow on Internet trends. Meeker was bullish once more, and it wasn't empty hype: the new generation of companies had better fundamentals, the market had matured, and the people operating and investing in them better understood what it took to make an Internet-based business succeed.[10]

The sages of Wall Street moved west, and the relationship between the Street and the Valley changed too. To help the economy recover from the housing bust and deep recession of 2008, the Federal Reserve had kept interest rates low, pumping liquidity into the market and leaving the monied classes looking for high-yield places to put their cash. The abundance and variety of investment capital—private equity, hedge funds, angels—lessened the need for IPOs in a company's early stages and increased SEC scrutiny of public companies further dampened start-ups' enthusiasm for a Wall Street offering. Overseas investment by the world's new mega-rich became another growing source of capital, particularly useful in the shaky days after the 2008 market crash. A cash-hungry Facebook entered into a lucrative deal in May 2009 with Russian financier Yuri Milner, a billionaire with close Kremlin ties; Milner ultimately ended up holding close to 9 percent of the company. "A number of firms approached us," Mark Zuckerberg said at the time, "but [Milner's] stood out because of the global perspective they bring." The Russian also made a large investment in Twitter.[11]

Lurking behind all of this cash, domestic and foreign, was the angel that had been there since the very beginning: the U.S. government. American financiers had so much to invest because of a U.S. tax code that—thanks to five decades of sustained lobbying—strongly advantaged those who made money from money. The capital gains tax stood at 15 percent. The carried-interest deduction stood firm, despite periodic attempts to abolish it. VCs, hedge fund managers, and private equity funds alike were able to rake in billions for managing other people's investments, and call all of it their "capital gains."

Then there were the taxes on corporate revenue. For more than fifty years, the U.S. had allowed American corporations operating overseas to defer taxation on profits earned in non-U.S. markets. In the early 2010s, this arrangement became a gold mine for software companies, which, because of their global reach and the ethereal nature of their product, were able to shift profits from high- to low-tax jurisdictions. (The United States, as tech giants and others would continually point out in their defense, had the second-highest corporate tax rate in the world.) Tech firms further reduced their tax bills by writing off stock options, depreciation of facilities, and expenditures on R&D.

The elaborate shell game—entirely permissible under IRS rules—made rich companies like Apple, Google, and Amazon even richer. Washington made spasmodic efforts to change the system in the Obama years, but it was hard to cast beloved tech brands as tax-dodging fat cats. "I love Apple!" rhapsodized Missouri Democrat Claire McCaskill at a 2013 Senate hearing where Tim Cook was supposedly being called to account for his company's creative accounting. Kentucky Republican Rand Paul berated his fellow Senators for "bullying" Cook and a company that was "one of America's greatest success stories."[12]

Facebook employees and alumni also joined the ranks of the breathtakingly wealthy. Napster co-founder and early Facebook leader Sean Parker became something of a latter-day Jerry Sanders, making headlines for the conspicuous consumption of the billions he had reaped from Facebook's 2012 IPO. Parker's wedding the following year in a Big Sur redwood forest was an homage to Silicon Valley's favorite fantasist, J. R. R. Tolkien, featuring custom-made medieval costumes for more than 350 wedding guests, a faux castle, and, as Facebook chronicler David Kirkpatrick reported, "a pen of bunnies . . . for anyone who needed a cuddle." Toward the end of the evening, Sting performed *a cappella*.[13]

The combination of Facebook stock and savvy investments in other Valley start-ups also made Chamath Palihapitiya a billionaire, and he co-founded a new venture operation called Social+Capital (later dropping the +). It was VC, Chamath-style, made possible by the immense amount of personal wealth generated in the Valley since 2000. Limited partners included not only outside investors but also a hand-picked group of very rich friends and one very rich corporation: Facebook. The goal was not only to further leverage the connective power of social platforms, but also to support a more diverse pool of entrepreneurs and build a "purpose-driven" portfolio.

Yet even in this new era of VC, the Valley's tight networks of friendship and familiarity still ruled: Social+Capital's first investment was in Yammer, a social network for business use, founded and led by *Diversity Myth* author-turned PayPal millionaire David Sacks. Palihapitiya became an observer on Yammer's board, whose members included Peter Thiel and Sean Parker.[14]

The new generation of money men wore designer T-shirts instead of sport coats, drove Teslas instead of Mercedes, and used rap lyrics as business metaphors. They had more flash and cash than their VC forebears, but the same relentlessness, elbowing past slower-moving East Coast competition to latch on to early-stage deals. They were brilliant and lucky, and they knew it. "This is not checkers," Ben Horowitz advised would-be entrepreneurs. "This is motherfuckin' chess."[15]

AN INCONVENIENT TRUTH

Among those made very wealthy in this era: Al Gore, who already was living one of the more extraordinary afterlives in American political history. Gore had first turned media mogul, embarking on a cable news venture called Current TV. He then had gained fame as a bearded prophet of climate change after a 2006 documentary in which he starred, *An Inconvenient Truth*, became an Oscar-winning smash. But it was his next act as a Silicon Valley advisor and venture capitalist that turned him into a multimillionaire.

Gore had hit the jackpot with his early stake in Google, and his net worth soared even higher after a few years on the board of Apple, which he joined in 2003. Four years later, the iPhone was a smash, the former veep was worth $100 million, and he accepted John Doerr's invitation to become a partner at Kleiner Perkins. The politician mocked for his earnest and robotic mien, who had longed for and very nearly achieved the presidency, had at last found his niche. "For whatever reason," he reflected to a reporter, "the business world rewards a long-term perspective more than the political world does."[16]

Doerr's decision to bring on Gore as a partner was motivated not only by friendship and politics. It was a business decision as well. Kleiner had exited the 1990s boom as one of the Valley's biggest names, and its investments in Google and Amazon had bolstered its portfolio as the 2000s began. Doerr was ready for a next act, and he was thinking bigger: not just consumer software, but global "grand challenges" that might, under the right conditions, provide huge market opportunity. Alternative energy—green tech—was the biggest challenge and opportunity of them all.

Like many others who saw Gore's *An Inconvenient Truth*, Doerr had become increasingly worried about the consequences of untrammeled consumption of fossil fuels. As he tooled around the Valley in his new Toyota Prius, Doerr realized that rising energy prices and turmoil in the Middle East were going to soon force a policy tipping point. The George W. Bush administration might be stocked with oil and gas men, but some limitations on carbon emissions and new renewable-energy mandates seemed unavoidable. Here was the public-sector push that the green-tech sector needed: just like with the integrated circuit and the Apollo program, government spending would allow an expensive and cutting-edge product to scale to market-altering proportions.

In the spring of 2007, Doerr went public with his new crusade, giving a heart-on-his-sleeve TED talk titled "Salvation (and profit) in greentech." The industry "is bigger than the Internet," he declared. "It could be the biggest opportunity of the twenty-first century." By November, Gore was talking an even bigger game about its potential impact. "What we are going to have to put in place is a combination of the Manhattan Project, the Apollo project, and the Marshall Plan," the former veep explained. "It'd be promising too much to say we can do it on our own, but we intend to do our part."[17]

By 2008, the Gore-Doerr team was again in full swing, lobbying the Bush Administration and encouraging presidential candidates to move forward on environmental treaties and other measures to reduce carbon emissions. By the fall, Kleiner's billion-dollar green-tech fund had backed forty different companies, and hundreds more entrepreneurs trekked to Sand Hill Road to deliver their pitches. Obama's election was a thrilling victory for the green team. An eco-friendly Democrat was back in the White House at last. The crash of the housing market that autumn was a setback, but Obama was promising a hefty infrastructure and spending program to bring the economy back—what better time to be building solar panels or developing electric cars?[18]

John Doerr always thought five years ahead, but this time politics made him fifteen or twenty years too early. Facing stiff opposition in a Republican-led Congress, Obama's big-spending stimulus plan was not as massive or market-disrupting as originally hoped. Carbon pricing went nowhere. Then the rapid scaling up of a different technological breakthrough transformed the economics of the U.S. energy market, and it wasn't green in the least. Most significantly of all, hydraulic fracking—which involved the high-velocity injection of millions of gallons of liquid into bedrock to release the natural gas within—vastly increased domestic energy production and drove down prices, taking away the market incentives to use alternative fuels.

On top of this came the cable-news-stoked political scandal of Solyndra, a solar energy company that collapsed after receiving $500 million in federal subsidies. (As staggering as the sum appeared, it was small potatoes in the world of green energy. Elon Musk's various ventures together received close to $5 billion in government subsidies by 2015.) Under fire, the Obama Administration scaled back their ambitions for a green-tech future, and Kleiner did as well.[19]

Doerr and Gore had made a gamble that fell far short of its promise, even though in another, less divided and less austerity-minded political moment it might have indeed been another successful moon shot. While Kleiner remained one of the Valley's biggest players, the firm's focus on alternative energy had come at the cost of missed opportunities in social and mobile, despite the fact that it had beefed up its consumer-Internet credentials by hiring on Mary Meeker. Then Doerr and his firm were hit by even worse news: a junior partner named Ellen Pao filed a sex bias suit against the company in the spring of 2012.

Doerr was extremely close to Pao. He had brought her into the firm as his chief of staff and remained her mentor and internal champion ever since. She was part of "Team JD," and her accusations that the firm was a hostile old-boys' club were a cutting personal blow. Over the previous decade, alarmed at the small numbers of women in venture investing, Doerr had pushed hard to recruit and promote women at Kleiner. "It is not easy to stand by as false allegations are asserted against the firm," Doerr wrote on the company's website soon after Pao filed her suit.[20]

The lawsuit and its attendant scandal dragged on for three more years, precipitating significant public discussion and soul-searching about the industry's endemic gender imbalances for the first time in Silicon Valley history. Kleiner's male-dominated culture was on trial, and so was that of the entire VC industry, a place where only 8 percent of investing partners were women and less than 5 percent of venture-backed companies had female founders.[21]

The low numbers weren't all that different from those of the rest of corporate America, the Valley's defenders would counter, where a comparably feeble percentage of Fortune 500 CEOs were women. But that argument didn't fly very far in the swelling public debate, and when the largest companies very reluctantly released data about the numbers of women and underrepresented minorities on their payrolls, the news went from bad to worse. Across the industry, women held only about 20 percent of technical roles. The percentage of women in computer science had been going down since the 1980s.[22]

Doerr agonized over what he could have done differently in Ellen Pao's case, he told author Emily Chang, feeling that if he had promoted his protégé earlier, "I don't think we would have gone through the trial." Pao lost her case in May of 2015, and while his allies publicly exulted at the win, Doerr was far more subdued. Coming right on the heels of green tech's challenges, the lawsuit had taken an obvious emotional and professional toll. While remaining Kleiner's chairman, he stepped back from active investing the following year. It was time for other masters of the universe to take over.[23]

PATTERN RECOGNITION

In the early 1960s, when Draper and Johnson trolled the prune sheds in search of places to invest, the Valley was a remote, lightly settled place, where cul-de-sacs abruptly gave way to orchards and grazing land, striped with lightly traveled highways of raw concrete. Its electronics industry was niche, its brands mostly unknown, its wealth a fraction of what was to come.

A half century later, the Valley was a pulsing sprawl of million-dollar bungalows and high-priced boutiques, with clusters of glass-sheathed office parks at every exit of its traffic-choked highways. San Francisco, once so remote from the Valley somnolence, now was a critical part of the Valley's high-tech realm. Twitter and enterprise-software giant Salesforce had their headquarters in the city. Private buses launched by Google and Facebook rolled up and down Highway 101, easing the grueling commutes of their many San Francisco–based employees and sparking protest from city folk who saw the vehicles as rolling symbols of tech gluttony. Consumers across the globe used the platforms of Bay Area and Seattle tech giants every day; everyone knew about Sergey and Larry and Mark and Steve and Jeff.

The tech industry had grown immensely large and powerful. Yet the networks of influence and investment were as tightly coiled and closely held as they had been in the days of "The Group's" steak lunches, so many decades earlier. The massively lucrative deals around Google and Facebook, not to mention other post-2000 rocket ships, involved a tiny number of people: all men, all wealthy, all pretty confident in their conviction that the industry was a marvelous meritocracy.

Some, like Peter Thiel and David Sacks, had translated their college networks into powerful instruments of wealth creation. Others, like Sean Parker and Ram Shriram, had parlayed early success with one company into enduring networks of influence and talent-spotting. Still others, like Marc Andreessen and Chamath Palihapitiya, followed the path from entrepreneurship to venture capital first laid by David Morgenthaler and Burt McMurtry. All thrived in a ferociously competitive environment of sharp elbows and unvarnished criticism, where working hard was matched by playing hard, and where business partners were like family.

As wealth grew, so did the mythos around how Silicon Valley was able to generate one innovative company after another. It was about allowing risks and not penalizing failure, they'd say. It was about putting engineering first—finding the best technical talent, with no bias about origin or pedigree. It was about that "pattern recognition" so fatefully identified by John Doerr, looking for the next Stanford or Harvard dropout with a wild but brilliant idea.

Of all those assertions, Doerr's slip-up came closest to the heart of the Valley's secret. "West Coast investors aren't bolder because they're irresponsible cowboys, or because the good weather makes them optimistic," wrote Paul Graham, founder of the Valley's most influential tech incubator, Y Combinator, in 2007. "They're bolder because they know what they're doing." The Valley power players knew the tech, knew the people, and knew the formula that worked.

They looked for "grade-A men" (who very occasionally were women) from the nation's best engineering and computer science programs, or from the most promising young companies, and who had validation from someone else they already knew. They sought out those exhibiting the competitive fire of a Gates or a Zuckerberg, the focus and design ascetism of Kapor or Andreessen or Brin and Page. They funded those who were working on a slightly better version of something already being attempted—a better search engine, a better social network. They surrounded these lucky entrepreneurs with support and seasoned talent; they got their names in the media and their faces on the stage at premier tech conferences. They picked winners, and because of the accumulated experience and connections in the Valley, those they picked often won.[24]

Keeping the networks tight and personal was a critical part of Silicon Valley's ability to keep the flywheel turning, to move from chips to micros to dot-com to the next Web without dropping the pace. VC had always been a men's world, but the post-2000 elite—with its overrepresentation from the overwhelmingly male worlds of Google, Facebook, and the enterprises founded by the PayPal Mafia—was even more so. The business of entrepreneurship and VC took place not only in boardrooms and cubicles but over beers and peanuts at Antonio's Nut House, breakfast at Hobee's or Buck's, late-night coding sessions and poker games, on forty-mile bicycle rides along Skyline Drive. It was a wonderful world if you were in it, and a tough place to hack into if you didn't have the time, the money, the poker skills, or the $10,000 bike.

In the flurry of public shaming and self-examination that ensued after the Pao case, the venture capitalists promised to try to do better. Subsequent revelations of sexual harassment and abuse in the industry added to the pressure to rectify tech's imbalances. The numbers started to budge, a little. It helped that some of those already inside the charmed circle were ready to point out its shortcomings. Because he once had been an outsider himself, Chamath Palihapitiya had come into the venture game determined to shake up its so-called meritocracy. The industry was filled with "rich douchebags," he told journalist Kara Swisher. This wasn't just bad for society, it was bad for innovation. "We've been overrun with too many people who don't understand what the real goal is," he said later. "Folks here have to reset some of that inequity," even though "we're not purpose built for that."[25]

The Chamath solution: take human bias out of the financing equation altogether, and let a computer algorithm make the investment decisions instead. Social Capital called the model "Capital as a Service," and began inviting early-stage ventures to pass on the traditional pitch meeting and simply submit their data instead—revenues, user base, costs. After running the numbers, the firm would invest or take a pass. "No hoops, no $7 artisanal coffee chats, no designer pitch decks, no bias, no politics, no bullshit," explained Palihapitiya's business partner Ashley Carroll. It was a classic Silicon Valley answer to a distinctively Silicon Valley problem. They'd hack their way out of it.[26]

Those who had been tackling tech's diversity problem for years remained skeptical. "Ten years, same damn conversation," reflected Trish Millines Dziko, as she looked back on the many times investors and executives had promised her that the industry would change its ways. The program she had started in 1996 had become wildly successful, a pathbreaking model that had inspired others to build similar schools for underrepresented kids. What had begun as a modest prep program for teenagers had grown to serve kindergarten through high school, bringing science and engineering education to more than 5,000 Seattle students and counting. But the faces inside those big tech companies looked much the same as they had when she started at Microsoft all those years ago. "Why are people of color not part of this movement, and why isn't anyone doing anything about it? Why is everyone continuing to give more to people that already have more?" she wondered. "There is no incentive for leaders in tech to do anything, no incentive for them to explore other communities for brainpower."[27]

Nonetheless, the mid-2010s conversation about tech diversity was unlike anything the Valley had seen before. Private rumblings that had seethed under the surface for decades burst out in loud, sustained, and very public conversation. Under fire, the largest tech companies released employee demographics, confirming what was obvious to anyone who set foot upon the bustling campuses of Google, Facebook, or Apple, not to mention the scores of other firms across the Valley: especially in technical and executive roles, the workforce was overwhelmingly male, young, and either white or Asian. As the Valley's biggest brands poured hundreds of millions into hiring efforts and launched slickly designed "diversity and inclusion" campaigns, the numbers began to bump up, but only slightly.

It was not just the twenty-first-century brogrammers and the arrogant-jerk CEOs who were responsible for limited opportunities the Valley presented to women and minorities. It was something of much deeper origin, rooted in a time when business was a man's world, when the Valley was young and remote, when girls and electronics didn't mix. With a culture this long in the making, meaningful change would take a while.

CHANGE THE WORLD

Although tech's shortcomings attracted increasing attention, it was hard not to be dazzled by the grand visions coming out of Northern California and Seattle. Government was enfeebled and polarized. Tech had flash, ambition, and billions to spend. The CEO of Google's advanced research laboratory was the Rollerblade-wearing grandson of H-bomb developer Edward Teller; his corporate title was "Captain of Moonshots." Elon Musk's Tesla produced roaringly fast roadsters that featured a button allowing drivers to enter "ludicrous mode." When he wasn't building cars or jokingly repurposing welding tools to sell as $500 flamethrowers to adoring fans, Musk was literally shooting the moon with his SpaceX commercial space venture, beating out Lockheed and others for prime contracts. Jeff Bezos joined the race, too, spending $1 billion a year on his space exploration company, Blue Origin. Its motto, emblazoned on a corporate crest, was *Gradatim Ferociter*: "Step by step, ferociously."

Beyond that, of course, was extraordinary philanthropy, led by Bill Gates, whose namesake foundation had a $40 billion endowment and had become the leading actor in global initiatives tackling public health and poverty. Gates, the *enfant terrible* turned elder statesman, became an inspiration to tech's younger generation. Mark Zuckerberg pumped $100 million into public education in the beleaguered school system of Newark, New Jersey, became an advocate for immigration reform, and announced that he and his wife would, like Bill and Melinda Gates, give away all their wealth during their lifetimes.

While the industry's longstanding lobbying work was as active as ever, tech's masters of the universe were becoming increasingly vocal about policy matters that went beyond the usual array of capital gains tax cuts or Internet sales taxes or net neutrality. Some came off as odd and self-serving: in 2014, third-generation Valley VC Tim Draper, son of Bill and grandson of William, spearheaded a campaign to break California into six states (one of which would be named, naturally, "Silicon Valley"). Others had a swagger only the Valley could bring: in 2016, Chamath Palihapitiya tried to persuade New York billionaire Michael Bloomberg to run as a third-party candidate for president, promising that he'd devote all of his firm's resources to the effort. "The same team that helped build Facebook to one billion users would do our best to activate the entire United States to put him in the White House," Palihapitiya promised. "I think we'd be successful." Nervous that his run would hand the presidency to someone like Donald Trump, Bloomberg declined to enter the race.[28]

That decision left the Valley's power players solidly in Hillary Clinton's corner, becoming a reliable source of campaign cash and policy advice. The singular exception to that trend was a highly visible one: Peter Thiel, who overcame his long disdain for the mess of electoral politics to come out publicly in favor of Trump's renegade bid for the White House. Thiel spoke at the Republican Convention. He penned op-eds slamming governmental inefficiency while wistfully invoking the glories of the Manhattan Project. Ironically, Thiel's version of making America great again involved a return to the big-spending era when the Valley first got its start. "When Americans lived in an engineering age rather than a financial one, they mastered far bigger tasks for far less money," he wrote. "We can't go back in time, but we can recover the common sense that guided our grandparents who accomplished so much." After the election, Thiel became the liaison between Trump Tower and the bewildered, more than slightly horrified leaders of big tech, who could not believe what November 2016 had wrought.[29]

And so very swiftly, sentiment swelled that the chief culprits in this electoral surprise were the big American tech platforms themselves. Their breezy confidence about connecting the world, their hubris about the power of engineering, their dazzlingly sophisticated thinking machines: all seemingly had opened the door for bad actors to come in, exploiting networks like Facebook and Twitter and YouTube and, really, the whole of the Internet, driving a divided America even further apart.

American politicians who had for so long regarded tech with gee-whiz amazement started calling Mark Zuckerberg into Capitol Hill hearing rooms for hostile face-offs about Facebook's business model. European lawmakers went one step further, forcing consumer Internet companies to adopt far stricter privacy rules. And with every week came a steady drip of revelation that the agents of all this social media mayhem came from overseas, orchestrated by the government of Vladimir Putin, the man who had pushed his protégé and placeholder Dmitry Medvedev aside only two short years after that earnest pilgrimage to Silicon Valley.

Since the end of the Cold War, the Valley's defense-centric origins had faded into the haze of collective memory. In the sun-dappled quads of Stanford and the polished-wood wine bars of University Avenue, the only history that seemed to matter was that of Sergey and Larry and Mark, or perhaps, if you wanted to get nostalgic, the two Steves in their Los Altos garage. The quiet land of Lockheed and Terman didn't seem to have much to do with the gloriously rich, pulsing center of global capitalism, a place that represented the resounding triumph of the agile new market economy over the lumbering bureaucracies of old.

The things the Valley now sold—their reach, their ubiquity, their intelligence—had an influence far greater than that of the hardware and software products that had originally put it on the map. Yet as the 2016 election showed, the social media platforms of the new era were some of the most powerful weapons the Valley had ever produced.

This was the place that had rejected the old politics, where titans like Steve Jobs became billionaires before bothering to vote, where belief systems often cleaved into either techno-libertarian rejection of the system or an impatient, technocratic "there's an app for that" belief that Silicon Valley could fix governmental failures. Heads down and engineering first, the boy wonders of the Valley had built extraordinary thinking machines without reckoning with how truly disruptive they might become. The Silicon Valley ethos was a product of the prosperous, rambunctious, and generally peaceful America of the late twentieth century, now gravely tested in the early twenty-first.

Those who had been present at the creation of the extraordinary creative explosion of the online era looked on in dismay at what all this freedom had wrought. "We were astoundingly naïve," Mitch Kapor remarked regretfully as he looked out at the upended political landscape. "We couldn't imagine what is now obvious: if people have bad motives and bad intentions they will use the Internet to amplify them."[30]

DEPARTURE

Into the Driverless Car

The driverless cars slid silently through Mountain View's eucalyptus-scented side streets, released into the wild during the quieter hiccup after the relentless morning rush. Compact and gleaming, all electric motor and spinning geolocators sprouting from the roof, the cars were not actually without a driver: if you peered closely, you could see the outline of one or two heads inside, belonging to the young Googlers whose job it was to test and monitor and take notes and seize the wheel if anything went wrong.

It wasn't just Google that was getting into the driverless car in those wealth-dazzled and tech-saturated years of the late 2010s. It was Apple and Uber and Tesla, too, all in a race to turn the car into a computer, a Shaky the Robot updated for the twenty-first century. Even if the vehicles weren't quite fully autonomous, they showed how far the American tech industry had come in its seventy-year pursuit of machines that think and computers that network—and how far its largest and richest companies intended to go.[1]

Cars were just part of it. With the billions made in search and social and mobile, the Valley's largest companies were plunging resources into AI and machine learning, as were its most far-seeing investors. "About every dozen years in the tech world there is a tsunami—a huge wave of disruption," declared John Doerr. First came the microchip and the PC, then the Internet, then mobile. AI was the next wave, and "it will be even larger." Chamath Palihapitiya was betting big on AI as well, partnering with several ex-Googlers in a stealth start-up that was developing a wholly new microchip built for machine learning. The next big thing, it seemed, would involve a return to the Valley's hardware roots.[2]

Travel north into the golden gleam of Palo Alto, and you'd pass other signs of prosperity and hustle: the luxury-car dealerships, the Crossfit gyms, the coffee chain where people lined up to wait ten minutes for a rare-bean pour-over. Cast a glance over to the commuter train screaming past as you drove, and you'd be reminded of the dark anxiety lurking just beneath the sunny surface. Those tracks had been the site of a spate of high school suicides, young people crushed by the college-prep pressure of growing up in the shadow of the world's most famously entrepreneurial university.

Bay Area people had forever complained about the high cost of real estate, but the home prices had gotten so sky-high that the carping was fully justified. Those modest bungalows built in the first postwar boom now cost upwards of $3 million, and that was before you updated the kitchen. Already facing criticism for the pressure their growth had placed on the housing market, Facebook and Google announced that they would build apartments and town houses to house their employees nearby, company towns in the mold of Pullman and Ford, updated for the twenty-first century. As if in call-and-response to the new welfare capitalism, campaigns to organize tech workers gained momentum, pushing to expand white-collar tech's perks and pay to its massive, contingent workforce, the shuttle-bus drivers and gig workers and coders for hire.

Drive beyond the sandstone-and-tile shimmer of Stanford, the obelisk of the Hoover Tower (still pointing straight up!), and the Stanford Shopping Center, developed on university land back in the 1950s and now home to every designer brand imaginable. Across the street, through a shaded drive, you'd find the Vi, the commodious and luxe retirement destination of the men and women who'd been in the Valley from the very start.

BURT AND DEEDEE MCMURTRY HAD THE MOST GENEROUSLY SIZED APARTMENT to be had in the Vi, on a high floor above the treetops, a view of Stanford's campus not too distant. There were a number of things with their names on them over on campus, most notably a gleaming new building for art and art history that opened in 2015. Philanthropy had been Burt McMurtry's main pursuit since stepping back from active investing two decades earlier, but even before that he'd been sharing his wealth with the two places that had given him his start: Stanford and Rice. "I had ten years of free education," he explained, "and it makes you realize how education is a tool for economic mobility, so it was very easy to think about giving back." He wondered whether the new generation would do the same, and if they had a similar sense of how success came not just from their talent, but from their circumstances and timing. "Many VCs," he observed, "take themselves way too seriously."[3]

David Morgenthaler agreed. He lived a couple of floors below the McMurtrys, and although he was in his mid-nineties his considerable intellect hadn't flagged. He'd kept in touch with all the friends he'd made from his earliest days in the venture business, from those many trips to Washington, from the many tech deals made since. Over tomato soup and club sandwiches in the Vi's lunchroom, they'd reminisce about the old days and talk about what the new generation of technologists might build next.

Morgenthaler seeded student engineering projects and kept an eye on the more interesting pieces of his firm's portfolio, and he worried about how much longer Silicon Valley could keep all of this going. The Detroit auto industry had a good sixty-year run, then it reached the outer limit of its S-curve. The Valley had been at it nearly as long, and although the place brimmed with talk of disruption, he couldn't quite see where this disruption might emerge. The silicon transistor had been its great enabler, but what would take its place? Where was the next Shockley? The next Jobs? Morgenthaler wondered whether the future of tech would be in the Valley at all.[4]

Ann Hardy didn't live at the Vi. She lived with her daughter and son-in-law, just across town, savoring the hours she got to spend with her young grandson. She kept busy, and life was wonderful. "I've never lived this well," she reflected happily. The spectacular growth of the Valley amazed her; the revelations of its persistent sexism didn't surprise her much at all. It was sad, after all this time, that it was still so tough to be a technical woman, and that so few men worked to make it otherwise. "Even if you're a nice guy going into a company, because it was started by jerks, it's hard to be a nice guy." She could see the shortcomings of Valley culture reflected in its products: now used by everyone, but all seemingly designed by and for twenty-year-old men. "There are so many trivial things you could do to make them more accessible," she mused, as much of a programmer as ever. "It is such a shame not to have better products."

If the current generation wasn't going to change, then she'd put her faith in the next one. It was going to be up to children like her grandson, and like my elementary-age daughters. "Tell your girls to play math games," she called to me across the sidewalk, as we parted after our last meeting.[5]

SAN DIEGO, 2018

The sun hit their faces as soon as they stepped off the plane. Seattle had been drearily chilly, spitting rain even in early June. It was nice to be back in San Diego's warmth and languor after a frenetic week in the city that once had been home. Seattle had become a city of cranes and construction sites, slow traffic and bikeshare-littered sidewalks, soaring rents and homeless encampments clustered along Interstate 5. Amazon was expanding like wildfire and all the other Valley giants had outposts there, too, leaving the city heaving with new arrivals and fraught with anxiety. The local tech scene had always prided itself on being a little saner and slower than the fake-it-'til-you-make-it Valley. Now Seattle seemed to be turning into San Francisco without the good weather.

Yaw Anokwa and Hélène Martin could have gotten great tech jobs anywhere—they had elite computer science degrees and gold-plated résumés—but they didn't want to be part of the rat race in either Seattle or the Bay Area. They knew they didn't have to. The two technologists already had spent three years as nomads, traveling around the globe, choosing their destinations based on good weather and availability of strong Internet connections.

They'd created a business as they roamed, one built on open-source software Anokwa had developed as a graduate student, which allowed for mobile data collection in places where the Internet couldn't reach. Their company helped its clients map Brazilian rainforests, track polio outbreaks in Somalia, and monitor crop outputs in Nigeria and Rwanda. They didn't have venture funding; operations were lean and revenue came from customers. "I wanted a business model I could explain to my parents," said Anokwa. The Gates Foundation was an important client and supporter, the irony of which was not lost on Martin, who once had been so much of an open-source disciple that a photo of Bill Gates getting a pie thrown in his face hung in her middle school locker.

The couple were both in their thirties and already had lived through several generations of technological change. Both were immigrants—Martin was French-Canadian and Anokwa was from Ghana—and had early exposure to the tech world. Anokwa was a professor's kid who had been hooked on computers ever since he was a nine-year-old newly arrived in Indiana, when his father brought a shiny new Mac home to their campus apartment. Martin had grown up in Palo Alto, the daughter of an electrical engineer, and spent her childhood tinkering, taking computers apart, and spending hours on the very early Internet. "Nobody told me that girls don't do this," she remembered. After earning her degree at the University of Washington, she started a computer science program at a local public high school, hoping to encourage girls and minorities in the same way. "You don't understand something," she liked to say, "until you've taught a teenager to teach a computer to do it."

The tech world of their years at the University of Washington was dominated by tech's Big Five, and they appreciated the resources those companies could bring. Anokwa's PhD advisor was on leave at Google, allowing Anokwa to build his software on a pre-release version of Android. Their company's interns learned open-source development during Google's Summer of Code. Microsoft's embrace of open source was a welcome plot twist, and the philanthropic accomplishments of Bill Gates had made the world a better place, at scale. These technologists and companies, they felt, understood the benefits they had reaped from the open-source infrastructure created over the decades by a global community of largely unsung programmers. But the breezy optimism of other giants could be unnerving. "If you haven't experienced a dictatorship," observed Anokwa, "you may be naïve to the dangers of amassing massive amounts of user data."

Both opted not to work for Google, or Microsoft, or the rest. They also didn't want to plunge the next several years of their lives into building the next Facebook. They didn't need to be billionaires or even millionaires; they just wanted to make a decent living. "How do we find balance?" Martin wondered. "How do we avoid getting into that grind?"

That's where San Diego came in. It was sunny. There was UCSD's computer science department, and biotech firms, but that really wasn't what drew them in. It was the presence of other people like them, people who were lucky enough to be able to choose their destination, people who were consciously opting out of places that had gotten crowded and competitive. "It's more and more possible," Martin observed, "to choose where I want to be not based on the job, but on the life I want to live." Silicon Valley still was a draw for many, but not for them.[6]

DRIVERLESS CARS ROAMED THE STREETS. SUPERCOMPUTERS BECAME PORTAble. The electronic brain was nearly as capable as the human brain. Yet the techno-optimism that had long propelled the industry had shifted. The change-the-world proclamations of its most famous leaders elicited smirks and scowls, even as the technologies themselves were more marvelous and full of awesome and fearsome possibility than ever. The cries of complaint rang out from Washington to Brussels, from major newspapers and (oh so ironically) all across Facebook and Twitter.

Yet branding big tech as the source of all society's problems was as problematic as those bold declarations, made at the dawn of the silicon age, that computers and their makers would fix it all. The story of the American tech revolution wasn't a binary code of heroes or villains; it was far messier and more interesting than that. All the connection had brought people together, but it had made it easier to drive them apart. All the openness had nurtured individual freedom, but at a cost to individual privacy. The concentration of talent and money in certain places had made the explosion of new ideas possible, but the financial benefits had flowed to a privileged few. Now these cities felt like millionaires' playgrounds, with too little room or opportunity for anyone else.

Facebook and Google hadn't even existed two decades earlier; Apple and Amazon had once been written off as failures. Now they were the largest and richest corporations on the planet, wildly successful even beyond their founders' considerably ambitious dreams. The regulatory scrutiny and fierce critiques came in the wake of this success, of these companies having so fully accomplished what they had set out to do—target ads, create ubiquitous platforms, change the world.

But now a new American generation was about to take over, and it included people like Yaw Anokwa and Hélène Martin. They wanted work-life balance instead of the next market-busting company. They wanted to be citizens of the world, and not just citizens of a small and very fortunate slice of western North America. The new entrepreneurs were taking the open-source platforms, the broadband networks, and the extraordinary hardware and software inventions that came out of America's seven-decade-long tech explosion, and they were building something new.

These entrepreneurs weren't necessarily doing it in Silicon Valley, nor were they doing it in "the next Silicon Valley." They were doing it everywhere, in places that had a different rhythm, were more affordable, more diverse in outlook and experience.

These technologists were thinking about a different kind of hacker ethic. It was one that built software to last rather than asking programmers to give over their lives to constant tweaks and updates. It was an ethic that brought in people from beyond the charmed circles that had dominated tech for so long, and one that blended engineering with humanism. It was one that postponed building colonies on Mars until technologists first tackled inequalities here on earth.

They were able to think this way because of what the tech revolution already had done. Silicon Valley wasn't just a place. It was a set of tools, a network of people, a bootstrapping sensibility. This was an only-in-America story of glorious accomplishments and unfinished business, made possible by the broader political and economic currents that shaped more than a half century of history. The latest generation of seekers and nomads and arrivals were ready to build on that past, and do things differently in the future.

You never knew what might happen next.

ACKNOWLEDGMENTS

It has been a thrill and a challenge to write a history of a place and industry as fast-moving as Silicon Valley. In the years I spent researching and writing *The Code*, Apple released five generations of the iPhone, Facebook added a billion users, Google's ad revenues doubled, and more than a few start-ups soared from obscurity to ten-figure valuations. Throughout this process, my colleagues, friends, and family inspired me and kept me going. My name is the only one on the cover, but I couldn't have gotten here without this amazing team.

First thanks go to the many people who spent so many hours talking with me, sharing their notes and personal papers, ground-truthing my analysis, and providing detail I could not have found anywhere else. Special thanks to Ann Hardy, Regis McKenna, Burt McMurtry and the McMurtry family, David Morgenthaler and the Morgenthaler family, and Ed Zschau for agreeing to have their lives and careers chronicled in this book, as well as connecting me to other friends and colleagues. Jennifer Jones and Gary Morgenthaler were the original connectors—thank you both for early conversations that led to so much more.

Thanks to the archivists and archives of Stanford University, the University of California, Harvard University, the University of Washington, the Computer History Museum, the Museum of History and Industry (MOHAI) Seattle, the Agilent History Center, the Palo Alto Historical Association, History San José, and the presidential libraries of the U.S. National Archives system for preserving and cataloguing the history of this relentlessly present- and future-tense industry and place. I also am indebted to the journalists in the Bay Area, Seattle, and nationally who covered the technology beat from the 1970s until today; your first drafts of history made my second draft possible.

The following institutions and fellowships provided time and resources that made this book possible: the American Council of Learned Societies' Frederick

Burkhardt Fellowship for Recently Tenured Scholars; the Stanford Center for Advanced Study in the Behavioral Sciences (CASBS) Fellowship; the Stanford Program for the History of Science; the Stanford University Department of History; the Walter Chapin Simpson Center for the Humanities at the University of Washington; Lenore Hanauer and the Hanauer History Funds of the University of Washington; and the Keller Fund of the University of Washington Department of History.

The year I spent at CASBS was the ideal launching pad for this project, and I am so grateful to my "fellow fellows" for their insights and feedback as I worked out early ideas. Special thanks to Fred Turner for a pep talk exactly when I needed it, Katherine Isbister for enlarging my thinking about technological possibility, Ann Orloff for our adventures in field research, and to CASBS's fearless leader, Margaret Levi, for her sustained support of this work. Thanks to others who helped make my time down south productive and memorable: Jennifer Burns, Jim Campbell, Paula Findlen, Zephyr Frank, Allyson Hobbs, and another visitor, Louis Hyman, with whom I always will agree about 1877. I presented working versions of parts of this book at Princeton, Stanford, Johns Hopkins, the University of California at Santa Barbara, the Miller Center at the University of Virginia, and the annual meetings of the American Historical Association and the Organization of American Historians. Many thanks to those who invited me and all who participated and commented in those sessions.

Thanks to my many wonderful colleagues at the University of Washington History Department for their sustained encouragement, support, and feedback on this work at various stages. Thank you to Ana Mari Cauce, Judith Howard, Lynn Thomas, and Anand Yang for ensuring I had the time and support needed to bring the book to completion. Much gratitude to those who provided expert research assistance at different points in this project: Kayla Schott-Bresler, Eleanor Mahoney, and Madison Heslop of the University of Washington; and Andrew Pope of Harvard University.

Thank you to Richard White for believing in this audacious venture from the start, and to David M. Kennedy and Lizabeth Cohen for the opportunity to think about how this story fits into the grand sweep of American history. Thanks to Thaïsa Way for helping me talk through thorny parts of this narrative as we hiked through evergreen forests and up Cascade peaks. Ed Lazowska introduced me to Seattle's tech world upon my arrival at the University of Washington fifteen years ago, and I have gained much from our collaboration and friendship since.

I was very lucky that two brilliant writers and dear friends, Leslie Berlin and Ingrid Roper, read early drafts and provided incisive comments. Ingrid has brought her keen editorial sensibility and nose for good storytelling to every book I have written, and I am so grateful for her sustained encouragement. And I could not have asked for a better reader on the semiconductor industry and the Valley of the 1970s than Leslie, Silicon Valley historian and biographer par excellence. Thanks to Bill Carr, Ryan Calo, Trish Millines Dziko, Bruce Hevly, Dan'l Lewin, Gary Morgenthaler, and Lissa Morgenthaler-Jones for critical vetting of key passages. Several very busy people generously agreed to read in full: Tom Alberg, Phil Deutch, Marne Levine, John Markoff, Brad Smith, Mark Vadon, and Ed Zschau. Their seasoned perspectives made this book better, and any remaining errors of fact or interpretation are mine alone.

Geri Thoma has been an unflagging advocate for this book and its author, going above and beyond the call of literary-agent duty as a trusted guide, sounding board, and friend. Thank you, thank you, Geri. Another great stroke of luck was teaming up with my editor, Scott Moyers, who immediately understood what I wanted to do with this project, and who gave me the essential guidance I needed to take it there. Heartfelt thanks to Scott and the rest of the world-class team at Penguin Press, especially Mia Council and all others who steered this so expertly through the production process and beyond.

Two people provided great inspiration and encouragement but did not live to see this project's completion. One was Michael B. Katz, my graduate mentor, collaborator, and friend. A historian of social policy and poverty who ably covered his bemusement when one of his advisees ended up specializing in the lives of high-tech billionaires, Michael was bullish on this project from the start. In one of our last e-mail exchanges, I told him about my ideas for this project; he responded with characteristic enthusiasm: "Get writing—we need this book!"

Another great champion was David Morgenthaler, who died in June 2016 at the age of 96, and whose appreciation for history's long arc came from having lived so long and so fruitfully. He not only spent many hours with me sharing his personal memories and judicious meditations on the past and future of the tech world, but he also very generously connected me to many of his friends and colleagues.

The professor and the venture capitalist were utterly different in their politics and in their chosen professions, but both of them firmly believed in the United

States' potential to be a land of opportunity, fairness, and bold ideas. I hope that I have served both of their legacies well.

Dear friends old and new made Palo Alto feel like home during our family's stay. Special thanks go to the world's best next-door neighbors, Monica Stemmle and Jamie Zeitzer. Alana Taube's work made mine possible, and I cannot wait to see where she goes next. Thanks as well to Katie Smith for the time she spent with our family, the teachers and parents of Lucile M. Nixon Elementary, and the wonder women of 9:00 a.m.

In Seattle and Mercer Island we are surrounded by friends who are like family, who watched and cheered as this book took shape. Thank you all. The talented and generous Alina Ostate kept our lives running smoothly while I was glued to the keyboard; I could not have done this without her. Love to far-flung family: my parents, Joel and Caroline Pugh; John Pugh and Liz Seklir-Pugh; my in-laws, Frank and Marge O'Mara; Erin O'Mara and Roger Aschbrenner. Thanks to Erin for summers of working solitude and happy family time in Harpswell.

My extraordinary daughters, Molly and Abby O'Mara, lived with this book for a good chunk of their lives, gamely adapting to two states, three houses, four schools, and one writing-distracted mother. They surround me with daily joy and necessary silliness, bring creative inspiration, and give me hope for our future.

Last, at every step on this long and winding road, there was Jeff O'Mara. After writing so many words, I now struggle to find ones that fully express my gratitude and love, and the best I can do is dedicate this book to you. Thank you for being my rock, my light, my home. I can't wait for the next chapter.

NOTE ON SOURCES

Like any work of history, *The Code* rests on a foundation of both primary and secondary materials, employing a variety of methods to tell a story that begins in the 1940s and ends in the late 2010s. My primary sources included corporate and governmental archives; personal papers; newspapers, magazines, and contemporaneous books; memoirs; corporate publications and financial prospectuses; published oral histories; and first-person interviews. Lists of the archives consulted and of individuals interviewed can be found at the end of this essay.

The secondary sources that inform this book cover a similarly wide swath of modern American technological, political, and economic history. When I was writing, I would joke that my subject was "everything about the high-tech revolution except the technology." Plenty of tech made it into the book, of course, and I sought to bring this vast subject into a straightforward, readable narrative that would be accessible to non-technologists. I was able to do this because many others have written the history of this technology so well and so extensively. I am grateful for the chroniclers of computer hardware, software, and telecommunications industries whose work informed my understanding of tech and technologists, as well as the scholars of science, technology, and society (STS) whose insights and interrogations have shaped the questions I ask and answer in these pages. The endnotes provide full bibliographic data on works consulted. For general readers interested in a deeper dive, I list some key titles here.

Historians of science and technology have woven the story of American invention into a broader context of social and structural change. A shaper of the field was Thomas P. Hughes, whose sweeping *American Genesis* (1989) is an excellent starting point for readers interested in this longer history. On computing and its related industries in particular, an indispensable synthesis is Martin Campbell-Kelly, William Aspray, Nathan Ensmenger, and Jeffrey R. Yost, *Computer: A*

History of the Information Machine (3rd ed., 2013). The latest edition contains discussion of mobile and social platforms and software as well as explanation of key landmarks and figures in computing from the nineteenth century to the present day. Another important overview is Paul E. Ceruzzi, *A History of Modern Computing* (2003). On the longer history of communication technologies, an inspiring, now-classic source is James R. Beniger, *The Control Revolution* (1986). A more recent and also valuable contribution is Tim Wu, *The Master Switch* (2011). A useful synthesis of information-technology policy in both the U.S. and Europe is Armand Mattelart, *The Information Society: An Introduction* (English translation, 2003).

A number of important studies trace the prominent role of women—the first "computers"—in early computer programming, and the discriminatory practices and cultural presumptions that gradually pushed them out of the spotlight and limited their opportunities in executive roles. On the U.S. case, see Nathan Ensmenger, *The Computer Boys Take Over* (2012); on Great Britain, see Marie Hicks, *Programmed Inequality* (2017). Also see Thomas J. Misa, ed., *Gender Codes: Why Women are Leaving Computing* (2010) and, on the longer history of gender and science, see Londa Schiebinger, *The Mind Has No Sex?* (1991).

Mainframe digital computers transformed the American industrial sector after World War II; for a comprehensive exploration of the many industries disrupted by electronic data processing, see James W. Cortada, *The Digital Hand* (2003). On consumer electronics and telecommunications devices, not a specialty of early Silicon Valley but important to the history of American technology in the mid-twentieth century, see Alfred D. Chandler Jr., *Inventing the Electronic Century* (2005), as well as Stephen B. Adams and Orville R. Butler's study of Western Electric, *Manufacturing the Future* (1999). On IBM, the dominant, market-defining company of the mainframe era, see Emerson W. Pugh, *Building IBM* (1995), as well as John Harwood's exploration of the company's branding and industrial design, *The Interface* (2011).

Much has been written on the U.S. government's World War II–era and immediate postwar investments in scientific research. Scott McCartney, *ENIAC* (1999) tells of the making of the world's first all-digital computer; G. Pascal Zachary, *Endless Frontier* (1997) remains the definitive biography of the influential and irrepressible Vannevar Bush. New insights into the Second World War's long economic shadow have come from scholars of American politics and capitalism such as James T. Sparrow, *Warfare State* (2011) and Mark R. Wilson, *Destructive Creation* (2016).

Bruce J. Schulman, *From Cotton Belt to Sunbelt* (1994) shows how the military-industrial complex remade America's economic geography. On Eisenhower and the Cold War, including nuanced discussion of Sputnik and the "missile gap," see William I. Hitchcock, *The Age of Eisenhower* (2018). For the fascinating history of DARPA, see Annie Jacobsen, *The Pentagon's Brain* (2015). On the foundational role of policy in technological development in both the U.S. and Europe, see Mariana Mazzucato's influential and important *The Entrepreneurial State* (2015).

A critical dimension of this government investment occurred in America's research universities, which became major political and economic actors in their own right. Important work here includes Stuart W. Leslie, *The Cold War and American Science* (1993); Roger L. Geiger, *Research and Relevant Knowledge* (1993); and Christopher P. Loss, *Between Citizens and the State* (2012). I also addressed this topic in my first book, *Cities of Knowledge* (2005). On the economic evolution of the university in a later period, see Elizabeth Popp Berman, *Creating the Market University* (2012). On the evolution of Stanford and its "steeples of excellence," see Rebecca S. Lowen, *Creating the Cold War University* (1997), and C. Stewart Gillmor, *Fred Terman at Stanford* (2004).

On the early history of the Santa Clara Valley electronics industry, especially important early firms like Ampex, Eitel-McCullough, and Varian, see Christophe Lécuyer, *Making Silicon Valley* (2005). David Beers, *Blue Sky Dream* (1996) is a vivid personal account of growing up in a Lockheed family in San Jose. Valuable essays on the growth of the regional ecosystem over time, including discussion of law firms, venture capital, and other specialized services, is found in Martin Kenney, ed., *Understanding Silicon Valley* (2000) and Chong-Moon Lee, William F. Miller, Marguerite Gong Hancock, and Henry S. Rowen, eds., *The Silicon Valley Edge* (2000).

On the transistor and the chip industry it spawned, see Michael Riordan and Lillian Hoddeson, *Crystal Fire* (1997); Leslie Berlin, *The Man Behind the Microchip* (2005); and Arnold Thackray, David C. Brock, and Rachel Jones, *Moore's Law* (2015). William Shockley, the man who brought the silicon to Silicon Valley, spent the later decades of his life as a vocal eugenicist and white supremacist; see Joel N. Shurkin's biography, *Broken Genius* (2006).

The Santa Clara Valley was of course about more than technology, and it has been the subject of several fine histories that interrogate the broader racial and social politics of the region: Glenna Matthews, *Silicon Valley, Women, and the*

California Dream (2002); Stephen J. Pitti, *The Devil in Silicon Valley* (2004); and Herbert G. Ruffin II, *Uninvited Neighbors* (2014). On the broader politics of California during this period, see Matthew Dallek, *The Right Moment* (2004); Kevin Starr, *Golden Dreams* (2011); Jonathan Bell, *California Crucible* (2012); and Miriam Pawel, *The Browns of California* (2018). On 1978's Proposition 13, a turning point in public financing in California, and the broader currents of race, property, and the politics of homeownership, see Robert O. Self, *American Babylon* (2003) and Isaac William Martin, *The Permanent Tax Revolt* (2008).

My discussion of Boston and its tech ecosystem drew on a number of studies. The "father of venture capital," Georges Doriot, is the subject of Spencer E. Ante, *Creative Capital* (2008). Tracy Kidder's now-classic *The Soul of a New Machine* (1981) captures the fever of the minicomputer product development cycle, and Steven Levy's *Hackers* (1984) captures the early MIT computer scene with vivid intensity. Lily Geismer, *Don't Blame Us* (2014) discusses the demographic and political dynamics propelled in part by the presence of the tech industry in and around Boston. Last but hardly least, Annalee Saxenian, *Regional Advantage* (1996) remains the definitive comparative study of Route 128 and the Valley more than two decades after its publication.

Several important histories of Silicon Valley have focused on the pivotal decade of the 1970s, tracing how countercultural ideas, technological inflection points, and market shifts contributed to the emergence of microcomputing and other industries. Fred Turner, *From Counterculture to Cyberculture* (2006) traces the intellectual lineage from cybernetics to personal computing to early online communities such as the WELL. On this generational and cultural confluence, also see John Markoff, *What the Dormouse Said* (2007), and David Kaiser, *How the Hippies Saved Physics* (2011). Michael A. Hiltzik tells the history of Xerox PARC in *Dealers of Lightning* (1999). Leslie Berlin, *Troublemakers* (2017) traces the lives and careers of seminal Valley entrepreneurs of the decade, and the industries they made. A comprehensive resource that helped inform my discussion of the early microcomputing era is Paul Freiberger and Michael Swaine, *Fire in the Valley*, 2nd ed. (1999).

Biotechnology was another important industry that grew in the Bay Area during this period, enabled by the venture capital ecosystem, university-based research institutions, and government money and regulation. While IT and biotech are often grouped together under a "high-tech" economic rubric (and many of the VCs I discuss in this book, including the Morgenthaler family, were important biotech investors), the timescale of developing and marketing drugs and medical devices, as well as the legal and regulatory context, is markedly different from computer hardware and software. The biotech sector is also more geographically dispersed; Boston may have ceded its once-formidable computer hardware and software lead to Silicon Valley, but it remains among the most important centers of the biotech industry. The contrast has sharpened even further in the post-2000 period as search, social, mobile, and cloud software companies grew to a scale and influence that dwarfed their IT predecessors.

For these substantive as well as narrative reasons, I chose to keep the book's focus on Silicon Valley's information-technology companies and industries. For readers interested in learning more about biotech's origins and evolution, a good place to start is Sally Smith Hughes's excellent *Genentech* (2011). In *Troublemakers*, Leslie Berlin discusses Genentech as well as the important role of Stanford's Office of Technology Licensing, an innovator in technology transfer, which helped turn medical research into a major profit center for the university. See also Elizabeth Popp Berman, *Creating the Market University* (2012).

Economic competition with Japan was one of the defining business and political stories of the 1980s, extending far beyond the electronics industry. To understand the remarkable evolution of the Japanese economy after World War II, I drew on Chalmers Johnson's classic *MITI and the Japanese Miracle* (1982) as well as Mark Metzler, *Capital as Will and Imagination* (2013). On the impact of Japanese consumer products, including the Sony Walkman, on American society, see Andrew C. McKevitt, *Consuming Japan* (2017). On the domestic political impacts of economic globalization, see Judith Stein, *Pivotal Decade* (2010); Jefferson Cowie, *Stayin' Alive* (2010); and Meg Jacobs, *Panic at the Pump* (2016). On countercultural capitalism, see Joshua Clark Davis, *From Head Shops to Whole Foods* (2017).

The defense buildup and cultural battles of the Reagan era had long-lasting effects on the Valley. On SDI, see Frances FitzGerald, *Way Out There in the Blue* (2001). An account of DARPA's strategic computing program, written by those who built it, is Alex Roland with Philip Shiman, *Strategic Computing* (2002). The environmental and social costs of the era's tech boom are explored in Lenny Siegel and John Markoff, *The High Cost of High Tech* (1985). On the culture wars on campus and beyond, see Andrew Hartman, *A War for the Soul of America* (2015). On broader cultural and political polarization, see Daniel T. Rodgers, *Age of Fracture* (2012).

The literature on artificial intelligence and "machines that think" is rich and engaging. Norbert Wiener's *Cybernetics* (1948) and its popularizing contemporary, Edmund Callis Berkeley's *Giant Brains, or, Machines That Think* (1949), remain fascinating and revealing reads. Secondary works that helped inform this part of the story include Daniel Crevier, *AI* (1993); John Markoff, *Machines of Loving Grace* (2015); and Thomas Rid, *Rise of the Machines* (2016). The impact of automation and robotics on work is a deservedly hot topic. For an optimistic take, see Erik Brynjolfsson and Andrew McAfee, *The Second Machine Age* (2016); for a more sobering one, see Martin Ford, *Rise of the Robots* (2015). Properly placing gig-economy phenomena in the context of a longer history of corporate restructuring and contingent work (in Silicon Valley and elsewhere) is Louis Hyman, *Temp* (2018).

On the venture capital industry in the Valley and elsewhere, useful sources are John W. Wilson, *The New Venturers* (1985); Udayan Gupta, *Done Deals* (2000); and William H. Draper III's memoir, *The Startup Game* (2011). Randall E. Stross, *eBoys* (2000) explores venture capital during the dot-com era. On investors' long-running battles against business taxation and regulation, see Monica Prasad, *The Politics of Free Markets* (2006); Julia C. Ott, *When Wall Street Met Main Street* (2011); and Isaac William Martin, *Rich People's Movements* (2015).

The changing regulatory and market environment on Wall Street made successive tech booms possible. For a comprehensive sweep, see B. Mark Smith, *A History of the Global Stock Market* (2004). On the exchange that became home to many of tech's biggest names, see Mark Ingebretsen, *NASDAQ* (2002). On another financial phenomenon of the late twentieth century, hedge funds, see Sebastian Mallaby, *More Money than God* (2010).

For more on the academic origins and commercial evolution of the Internet, a place to start is Janet Abbate's careful survey, *Inventing the Internet* (1999). For more about the people and seminal technologies of the early, noncommercial network, see Katie Hafner and Matthew Lyon, *Where Wizards Stay Up Late* (1996) as well as the account of the Web's creation by the man who invented it, Tim Berners-Lee with Mark Fischetti, *Weaving the Web* (1999).

Once we reach the 1990s, the academic histories become scarcer; instead, we have a torrent of book-length profiles of tech's major companies and the people who led them, many written within months or a few years of the events depicted. An important source is the longform reporting and books produced by journalists associated with *Wired* magazine, which among other things paid attention to the Washington, D.C., story long before others did. Of particular relevance: Sara Miles's account of the Democrats' wooing of the Valley, *How to Hack a Party Line* (2001); Paulina Borsook's wild ride through techno-libertarian thinking, *Cyberselfish* (2000); and John Heilemann's riveting account of the Microsoft antitrust saga, *Pride Before the Fall* (2001).

For my discussion of Bill Gates and Microsoft, I also drew on Stephen Manes and Paul Andrews, *Gates* (1993), as well as Gates's own Internet-era account, *The Road Ahead* (1996). G. Pascal Zachary documents the race to build a better Windows in *Showstopper!* (1994); Ken Auletta, *World War 3.0* (2001) and David Bank, *Breaking Windows* (2001) further document the turmoil of the late 1990s within the company.

On Apple, Walter Isaacson's *Steve Jobs* (2011) explores this complicated leader in all his irascible and brilliant glory—a portrayal with which, it should be added, some of those closest to Jobs disagree. On the Mac, see Steven Levy, *Insanely Great* (1994) as well as the firsthand account by one of the Mac's creators, Andy Hertzfeld, *Revolution in the Valley* (2011). Randall Stross, *Steve Jobs and the NeXT Big Thing* (1993) tells the story of Jobs's short-lived but long-influential venture. On latter-day Apple, see Adam Lashinsky, *Inside Apple* (2012). On the iPhone and its ecosystem, as well as an account of the mobile devices that preceded and enabled it, see Brian Merchant, *The One Device* (2017).

Katie Hafner, *The Well* (2001), tells the history of the influential online network and its denizens. On America Online and the dial-up networking era, see Kara Swisher, *AOL.com* (1998); on AOL's merger with Time Warner and the dot-com euphoria that accompanied it, see Swisher's aptly titled *There Must Be a Pony in Here Somewhere* (2003). Michael Lewis tells the story of Netscape and the irrepressible Jim Clark in *The New New Thing* (2000).

The rise of a new generation of companies from the ashes of the dot-com bust is the subject of Sarah Lacy, *Once You're Lucky, Twice You're Good* (2008). John Battelle explores the technologies and technologists behind the first wave of search engines, and Google's rise above them all, in *The Search* (2005). An excellent study of the original smash-hit social network is Julia Angwin, *Stealing MySpace* (2009).

For my discussion of Google, I drew on Ken Auletta, *Googled* (2009). On Facebook's early years, I consulted David Kirkpatrick, *The Facebook Effect* (2010) as well as Katherine Losse's vivid and reflective first-person account, *The Boy Kings* (2012). On Amazon, see Brad Stone's deep dig into *The Everything Store* (2013) as well as Robert Spector's chronicle of the company in the dot-com era, *Get Big Fast* (2000).

As the biggest tech companies have swelled in influence and wealth, there have been a number of carefully researched works by scholars examining the limitations and biases embedded within the products these companies have built (and whose titles alone reveal much about the current mood). They include Frank Pasquale, *The Black Box Society* (2015); Sara Wachter-Boettcher, *Technically Wrong* (2017); Safiya Umoja Noble, *Algorithms of Oppression* (2018); Virginia Eubanks, *Automating Inequality* (2018); Siva Vaidhyanathan, *Antisocial Media* (2018); and Meredith Broussard, *Artificial Unintelligence* (2018).

ARCHIVES

I am grateful to the archives whose collections informed my research and the archivists who manage and lead them. Deep thanks also go to the technologists, executives, and companies who have recognized that a history-making industry and place needs to preserve its past, and who have donated their papers and artifacts to archival repositories. (Note to current leaders of the tech world: please do this too!). Digital archives were another huge boon to me as a researcher, and special thanks go to three digitized archival oral history collections on which I relied extensively: Stanford's Silicon Genesis Project; the oral histories conducted by the Computer History Museum; and the "Early Bay Area Venture Capitalists" project of the Regional Oral History Office of the Bancroft Library at the University of California, Berkeley.

Other archives consulted, including abbreviations used in the notes:

CA Carl Albert Center Congressional and Political Collections, Norman, Okla.
CHM Computer History Museum, Mountain View, Calif.
HH Hoover Institution Library & Archives, Stanford, Calif.
HP Agilent (Hewlett Packard) History Center, Palo Alto, Calif.
HV Harvard University Archives, Cambridge, Mass.

MO Museum of History and Industry (MOHAI), Seattle, Wash.

NA U.S. National Archives, College Park, Md.

PA Palo Alto Historical Association, Palo Alto, Calif.

PT Paul E. Tsongas Collection, University of Massachusetts Lowell, Lowell, Mass.

RMN Richard M. Nixon Presidential Library, Yorba Linda, Calif.

SJ History San José, San Jose, Calif.

SU Stanford University Special Collections and University Archives, Stanford, Calif.

UW University of Washington Special Collections and Archives, Seattle, Wash.

WJC William J. Clinton Presidential Library, Little Rock, Ark.

INTERVIEWS

If one secret of Silicon Valley was the people, then one secret of this book about Silicon Valley was the opportunity to talk to so many of those who lived this history. A list of interviews I conducted between 2014 and 2018 is below; additional interviewees requested anonymity. I am deeply grateful for the insights and contributions of all who spoke with me.

Yaw Anokwa, June 7, 2018

Pete Bancroft, November 3, 2015

John Seely Brown, December 16, 2014, March 22, 2018

Luis Buhler, February 8, 2016

Tom Campbell, February 17, 2016

Jim Cunneen, February 1, 2016

Reid Dennis, May 26, 2015

Bill Draper, June 23, 2015

Trish Millines Dziko, April 3, 2018

James Gibbons, November 14, 2015

Stewart Greenfield, May 19, 2015

Ken Hagerty, September 9, 2015

Kip Hagopian, February 8, 2016

Ann Hardy, April 20, 2015, September 19, 2017, August 28, 2018

Pitch Johnson, May 26, 2015

Jennifer Jones, November 14, 2014

Tom Kalil, August 7, 2017

Mitch Kapor, September 19, 2017

Roberta Katz, November 12, December 10, 2014

Guy Kawasaki, January 26, February 12, 2015

Chop Keenan, March 17, 2016

Floyd Kvamme, February 16, 2016

Arthur Levitt, May 7, July 10, 2015

Dan'l Lewin, November 21, 2017

Audrey Maclean, May 14, 2015

Hélène Martin, June 4, 2018

Bob Maxfield, May 28, 2015

Kathie Maxfield, May 28, 2015

Pete McCloskey, February 18, 2016

Tom McEnery, February 2, March 9, 2016

Regis McKenna, December 3, 2014, April 21, 2015, May 31, 2016

Burt McMurtry, January 15, 2015, October 2, 2017

Bob Miller, December 16, 2014

William F. Miller, February 27, 2015

Becky Morgan, May 13, 2016

David Morgenthaler, February 12, May 19, June 23, November 3, 2015

Gary Morgenthaler, November 24, 2014

Chamath Palihapitiya, December 5, 2017

Paul Saffo, March 24, 2017

Allan Schiffman, March 22, 2018

Charles Simonyi, October 4, 2017

Larry Stone, April 7, 2015

Marty Tenenbaum, February 9, February 21, March 16, 2018

Avie Tevanian, December 13, 2017

Andy Verhalen, November 18, 2014

Ed Zschau, April 9, June 24, 2015, January 19, 2016

NOTES

1. *Night Shift*, directed by Ron Howard, written by Lowell Ganz and Babaloo Mandel (Burbank, Calif.: Warner Brothers Pictures, 1982). Reproduced with permission.
2. John Perry Barlow, "A Declaration of the Independence of Cyberspace," Electronic Frontier Foundation, February 8, 1996, https://projects.eff.org/~barlow/Declaration-Final.html.
3. Ellen Ullman, *Life in Code: A Personal History of Technology* (New York: Farrar, Straus and Giroux, 2017), 47.

INTRODUCTION: THE AMERICAN REVOLUTION

1. Associated Press, "Apple, Amazon, Facebook, Alphabet, and Microsoft Are Collectively Worth More Than the Entire Economy of the United Kingdom," April 27, 2018, https://www.inc.com/associated-press/mindblowing-facts-tech-industry-money-amazon-apple-microsoft-facebook-alphabet.html, archived at https://perma.cc/HY68-RJYG.
2. Reyner Banham, "Down in the Vale of Chips," *New Society* 56, no. 971 (June 25, 1981): 532–33.
3. John Doerr, "The Coach," interview by John Brockman, 1996, Edge.org, https://www.edge.org/digerati/doerr/, archived at https://perma.cc/9KWX-GLWK.
4. Marc Andreessen, "Why Software Is Eating the World," *The Wall Street Journal*, August 20, 2011, C2. Billions of dollars of public investment later, many of the would-be Silicon Somethings have fallen short of original expectations; see Margaret O'Mara, "Silicon Dreams: States, Markets, and the Transnational High-Tech Suburb," in *Making Cities Global: The Transnational Turn in Urban History*, ed. A. K. Sandoval-Strausz and Nancy H. Kwak (Philadelphia: University of Pennsylvania Press, 2017), 17–46.
5. Chiat/Day, "Macintosh Introductory Advertising Plan FY 1984," November 1983, Apple Computer Records, Box 14, FF 1, SU.
6. Ronald Reagan, "Remarks and Question-and-Answer Session with Students and Faculty at Moscow State University," May 31, 1988, posted by John T. Woolley and Gerhard Peters, *The American Presidency Project*, https://www.presidency.ucsb.edu/node/254054.
7. Steven Levy, *Hackers: Heroes of the Computer Revolution* (New York: Anchor Press/Doubleday, 1984); Reagan, "Remarks and Question-and-Answer Session with Students and Faculty at Moscow State University." Also see Fred Turner, *From Counterculture to Cyberculture: Stewart Brand, the Whole Earth Network, and the Rise of Digital Utopianism* (Chicago: The University of Chicago Press, 2006); John Markoff, *What the Dormouse Said: How the Sixties Counterculture Shaped the Personal Computer Industry* (New York: Penguin, 2005); David Kaiser, *How the Hippies Saved Physics: Science, Counterculture, and the Quantum Revival* (New York: W. W. Norton, 2011).
8. Guy Kawasaki, interview with the author, January 26, 2015, Menlo Park, Calif.

ACT ONE

1. Fred Terman, interview by Jane Morgan for the 75th Palo Alto Anniversary, Palo Alto Historical Association, 1969, https://www.youtube.com/watch?v=Jwk2Y4mi87w, archived at https://perma.cc/5FSW-SXBF.

ARRIVALS

1. David T. Morgenthaler, interview with the author, May 26, 2015, Palo Alto, Calif.; David T. Morgenthaler, oral history interview by John Hollar, December 2, 2011, Computer History Museum, Mountain View, Calif., CHM Ref. X6305.2012, 21, http://archive.computerhistory.org/resources/access/text/2013/11/102746212-05-01-acc.pdf.
2. Ann Hardy, interview with the author, April 20, 2015, Stanford, Calif.; Hardy, phone conversation, August 28, 2018; John Harwood, *The Interface: IBM and the Transformation of Corporate Design, 1945–1976* (Minneapolis: University of Minnesota Press, 2011).
3. Burton J. McMurtry, interview with the author, January 15, 2015, Palo Alto, Calif.; McMurtry, oral history interview by Sally Smith Hughes, 2009, "Early Bay Area Venture Capitalists: Shaping the Economic and Business Landscape," Regional Oral History Office, The Bancroft Library, University of California, Berkeley, 12, http://digitalassets.lib.berkeley.edu/roho/ucb/text/mcmurtry_burt.pdf.
4. Alfred R. Zipser Jr., "Microwave Relay Replacing Cables," *The New York Times*, March 21, 1954, F1. On the Stanford microwave laboratory, see Christophe Lécuyer, *Making Silicon Valley: Innovation and the Growth of High Tech* (Cambridge, Mass.: The MIT Press, 2007) and Rebecca S. Lowen, *Creating the Cold War University: The Transformation of Stanford* (Berkeley: University of California Press, 1997). The symbiotic relationship between Stanford and the early electronics industry in the Valley is described by Robert Kargon, Stuart W. Leslie, and Erica Schoenberger, "Far Beyond Big Science: Science Regions and the Organization of Research and Development," in *Big Science: The Growth of Large-Scale Research*, ed. Peter Galison and Bruce Hevly (Stanford, Calif.: Stanford University Press, 1992), 334–54.
5. Burt McMurtry, interview with the author, January 15, 2015, Palo Alto, Calif.; interview with the author, October 2, 2017, by phone.

CHAPTER 1: ENDLESS FRONTIER

1. Harold D. Watkins, "Hometown, U.S.A.: High IQ, High Income Help Palo Alto Grow," *The Wall Street Journal*, August 10, 1956, 1. Adjusted for inflation, the median 1956 home price was roughly equivalent to $180,000 in 2018.
2. *The Founding Grant with Amendments, Legislation, and Court Decrees* [1885] (Stanford, Calif.: Stanford University, 1987), 4; Jane Stanford, Speech at Opening Ceremony of Stanford University, October 1, 1891, https://sdr.stanford.edu/uploads/rr/050/nb/1367/rr050nb1367/content/sc0033b_s5_b2_f04.pdf, archived at http://perma.cc/6JXE-A3U6.
3. "Crosses U.S. to Shop: San Jose Woman Finds New Yorkers Courteous—Tells of Prune Crops," *The New York Times*, July 1, 1923, 20; "San Jose Campaign for Prune Week," *The Los Angeles Times*, November 25, 1916, I3; "Prune Week in United States and Canada Begins February 27th," *Western Canner and Packer* 13, no. 10 (February 1922): 116; E. Alexander Powell, "Valley of Heart's Delight," *Sunset* 29 (August 1912): 115–25.
4. Postwar regional planning acknowledged this strength. As in many U.S. cities after the war, the San Francisco civic and business elite sat down in 1945 and mapped out a plan for

postwar regional development to ensure that the dismal Depression-era economy would not return once the hyperactive war machine wound down. Heavy industry would remain on the East Bay, finance in San Francisco, and the peninsula would be the hub of "light industry," zoned accordingly. See Margaret Pugh O'Mara, *Cities of Knowledge: Cold War Science and the Search for the Next Silicon Valley* (Princeton, N.J.: Princeton University Press, 2004). On California and defense mobilization during and after World War II, see Roger W. Lotchin, *Fortress California, 1910–1961: From Warfare to Welfare* (Champaign: University of Illinois Press, 2002); Kevin Starr, *Embattled Dreams: California in War and Peace, 1940–1950* (Oxford, U.K.: Oxford University Press, 2002); and Starr, *Golden Dreams: California in an Age of Abundance, 1950–1963* (Oxford, 2009). Seattle and the Pacific Northwest were another important node of defense production; see Richard S. Kirkendall, "The Boeing Company and the Military-Metropolitan-Industrial Complex, 1945–1953," *Pacific Northwest Quarterly* 85, no. 4 (October 1994): 137–49.

5. HP's startup capital from William Bates, "HP tips Toward Computers," *The New York Times*, July 2, 1978, p. F1. These seminal companies and entrepreneurs are explored in detail in Christophe Lécuyer, *Making Silicon Valley: Innovation and the Growth of High Tech* (Cambridge, Mass.: The MIT Press, 2007).
6. Floyd J. Healey, "Dirigible Base North's Dream," *The Los Angeles Times*, October 31, 1929, 6.
7. Larry Owens, "The Counterproductive Management of Science in the Second World War: Vannevar Bush and the Office of Scientific Research and Development," *Business History Review* 68, no. 4 (1994): 515–76; *Time* magazine, April 3, 1944, cover. Also see G. Pascal Zachary, *Endless Frontier: Vannevar Bush, Engineer of the American Century* (New York: Free Press / Simon and Schuster, 1997). The idea of marshaling America's scientists in the cause of war hadn't been Bush's alone—he shared credit with MIT President Karl Compton and Harvard President James Conant—but Bush was both the public face and the operational mind that put the idea into action.
8. Vannevar Bush, "As We May Think," *The Atlantic*, July 1, 1945, reprinted in *Interactions* 3, no. 2 (March 1996): 35–46.
9. *The New York Times*, January 3, 1943, quoted in Owens, "Vannevar Bush and the OSRD." On the war's effect on state building and business-government relationships, see Mark R. Wilson, *Destructive Creation: American Business and the Winning of World War Two* (Philadelphia: University of Pennsylvania Press, 2016) and James T. Sparrow, *Warfare State: World War II Americans and the Age of Big Government* (Oxford, 2011).
10. C. Stewart Gillmor, *Fred Terman at Stanford: Building a Discipline, a University, and Silicon Valley* (Stanford, Calif.: Stanford University Press, 2004); Mitchell Leslie, "The Vexing Legacy of Lewis Terman," *Stanford Alumni Magazine* (July/August 2000), https://alumni.stanford.edu/get/page/magazine/article/?article_id=40678, archived at https://perma.cc/YFZ3-HJD4.
11. Carolyn Caddes, *Portraits of Success: Impressions of Silicon Valley Pioneers* (Wellsboro, Penn.: Tioga Publishing, 1986), 30; Gillmor, *Fred Terman at Stanford.*
12. Sybil Terman, quoted in Gillmor, *Fred Terman at Stanford*, 210.
13. Rebecca S. Lowen, *Creating the Cold War University: The Transformation of Stanford* (Berkeley: University of California Press, 1997); Paul H. Mattingly, *American Academic Cultures: A History of Higher Education* (Chicago: The University of Chicago Press, 2017).
14. Owens, "Vannevar Bush and the OSRD."

15. Franklin Delano Roosevelt, Letter to Vannevar Bush, November 17, 1944, reprinted in *Science, the Endless Frontier, A Report to the President by Vannevar Bush, Director of the Office of Scientific Research and Development* (Washington, D.C.: U.S. Government Printing Office, 1945), vii.
16. *Science, the Endless Frontier,* vi, 6, 34.
17. Joint Committee on the Economic Report, *National Defense and the Economic Outlook,* 82nd Congress, 1st Session (USGPO, 1951), 3.
18. Ibid., 31–38.
19. National Science Foundation, *First Annual Report* (USGPO, 1951), 8.
20. W. C. Bryant, "Electronics Industry: It's Due for a Vast Expansion," *The Wall Street Journal,* March 26, 1951, 3; N. E. Edlefsen, "Supersonic Era Pilots Need Help," *The Los Angeles Times,* June 17, 1951, 27.
21. Michael Amrine, "To Mobilize Science Without Hobbling It," *The New York Times,* December 3, 1950, 13; National Science Foundation, "Scientific Manpower and the Graduate Fellowship Program," *Annual Report* (1952), 25; "Electronic Sight: $20 Billon by '56," *The New York Times,* August 25, 1955, 34; Charles E. Wilson, *Three Keys to Strength: Production, Stability, Free-World Unity, Third Quarterly Report to the President by the Director of the Defense Mobilization Board* (October 1951), 11.
22. *The New York Times,* November 6, 1955, F11–13.
23. Amrine, "To Mobilize Science."
24. Benjamin A. Collier, "Wanted: Specialist in Electronics!," *The Chicago Defender,* December 17, 1949, 4; Frank E. Bolden, "Prober of Electronic Secrets," *The Pittsburgh Courier,* September 25, 1954, SM7.
25. Nathan Ensmenger, *The Computer Boys Take Over: Computers, Programmers, and the Politics of Technical Expertise* (Cambridge, Mass.: The MIT Press, 2010), 35. Great Britain, the other great computing power of wartime and after, treated its female programmers similarly; see Marie Hicks, *Programmed Inequality: How Britain Discarded its Women Technologists and Lost its Edge in Computing* (MIT, 2017). Also see Jennifer S. Light, "When Computers Were Women," *Technology and Culture* 40, no. 3 (July 1999): 455–83; Thomas J. Misa, ed., *Gender Codes: Why Women are Leaving Computing* (Hoboken, N.J.: Wiley / IEEE Computer Society Press, 2010). On the longer history, see Londa Schiebinger, *The Mind Has No Sex? Women in the Origins of Modern Science* (Cambridge: Harvard University Press, 1989).
26. Harwood G. Kolsky, Notes of Meeting with IBM at Los Alamos, 14 March 1956, IBM Project Stretch Collection, Lot No. X3021.2005, Computer History Museum, Mountain View, Calif. (CHM); Ann Hardy, interview with the author, April 20, 2015, Palo Alto, Calif.
27. Ann Hardy, interviews with the author, April 20, 2015, and September 19, 2017, Palo Alto, Calif.
28. Frederick E. Terman, Letter to Paul Davis, 29 December 1943, FF2, Box 1, Series I, SC 160, SU. Quoted in Stuart W. Leslie, *The Cold War and American Science: The Military-Academic-Industrial Complex at MIT and Stanford* (New York: Columbia University Press, 1993), 44; Terman to Donald Tresidder, April 25, 1947, quoted in Robert Kargon, Stuart W. Leslie, and Erica Schoenberger, "Far Beyond Big Science: Science Regions and the Organization of Research and Development," in *Big Science: The Growth of Large-Scale Research,* ed. Peter Galison and Bruce Hevly (Stanford, Calif.: Stanford University Press, 1992), 341.
29. Clark Kerr, *The Uses of the University* (Harvard, 1963).

CHAPTER 2: GOLDEN STATE

1. "Board Selects New President," *The Stanford Daily*, November 19, 1948, 1. On "steeples of excellence," see Stuart W. Leslie, *The Cold War and American Science: The Military-Academic-Industrial Complex at MIT and Stanford* (New York: Columbia University Press, 1993), and Rebecca S. Lowen, *Creating the Cold War University: The Transformation of Stanford* (Berkeley: University of California Press, 1997).
2. For discussion of the Stanford Research Park see Margaret Pugh O'Mara, *Cities of Knowledge: Cold War Science and the Search for the Next Silicon Valley* (Princeton, N.J.: Princeton University Press, 2004), 97–141, and John Findlay, *Magic Lands: Western Cityscapes and American Culture After 1940* (Berkeley: University of California Press, 1992), 117–59.
3. Frederick E. Terman, "The University and Technology Utilization" (speech, NASA—University Conference, Kansas City, Mo., March 3, 1963), SU; David Packard, "Electronics and the West" (speech, Stanford Research Institute, San Francisco, Calif., November 23, 1954), Box 2, Folder 27, Packard Speeches, Agilent Archives, Mountain View, Calif. (HP).
4. Stephen B. Adams, "Growing Where You Are Planted: Exogenous Firms and the Seeding of Silicon Valley," *Research Policy* 40, no. 3 (April 2011): 368–79; "Electronic Sight: $20 Billon by '56," *The New York Times*, August 25, 1955, 34.
5. Burton J. McMurtry, interview with the author, October 2, 2017, by phone.
6. "Bet with a Multiple Payoff," *BusinessWeek*, December 14, 1957, 107; "Hewlett Co-Founder Retiring from Post," *The New York Times*, Jan 20, 1987, D2; Ted Sell, "Defense's Packard—Low-Key Titan," *The Los Angeles Times*, May 3, 1970, N1.
7. Jim Collins, Foreword to David Packard, *The HP Way: How Bill Hewlett and I Built Our Company* (New York: HarperBusiness, 2005). Also see Findlay, *Magic Lands*, 138–39.
8. A decade later, this political engagement and sense of public service persuaded Packard to spend more than two years as Deputy Secretary of Defense in the Nixon Administration, where he applied his management ethos to the sprawling, war-consumed Pentagon. Packard's life as "Mr. Inside" to Defense Secretary Mel Laird's "Mr. Outside" is discussed briefly in Packard's memoir and more fully in Dale Van Atta, *With Honor: Melvin Laird in War, Peace, and Politics* (Madison: The University of Wisconsin Press, 2008).
9. David Packard, "Electronics—Glamour or Substance?" (speech to Purchasing Agents Association, February 13, 1958), Packard Speeches – 1958, Box 2, FF 31, HP.
10. Packard, "Acceptance of 'The American Way of Life' Award" (speech, Sertoma Club, Pueblo, Colorado, April 19, 1963), Box 2, FF 51, HP; Packard, "Business Management and Social Responsibility" (speech, Children's Home Society of California, Palo Alto, Calif., May 17, 1965) Packard Speeches, Box 2, FF 30, HP.
11. On the importance of Sunbelt business leaders to the emerging modern conservative movement as well as emerging ideas of free enterprise and entrepreneurship, see Elizabeth Tandy Shermer, *Sunbelt Capitalism: Phoenix and the Transformation of American Politics* (Philadelphia: University of Pennsylvania Press, 2013); Kathryn S. Olmsted, *Right Out of California: The 1930s and the Big Business Roots of Modern Conservatism* (New York: The New Press, 2015). On business conservatism more broadly, see Angus Burgin, *The Great Persuasion: Reinventing Free Markets Since the Depression* (Cambridge, Mass.: Harvard University Press, 2012); Kim Phillips-Fein, *Invisible Hands: The Businessmen's Crusade Against the New Deal* (New York: W. W. Norton, 2010).
12. Hoover chose Campbell on the recommendation of one of the fiercest critics of the New Deal, ex-Roosevelt aide Raymond Moley. See Gary Atkins, "Attacked for Politics, Policies;

Critics Center on Hoover Boss," *The Stanford Daily*, January 7, 1972, 1. On Campbell at Hoover, see Mary Yuh, "Governance, bias: enduring controversies," *The Stanford Daily Magazine*, April 18, 1986, 7; Thomas Sowell, "W. Glenn Campbell, 1924–2001," *Hoover Digest* No. 1, 2002, January 30, 2002.

13. H. Myrl Stearns, Varian Associates, quoted in Charles Elkind, "Riding the High-Tech Boom: The American Electronics Association Story, 1945–1990," unpublished manuscript, c. 1991, MISC 333, FF 1, SU, 17.
14. McMurtry, interview with the author, January 15, 2015.
15. David W. Kean, *IBM San Jose: A Quarter Century of Innovation* (New York: IBM, 1977), 47–48.
16. Adams, "Growing Where You Are Planted." On dispersion policy, see O'Mara, *Cities of Knowledge*, 36–54.
17. Robert Kargon, Stuart W. Leslie, and Erica Schoenberger, "Far Beyond Big Science: Science Regions and the Organization of Research and Development," in *Big Science: The Growth of Large-Scale Research*, ed. Peter Galison and Bruce Hevly (Stanford, Calif.: Stanford University Press, 1992), 348; "Fire in Locked Vault Destroys Missile Data," *The Washington Post*, December 22, 1957, 2; "Electronic Sight." In 1986, well into the era of the personal computer, Lockheed had a head count of 24,000. Only Hewlett Packard had more employees. See "Companies with over 500 Employees (Nov. 1986)," Silicon Valley Ephemera Collection, Series 1, Box 5, FF 12, SU.
18. "Bias Suit at Lockheed Unit," *The Wall Street Journal*, November 14, 1973, 22. Also see Herbert G. Ruffin II, *Uninvited Neighbors: African Americans in Silicon Valley, 1769–1990* (Norman: University of Oklahoma Press, 2014).
19. Peter J. Brennan, "Advanced Technology Center: Santa Clara Valley, California," Silicon Valley Ephemera Collection, Series 1, Box 1, FF 17, SU.
20. James Gibbons, interview with the author, November 4, 2015, Stanford, Calif.
21. Shockley was able to set up his own shop because the U.S. government had mandated that Bell Labs—a division of telecommunications monopsony AT&T—allow free licensing of the transistor patent originally developed by Shockley in its facilities. The fact that the transistor was not proprietary technology was instrumental in its wide adaptation and iteration, notably by Texas Instruments, which morphed from an oilfield instrumentation company into a leading transistor and microchip maker, and home to a team that co-invented the integrated circuit, led by Jack Kilby.
22. Shockley's fixation on IQ tests, it turned out, masked an unrestrained and unapologetic belief in white supremacy. After the disintegration of his company, the Nobel winner spent the last twelve years of his career at Stanford, focusing chiefly on the pseudoscience of eugenics. In doing so, he followed in the infamous footsteps of several prominent Stanford faculty members (including, to a certain degree, Fred Terman's father Lewis) who had given the field academic legitimacy a half century earlier. By the end of his life, Shockley had come to consider his work on eugenics more significant than his discovery of the transistor. See Wolfgang Saxon, "William B. Shockley, 79, Creator of Transistor and Theory on Race," *The New York Times*, August 14, 1989, D9; Joel N. Shurkin, *Broken Genius: The Rise and Fall of William Shockley, Creator of the Electronic Age* (New York: Macmillan Science, 2006). Shockley's white supremacism often goes unmentioned in discussions of his role in the genesis of Silicon Valley. A monument and historical plaque installed in August 2018 on the site where

Shockley Semiconductor once stood made no mention of the founder's eugenics research, nor did the several speakers who celebrated Shockley's legacy at the marker's unveiling ceremony. See Sam Harnett, "Mountain View Commemorates Lab of William Shockley, Acclaimed Physicist and Vocal Racist," The California Report, KQED Radio, August 21, 2018, https://www.kqed.org/news/11687943/mountain-view-commemorates-lab-of-william-shockley-acclaimed-physicist-and-vocal-racist, archived at https://perma.cc/M9CK-NZ25.

23. Gibbons, interview with the author. Also see James Gibbons, oral history interview by David Morton, May 31, 2000, IEEE History Center, https://ethw.org/Oral-History:James_Gibbons, archived at http://perma.cc/6Z4M-MHMG. The story of the transistor, Shockley Semiconductor, and the "Traitorous Eight" has been explored by a number of authors, most originally and notably in two biographies: Leslie Berlin, *The Man Behind the Microchip: Robert Noyce and the Invention of Silicon Valley* (Oxford, U.K.: Oxford University Press, 2005), and Arnold Thackray, David C. Brock, and Rachel Jones, *Moore's Law: The Life of Gordon Moore, Silicon Valley's Quiet Revolutionary* (New York: Basic Books, 2015).
24. Arthur Rock, interviews by Sally Smith Hughes, 2008 and 2009, "Early Bay Area Venture Capitalists: Shaping the Economic and Business Landscape," Regional Oral History Office, The Bancroft Library, University of California, Berkeley, California; Arthur Rock, "Strategy Versus Tactics from a Venture Capitalist," 1992, in *The Book of Entrepreneurs' Wisdom: Classic Writings by Legendary Entrepreneurs,* ed. Peter Krass (New York: John Wiley and Sons, 1999), 131–41.
25. On the importance of federal contracts to Fairchild's early business, see Daniel Holbrook, "Government Support of the Semiconductor Industry: Diverse Approaches and Information Flows," *Business and Economic History* 24, no. 2 (Winter 1995): 133–77. On Fairchild's founding, see Berlin, *The Man Behind the Microchip,* 75–96.

CHAPTER 3: SHOOT THE MOON

1. "Soviet Fires Earth Satellite into Space," *The New York Times,* October 5, 1957, 1; U.S. Naval Research Laboratory, "Orbits of USSR Satellite," released October 8, 1957; "Presidency is Filled by Foundry Services," *The New York Times,* October 10, 1957, 51.
2. Dwight D. Eisenhower, *Waging Peace, 1956-1961: The White House Years* (New York: Doubleday, 1965), excerpted in "Eisenhower Describes Repercussions Over Launching of 1st Soviet Sputnik," *The Washington Post,* September 21, 1965, 1; "Moscow Denounces Dog-Lover Protests," *The Washington Post,* November 6, 1957, 3. Adding to Eisenhower's political heartburn was the fact that (against the counsel of some of his science advisors) he had opted in 1955 to prioritize missile development over space-satellite research, reasoning that missiles were more important to national security. On Sputnik, the "missile gap," and the political sea change the events of October 1957 precipitated, see William I. Hitchcock, *The Age of Eisenhower: America and the World in the 1950s* (New York: Simon & Schuster, 2018), 376–406.
3. United States, President's Science Advisory Committee, Security Resources Panel, *Deterrence and Survival in the Nuclear Age (the "Gaither report" of 1957)* (Washington, D.C.: U.S. Government Printing Office, 1976); David L. Snead, *The Gaither Committee, Eisenhower, and the Cold War* (Columbus: Ohio State University Press, 1999). Many years and many billions in spending later, it became clear that the Gaither Committee was operating on some bad intelligence, and its projection that Russians would soon possess thousands of intercontinental

ballistic missiles turned out to be far off the mark. Soviet missile capabilities were, in fact, far less than the Americans believed at the time. See Annie Jacobsen, *The Pentagon's Brain: An Uncensored History of DARPA, America's Top Secret Research Agency* (Boston: Back Bay Books / Little, Brown, 2015), 46–54.

4. Neil H. McElroy, testimony in Subcommittee on Department of Defense Appropriations; Committee on Appropriations, House of Representatives, "Department of Defense: Ballistic Missile Program," Hearings, November 20 and 21, 1957, 7.
5. Don Shannon, "U.S. Missile Czar Appointed by Ike: MIT President Will Lead Drive to Speed Rockets, Satellites," *The Los Angeles Times,* November 8, 1957, 1; Richard V. Damms, "James Killian, the Technological Capabilities Panel, and the Emergence of President Eisenhower's 'Scientific-Technological Elite,'" *Diplomatic History* 24, no. 1 (January 1, 2000): 57–78. Also see James R. Killian, *Sputnik, Scientists, and Eisenhower: A Memoir of the First Special Assistant to the President for Science and Technology* (Cambridge, Mass.: The MIT Press, 1977).
6. "Scientific Progress, the Universities, and the Federal Government," statement by the President's Science Advisory Committee, November 15, 1960, 11.
7. Richard Witkin, "Missiles Program Dwarfs First Atom Bomb Project," *The New York Times,* April 7, 1957, 1.
8. John F. Kennedy, "Address at Rice University in Houston on the Nation's Space Effort," September 12, 1962, Houston, Texas, posted by Gerhard Peters and John T. Woolley, *The American Presidency Project*, https://www.presidency.ucsb.edu/node/236798.
9. National Science Foundation, *Federal Funds for Research, Development, and Other Scientific Activities* (1972), 3; Edwin Diamond, "That Moon Trip: Debate Sharpens," *The New York Times,* July 28, 1963, 150; Harold M. Schmeck Jr., "Scientists Riding Wave of Future," *The New York Times,* January 3, 1963, 20, 22. On Southern industrialization, see Bruce J. Schulman, *From Cotton Belt to Sunbelt: Federal Policy, Economic Development, and the Transformation of the South, 1938–1980* (Durham, N.C.: Duke University Press, 1994).
10. ". . . And in the Meantime at LMSC," *Lockheed MSC Star* (Sunnyvale, Calif.), July 12, 1974, 4, Box 3, FF "STC Historical Data," Joseph D. Cusick Papers, SU; Leif Erickson, "B58's Electronic Shield Guards Against Missiles," *The Washington Post*, January 25, 1958, C11.
11. "Discoverer XIII Life Cycle," c. 1961, 59–60, Box 3, FF "STC Historical Data," Joseph D. Cusick Papers, SU; Walter J. Boyne, *Beyond the Horizons: The Lockheed Story* (New York: Thomas Dunne / St. Martin's Press, 1998), 275.
12. Terman added to the momentum by aggressively recruiting star professors. By 1963, Stanford had five Nobel laureates on its faculty. The University of California at Berkeley had eleven. California edged out Massachusetts in having the most residents who were members of the august National Academy of Sciences. And once these scientists got to the sunshine, they tended to stay. See Rebecca S. Lowen, *Creating the Cold War University: The Transformation of Stanford* (Berkeley: University of California Press, 1997), 180–81; Schmeck, "Scientists Riding Wave of Future."
13. *Stanford University Bulletin*, May 15, 1958, FF "Palo Alto History," SC 486, 90-052, SU.
14. "Attempts to Stir Up Union Trouble on San Francisco Visit: Touring Premier Picks Up a Cap on Surprise Visit to Union Hall," *Los Angeles Times*, September 22, 1959, 1; David W. Kean, *IBM San Jose: A Quarter Century of Innovation* (New York: IBM, 1977), 59. Also see Harry McCracken, "Khrushchev Visits IBM: A Strange Tale of Silicon Valley History," *Fast*

Company, October 31, 2014, https://www.fastcompany.com/3037598/khrushchev-visits-ibm-a-strange-tale-of-silicon-valley-history, archived at https://perma.cc/X96L-UZAK.

15. Lawrence E. Davies, "De Gaulle Hailed by San Francisco," *The New York Times*, April 28, 1960, 2; John Markoff, correspondence with the author, September 21, 2018; audience conversation with the author, Stanford Historical Society, Stanford, Calif., January 22, 2004. Waverly Street, already famous for the HP garage, later became known for the tech billionaires who lived there: Google's Larry Page would one day live directly across from the home of the teenage Napoleon; Steve Jobs and his family were down the road. Palo Alto was, and is, a small town.
16. Margaret O'Mara, "Silicon Dreams: States, Markets, and the Transnational High-Tech Suburb," in *Making Cities Global: The Transnational Turn in Urban History*, ed. A. K. Sandoval-Strausz and Nancy H. Kwak (Philadelphia: University of Pennsylvania Press, 2017); Stuart W. Leslie and Robert H. Kargon, "Selling Silicon Valley: Frederick Terman's Model for Regional Advantage," *Business History Review* 70, no. 4 (Winter 1996): 435–72.
17. Leslie Berlin, *The Man Behind the Microchip: Robert Noyce and the Invention of Silicon Valley* (Oxford, U.K.: Oxford University Press, 2005), 130.
18. Robert N. Noyce, "Semiconductor Device-and-Lead Structure," *U. S. Patent 2981877* (filed July 30, 1959; issued April 25, 1961); Berlin, *The Man Behind the Microchip*, 108–10, 138–40; Jonathan Weber, "Chip Industry's Leaders Begin Bowing Out," *The Los Angeles Times*, February 4, 1991, D1.
19. Charles Elkind, "Riding the High-Tech Boom: The American Electronics Association Story, 1945–1990," unpublished manuscript, c. 1991, MISC 333, FF 1, SU, 19.
20. "Johnson and McNamara Letters on Defense Costs," *The New York Times*, December 2, 1963, 16; Ronald J. Ostrow, "Defense Cost-Cutting Procedures Outlined," *The Los Angeles Times*, January 31, 1963, B7.
21. Important discussion of the many changes to the contracting system and the shakeup they precipitated is in Christophe Lécuyer, *Making Silicon Valley: Innovation and the Growth of High Tech* (Cambridge, Mass.: The MIT Press, 2007), 171–75. Also see Jonathan D. Kowalski, "Industry Location Shift Through Technological Change—A Study of the U.S. Semiconductor Industry (1947–1987)," PhD dissertation, Carnegie Mellon University, 2012.
22. Gordon E. Moore, "Cramming more components onto integrated circuits," *Electronics* 38, no. 8 (April 1965): 114–17. For discussion of this technological transition, see Paul E. Ceruzzi, *A History of Modern Computing* (Cambridge, Mass.: MIT, 1998).
23. Martin Campbell-Kelly, William Aspray, Nathan Ensmenger, and Jeffrey R. Yost, *Computer: A History of the Information Machine* (3rd ed.; Boulder, Colo.: Westview Press, 2013), 221–23.

CHAPTER 4: **NETWORKED**

1. *Digital Equipment Corporation: Nineteen Fifty-Seven to the Present* (Maynard, Mass.: Digital, 1978); AnnaLee Saxenian, *Regional Advantage: Culture and Competition in Silicon Valley and Route 128* (Cambridge, Mass.: Harvard University Press, 1994), 59–82; Gene Bylinsky, *The Innovation Millionaires: How they Succeed* (New York: Charles Scribner's Sons, 1976).
2. Quoted in "About Ken Olsen," Gordon College, Ken Olsen Science Center, https://www.gordon.edu/kenolsen, *archived at* https://perma.cc/RF3V-U8MJ; Martin Campbell-Kelly, William Aspray, Nathan Ensmenger, and Jeffrey R. Yost, *Computer: A History of the Information Machine* (3rd ed.; Boulder, Colo.: Westview Press, 2013), 216–19.

3. John McCarthy, "What is Artificial Intelligence?," Computer Science Department, Stanford University, revised November 12, 2007; David Walden, "50th Anniversary of MIT's Compatible Time-Sharing System," *IEEE Annals of the History of Computing* 33, no. 4 (October–December 2011): 84–85; Daniel Crevier, *AI: The Tumultuous History of the Search for Artificial Intelligence* (New York: Basic Books, 1993); John Markoff, *Machines of Loving Grace: The Quest for Common Ground Between Humans and Robots* (New York: HarperCollins, 2015).
4. McCarthy, "Memorandum to P. M. Morse Proposing Time-Sharing," January 1, 1959, collection of Professor John McCarthy, http://jmc.stanford.edu/computing-science/timesharing-memo.html, archived at https://perma.cc/QU7M-7CM4.
5. Kent C. Redmond and Thomas M. Smith, *From Whirlwind to MITRE: The R&D Story of the SAGE Air Defense Computer* (Cambridge, Mass.: The MIT Press, 2000).
6. J. C. R. Licklider, "Man-Computer Symbiosis," *IRE Transactions on Human Factors in Electronics* HFE-1 (March 1960), 4–11.
7. William F. Miller, interview with the author, February 27, 2015, Stanford, Calif.; Miller, Interview by Patricia L. Devaney, August 9, 2009, Stanford Oral History Program, SU.
8. Quoted in Martin Campbell-Kelly, William Aspray, Nathan Ensmenger, and Jeffrey R. Yost, *Computer: A History of the Information Machine* (3rd ed., 2013), 211. Also see Martin Greenberger, *Computers, Communications, and the Public Interest* (Baltimore: Johns Hopkins University Press, 1971); Manley R. Irwin, "The Computer Utility: Competition or Regulation?" *Yale Law Journal* 76, no. 7 (1967): 1299–1320; Fred Gruenberger, ed., *Computers and Communications—Toward a Computer Utility* (Englewood Cliffs, N.J.: Prentice-Hall, 1967).
9. The founder of SDS, Max Palevsky, later sold his company to Xerox in a $100 million deal and became a major Democratic campaign donor, shoveling money and unsolicited policy advice toward liberals like George McGovern and Los Angeles Mayor Tom Bradley (Bill Boyarsky, "Palevsky Dives into New Political Waters," *The Los Angeles Times*, February 4, 1973, F1).
10. Ann Hardy, interviews with the author, April 20, 2015, and September 19, 2017; "Tymshare Reunion," Collection Item #102721147, CHM. For statistics, see Martin Campbell-Kelly and Daniel D. Garcia-Swartz, "Economic Perspectives on the History of the Computer Time-Sharing Industry, 1965–1985," *IEEE Annals of the History of Computing* 30, no. 1 (January–March 2008): 16–36.
11. Ann Hardy, phone conversation with the author, August 28, 2018.
12. LaRoy Tymes, interview by George A. Michael, July 1, 2006, in "Stories of the Development of Large Scale Scientific Computing at Lawrence Livermore National Laboratory," http://www.computer-history.info/Page1.dir/pages/Tymes.html, archived at https://perma.cc/6KFZ-UMA7.
13. LaRoy Tymes, oral history interviews by Luanne Johnson and Ann Hardy, June 11, 2004, Cameron Park, Calif., CHM. Also see Nathan Gregory, *The Tym Before: The Untold Origins of Cloud Computing* (Independently published, 2018).
14. "Tymshare Offer Sold Out," *The Wall Street Journal*, September 25, 1970, p. 18; Hardy, interview with the author, April 20, 2015, Stanford, Calif.
15. Andrew Pollack, "The Man Who Beat AT&T," *The New York Times*, July 14, 1982, D1; Wayne E. Green, "Tiny Firm Faces AT&T, General Telephone In Battle Over Telephone-Radio Connector," *The Wall Street Journal*, March 1, 1968, 30; Peter Temin with Louis Galambos, *The Fall of the Bell System: A Study in Prices and Politics* (Cambridge, U.K.:

Cambridge University Press, 1987); Katherine Maxfield, *Starting Up Silicon Valley: How ROLM became a Cultural Icon and Fortune 500 Company* (Austin, Tex.: Emerald Book Co., 2014). The precedent at work in the *Carterfone* decision was the 1956 *Hush-A-Phone* case, which ruled in support of a company that manufactured popular telephone receiver attachments that kept a user from being overheard. The Hush-A-Phone was a nonmechanical attachment to a telephone handset, however, not a self-powered electronic device like the Carterfone. See Nicholas Johnson, "Carterfone: My Story," *Santa Clara Computer & High Technology Law Journal* vol. 25, no. 3 (2008): 677–700. On the AT&T monopsony and more, see Tim Wu, *The Master Switch: The Rise and Fall of Information Empires* (New York: Alfred A. Knopf, 2010).

16. Herbert F. Mitchell, Unsolicited Proposal for Technical Assistance to NASA Goddard Space Flight Center, No. 5656-926 (Silver Spring, MD: Bunker-Ramo Eastern Technical Center, 1965).
17. "Bunker-Ramo Formed to do System Work," *The Washington Post*, January 24, 1964, B6.
18. Hardy, phone conversation with the author, August 28, 2018. For a historical overview of this regulatory shift see Gerald W. Brock, *Telecommunication Policy for the Information Age: From Monopoly to Competition* (Cambridge, Mass.: Harvard University Press, 1994); for discussion of the FCC policymaking process see Michael J. Zarkin, "Telecommunications Policy Learning: The Case of the FCC's Computer Inquiries," *Telecommunications Policy* 27, nos. 3–4 (April–May 2003): 283–99. Prodigy originally was named Trintex (1984–88).
19. By the time the ARPANET became the commercial Internet in the early 1990s, the OTC broker-traders had indeed smartened up their backroom image. But they'd kept barriers to entry friendly and low, making their board accessible to young firms with little equity. Their market now went by a new name: the NASDAQ. Computer hardware and software firms made up more than half of its listings. See Mark Ingebretsen, *NASDAQ: A History of the Market That Changed the World* (Roseville, Calif.: Forum / Random House, 2002).
20. On Licklider, ARPA, and Bob Taylor's work to build the ARPANET, see Leslie Berlin, *Troublemakers: Silicon Valley's Coming of Age* (New York: Simon & Schuster, 2017), 6–31.
21. Bolt Beranek and Newman Inc., "A History of the ARPANET: The First Decade" (Arlington, Va.: Defense Advanced Research Projects Agency, April 1981).

CHAPTER 5: THE MONEY MEN

1. AnnaLee Saxenian, *Regional Advantage: Culture and Competition in Silicon Valley and Route 128* (Cambridge, Mass.: Harvard University Press, 1994), 12. Saxenian's study is a notable part of a large body of work in economic geography that investigates the location of high-tech industrial districts, following on the concept developed by Alfred Marshall in *Principles of Economics* in 1890. See, for example, Timothy Bresnahan and Alfonso Gambardella, eds., *Building High-tech Clusters: Silicon Valley and Beyond* (Cambridge, U.K.: Cambridge University Press, 2004); Maryann P. Feldman, *The Geography of Innovation* (Dordrecht, Netherlands: Kluwer Academic, 1994); Ann Markusen, Peter Hall, Scott Campbell, and Sabina Deitrick, *The Rise of the Gunbelt: The Military Remapping of Industrial America* (Oxford, U.K.: Oxford University Press, 1991); Edward J. Malecki, *Technology and Economic Development: The Dynamics of Local, Regional, and National Change* (New York: Longman Scientific & Technical, 1991).
2. Marty Tenenbaum, interview with the author, February 9, 2018, by phone; Stewart Greenfield, interview with the author, May 19, 2015, by phone.

3. Bill Draper, interview with the author, June 23, 2015, Palo Alto, Calif.
4. William H. Draper III, interview by John Hollar, Computer History Museum, April 14, 2011, Mountain View, Calif., 5. On Doriot, see Gene Bylinsky, *The Innovation Millionaires: How They Succeed* (New York: Charles Scribner's Sons, 1976), 3–23; Christina Pazzanese, "The Talented Georges Doriot," *The Harvard Gazette*, February 24, 2015, archived at https://perma.cc/U7JL-KD2T; Spencer E. Ante, *Creative Capital: Georges Doriot and the Birth of Venture Capital* (Cambridge, Mass.: Harvard Business School Publishing, 2008).
5. Bylinsky, *The Innovation Millionaires*, 9.
6. Ernest A. Schonberger, "Inside the Market," *The Los Angeles Times*, November 2, 1969, L1.
7. Arthur Rock, interviews by Sally Smith Hughes, 2008 and 2009, "Early Bay Area Venture Capitalists: Shaping the Economic and Business Landscape," Regional Oral History Office, The Bancroft Library, University of California, Berkeley, California, 20–21. Also see Martin Kenney and Richard Florida, "Venture Capital in Silicon Valley: Fueling New Firm Formation," in Martin Kenney, ed., *Understanding Silicon Valley: The Anatomy of an Entrepreneurial Region* (Stanford, Calif.: Stanford University Press, 2000), 98–123.
8. Leslie Berlin, "The First Venture Capital Firm in Silicon Valley: Draper, Gaither & Anderson," in *Making the American Century: Essays on the Political Culture of Twentieth Century America*, ed. Bruce J. Schulman (Oxford, U.K.: Oxford University Press, 2014), 155–70; William H. Draper III, interview by John Hollar, 9.
9. Dwight D. Eisenhower, "Letter to Jere Cooper, Chairman, House Committee on Ways and Means, Regarding Small Business," July 15, 1957, posted by Gerhard Peters and John T. Woolley, *The American Presidency Project*, https://www.presidency.ucsb.edu/node/23; Robert Caro, *Master of the Senate*, vol. 3, *The Years of Lyndon Johnson* (New York: Alfred A. Knopf, 2002).
10. John W. Wilson, *The New Venturers: Inside the High-Stakes World of Venture Capital* (Reading, Mass.: Addison-Wesley, 1985), 21–24.
11. William John Martin Jr. and Ralph J. Moore Jr., "The Small Business Investment Act of 1958," *California Law Review* 47, no. 1 (March 1959): 144–70; Richard L. VanderVeld, "Small Business Symposium Set," *The Los Angeles Times*, September 11, 1960, D13.
12. Pitch Johnson, interview with the author, June 23, 2015, Palo Alto, Calif.
13. Reid Dennis, interview with the author, May 26, 2015, Palo Alto, Calif.; William H. Draper III, interview by John Hollar, 5; Wilson, *The New Venturers*, 49.
14. Franklin P. Johnson testimony, "Climate for Entrepreneurship and Innovation in the United States," Hearings before the Joint Economic Committee, Congress of the United States, Ninety-eighth Congress, Second Session, Part 2, August 27 and 28, 1984—A Silicon Valley Perspective, 167; U.S. Census, "Educational Attainment, by Race and Hispanic Origin: 1960 to 1998," *Statistical Abstract of the United States* (Washington, D.C.: United States Census Bureau, 1999), 160; William D. Bradford, "Business, Diversity, and Education," in James A. Banks, ed., *Encyclopedia of Diversity in Education* (Los Angeles: Sage Publications, 2012).
15. San Francisco's University Club, one of The Group's meeting places, did not open its dining room to women until 1988. For more on the "disappearance" of women from tech, see Nathan Ensmenger, *The Computer Boys Take Over: Computers, Programmers, and the Politics of Technical Expertise* (Cambridge, Mass.: The MIT Press, 2010), and Marie Hicks, *Programmed Inequality: How Britain Discarded its Women Technologists and Lost its Edge in Computing* (MIT, 2017).

16. John Doerr, in conversation with Michael Moritz, National Venture Capital Association Annual Meeting, Santa Clara, Calif., May 2008, quoted in Scott Austin, "Doerr and Moritz Stir VCs in One-on-One Showdown," *The Wall Street Journal*, May 8, 2008, http://www.wsj.com/articles/SB121025688414577219, archived at https://perma.cc/FM7F-CUSS.
17. Wilson, *The New Venturers*, 31–34; Maochun Yu, *OSS in China: Prelude to Cold War* (Annapolis, Md.: Naval Institute Press, 1996).
18. Rock interview, "Early Bay Area Venture Capitalists"; Wilson, *The New Venturers*, 31–40.
19. Bylinsky, *The Innovation Millionaires*.
20. Burton J. McMurtry, interview with the author, January 15, 2015, Palo Alto, Calif.
21. McMurtry, interview with the author, October 2, 2017, by phone.
22. McMurtry, "Evolution of High Technology Entrepreneurship and Venture Capital in Silicon Valley," Presentation to the Houston Philosophical Society, April 21, 2005. Manuscript in possession of the author.
23. "John Wilson," in Carolyn Caddes, *Portraits of Success: Impressions of Silicon Valley Pioneers* (Wellsboro, Penn.: Tioga Publishing, 1986); "Law firm founder John Arnot Wilson dies at 83," *The Almanac* (Menlo Park, Calif.), December 22, 1999, https://www.almanacnews.com/morgue/1999/1999_12_22.oawilson.html, archived at https://perma.cc/UQM3-YTM9; Paul "Pete" McCloskey and Helen McCloskey, interview with the author, February 18, 2016, Rumsey, Calif.
24. Patrick McNulty, "They Shrugged When Pete McCloskey Challenged the President," *The Los Angeles Times*, May 23, 1971, O24; Pete McCloskey, interview with the author.
25. Lawrence R. Sonsini, oral history interview by Sally Smith Hughes, 2011, "Early Bay Area Venture Capitalists: Shaping the Economic and Business Landscape," Regional Oral History Office, The Bancroft Library, University of California, Berkeley, 2011; Mark C. Suchman, "Dealmakers and Counselors: Law Firms as Intermediaries in the Development of Silicon Valley," in Kenney, ed., *Understanding Silicon Valley*, 71–97.
26. Roberta Katz, interview with the author, December 10, 2014, Stanford, Calif.; *Howard D. Hendrickson & Another v. Clark S. Sears*, 365 Mass. 83, 91 (1974); Therese H. Maynard, "Ethics for Business Lawyers Representing Start-Up Companies," *Wake Forest Journal of Business and Intellectual Property Law* 11, no. 3 (2010–11): 401–31. The "no conflict, no interest" remark also has been credited to venture capitalist John Doerr, but no record exists of when and where Doerr may have said this.
27. J. P. Mangalindan, "The Secretive Billionaire who Built Silicon Valley," *Fortune*, July 7, 2014, http://fortune.com/2014/07/07/arrillaga-silicon-valley/, archived at https://perma.cc/B347-42NW.
28. Chop Keenan, interview with the author, March 17, 2016, Palo Alto, Calif.; Tom McEnery, interview with the author, March 9, 2016, San Jose, Calif.; Pete McCloskey, e-mail correspondence with the author, February 3, 2016.
29. On postwar California politics see Jonathan Bell, *California Crucible: The Forging of Modern American Liberalism* (Philadelphia: University of Pennsylvania Press, 2012), and Miriam Pawel, *The Browns of California: The Family Dynasty that Transformed a State and Shaped a Nation* (New York: Bloomsbury, 2018).
30. Mary Soo and Cathryn Carson, "Managing the Research University: Clark Kerr and the University of California," *Minerva* 42, no. 3 (September 2004): 215–36; Margaret O'Mara, "The Uses of the Foreign Student," *Social Science History* 36, no. 4 (Winter 2012): 583–615.

31. Ronald J. Gilson, "The Legal Infrastructure of High Technology Industrial Districts: Silicon Valley, Route 128, and Covenants Not to Compete," *New York University Law Review* 74, no. 3 (June 1999): 575–629. By the early twenty-first century, usage of non-compete clauses spread to sectors far beyond tech and typical "knowledge work," prompting widespread calls for reform; Matt Marx, "Reforming Non-Competes to Support Workers," Policy Proposal 2018-04, The Hamilton Project, Brookings Institution, Washington, D.C., February 2018.

ARRIVALS

1. Lyndon B. Johnson, Remarks at the Signing of the Immigration and Nationality Act of 1965, Liberty Island, New York City, October 3, 1965.
2. "Ervin Challenges Immigration Bill," *The New York Times*, February 26, 1965, 9. Also see Tom Gjelten, "The Immigration Act that Inadvertently Changed America," *The Atlantic*, October 2, 2015, https://www.theatlantic.com/politics/archive/2015/10/immigration-act-1965/408409/, archived at https://perma.cc/Z6YP-KFUD.
3. AnnaLee Saxenian, "Silicon Valley's New Immigrant Entrepreneurs" (Public Policy Institute of California, 1999); Vivek Wadhwa, AnnaLee Saxenian, Ben Rissing, and Gary Gereffi, "America's New Immigrant Entrepreneurs," Master of Engineering Management Program, Duke University; School of Information, University of California, Berkeley, January 4, 2007.
4. Tim Larimer, "It's Still Anglo at the Top: Industry's Rainbow Coalition is Diverse," *The San Jose Mercury News*, October 1, 1989, A1.

CHAPTER 6: BOOM AND BUST

1. "Digital Equipment Offer of $8,250,000 Marketed," *The Wall Street Journal*, August 19, 1966; "Digital Equipment's Joint Offering Sells Out," *The Wall Street Journal*, August 30, 1968.
2. William D. Smith, "Wang Stock Makes Lively Debut," *The New York Times*, August 24, 1967, 51; An Wang, *Lessons: An Autobiography* (New York: Da Capo Press, 1986), 77, 149.
3. Adam Osborne, "From the Fountainhead: Wall Street Embraces Micros," *InfoWorld* 3, no. 3 (February 16, 1981): 16; Leslie Berlin, *The Man Behind the Microchip: Robert Noyce and the Invention of Silicon Valley* (Oxford, U.K.: Oxford University Press, 2005), 125.
4. David Morgenthaler, interviews with the author, January 13, February 12, and May 19, 2015.
5. Jean-Jacques Servan-Schreiber, *Le Défi Américain* (Paris: Éditions Denoël, 1967).
6. Henry R. Lieberman, "Technology Gap Upsets Europe: U.S. Lead Is Putting Strains on Ties of Atlantic Alliance," *The New York Times*, March 12, 1967, 1. Evidence of European anxieties came through in the plethora of publications and conferences about science and policy emerging during these years from European-led supranational organizations like the Organization for Economic Co-operation and Development (OECD); see for example Joseph Ben-David, *Fundamental Research and the Universities: Some Comments on International Differences* (Paris: OECD, 1968); *Problems of Science Policy: Seminar Held at Jouy-en-Josas (France) 19th–25th February 1967* (OECD, 1968).
7. United States Congress, House Committee on Science and Astronautics, Subcommittee on Science, Research, and Development, *Science, Technology, and the Economy: Hearings*, Ninety-second Congress, First Session, July 27, 28, 29, 1971 (1971); William Barry Furlong, "For the Class of '71, the Party's Over; A Report from the University of Chicago Suggests the

Nation's June Graduates Are Facing Some Sobering Facts of Life," *The New York Times*, June 6, 1971, SM35.

8. Herbert G. Lawson, "In a Stunned Seattle, Only Radicals See Good in Rejection of SST," *The Wall Street Journal*, December 7, 1970, 1; Sharon Boswell and Lorraine McConaghy, "Lights Out, Seattle," *The Seattle Times*, November 3, 1996, 1. The brokers behind the famous billboard later averred that it was a prankish sendup of the doom-and-gloom mood in town, as Seattle's downtown real estate market actually was booming due to a growing financial and white-collar services sector. As odd as the joke might have been (nearly no one seemed to get it at the time or afterward), these sectors—along with high-technology firms—soon came to dominate and define Seattle's regional economy. See Erik Lacitis, "Iconic 'Will the Last Person' Billboard Bubbles Up Again," *The Seattle Times*, February 2, 2009, https://www.seattletimes.com/seattle-news/iconic-will-the-last-person-seattle-billboard-bubbles-up-again/, archived at https://perma.cc/3LDM-6PK4.
9. U.S. Department of Commerce, Bureau of Economic Analysis, "Private Nonfarm Employment by Metropolitan Statistical Area: San Jose-Sunnyvale, 1969–2000."
10. Burton J. McMurtry, "Evolution of High Technology Entrepreneurship and Venture Capital in Silicon Valley," Presentation to the Houston Philosophical Society, April 21, 2005. Manuscript in possession of the author.
11. William F. Miller, interview with the author; Miller interviews, Stanford Oral History Program, SU; Committee on Innovations in Computing and Communications, National Academy of Sciences, *Funding a Revolution: Government Support for Computing Research* (Washington, D.C.: National Academies Press, 1999). As the National Academy report notes, corporate investment in computer science research may have been higher overall, but 70 percent of all funds flowing to academic computer science came from the federal government, funding development of software code and graduate education that was foundational to the marquee companies and products of the Valley of the 1990s and beyond.
12. Brad Darrach, "Meet Shaky, the First Electronic Person—The Fearsome Reality of a Machine with a Mind of Its Own," *Life* Magazine, November 20, 1970, 58B–68; John Markoff, *Machines of Loving Grace: The Quest for Common Ground Between Humans and Robots* (New York: HarperCollins, 2015), 7–8, 95–131.

ACT TWO

1. Floyd Kvamme, interview with the author, February 16, 2016, Stanford, Calif.

ARRIVALS

1. Ed Zschau, interview with the author, January 19, 2016, Stanford, Calif.; John Balzar, "A Portrait of Serendipity: Ed Zschau: An Unknown Grabs for the Brass Ring," *The Los Angeles Times*, September 7, 1986, 1.
2. Regis McKenna, interview with the author, May 31, 2016, Menlo Park, Calif.; "CHM Revolutionaries: Regis McKenna in Conversation with John Markoff," video, The Computer History Museum, February 6, 2014; Jaime González-Arintero, "Digital? Every Idiot Can Count to One," *Elektor*, May 27, 2015; Harry McCracken, "Regis McKenna's 1976 Notebook and the Invention of Apple Computer, Inc.," *Fast Company*, April 1, 2016, https://www.fastcompany.com/3058227/regis-mckennas-1976-notebook-and-the-invention-of-apple-computer-inc, archived at https://perma.cc/P4JC-NWU8.

CHAPTER 7: THE OLYMPICS OF CAPITALISM

1. Don C. Hoefler, "Silicon Valley, U.S.A.," *Electronic News*, January 11, 1971, 1.
2. "Don C. Hoefler," *Datamation* 32, no. 5 (May 15, 1986); David Laws, "Who Named Silicon Valley?" CHM, January 7, 2015, http://www.computerhistory.org/atchm/who-named-silicon-valley/, archived at https://perma.cc/EMT2-KUCG.
3. James J. Mitchell, "Curtain to Fall on Valley Era," *The San Jose Mercury News*, October 2, 1988, Silicon Valley Ephemera Collection, MISC 33, FF 2, SU; Regis McKenna, interview transcript, August 22, 1995, Silicon Genesis Project, SU.
4. Jonathan Weber, "Chip Industry's Leaders Begin Bowing Out," *The Los Angeles Times*, February 4, 1991, D1.
5. Gordon Moore, transcript of video history interview by Daniel S. Morrow, March 28, 2000, Santa Clara, Calif., Computerworld Honors Program International Archives, 32. The design breakthrough of the 4004 became a case study familiar to generations of MBAs to come. Faced with the daunting task of building a custom-designed chip for each calculator, designers Ted Hoff and Federico Faggin instead built one chip that could be programmed to adapt to the different functions. See Gary P. Pisano, David J. Collis, and Peter K. Botticelli, *Intel Corporation: 1968–1997*, Harvard Business School Case 797-137, May 1997.
6. Regis McKenna, *The Regis Touch: Million-Dollar Advice from America's Top Marketing Consultant* (Reading, Mass.: Addison-Wesley, 1985), 23–24; McKenna, correspondence with the author, September 6, 2018.
7. Quoted in Gene Bylinsky, *The Innovation Millionaires: How They Succeed* (New York: Charles Scribner's Sons, 1976), 145.
8. Grove quoted in Bylinsky, *The Innovation Millionaires*, 156.
9. Victor K. McElheny, "An 'Industrial Innovation Crisis' Is Decried at MIT Symposium," *The New York Times*, December 10, 1976, 85.
10. Robert Lloyd quoted in Victor K. McElheny, "There's A Revolution in Silicon Valley," *The New York Times*, June 20, 1976, 11; Bylinsky, "California's Great Breeding Ground for Industry," *Fortune*, June 1974, 128–35; Don Hoefler, "He's on Their List," *Microelectronics News*, November 27, 1975, 4, Catalog #102714139, CHM.
11. Steven Brandt, quoted in Bylinsky, "California's Great Breeding Ground for Industry," reprinted in Bylinsky, *The Innovation Millionaires*, 55.
12. David P. Angel, "High-Technology Agglomeration and the Labor Market: The Case of Silicon Valley," in Martin Kenney, ed., *Understanding Silicon Valley: The Anatomy of an Entrepreneurial Region* (Stanford, Calif.: Stanford University Press, 2000), 131; "Salesforce: 100 Best Companies to Work For 2015," *Fortune*, September 21, 2015, http://fortune.com/best-companies/2015/salesforce-com-8/, archived at https://perma.cc/96UG-X9LH.
13. Bylinsky, *The Innovation Millionaires*, 160.
14. Judy Vadasz to Leslie Berlin, *The Man Behind the Microchip: Robert Noyce and the Invention of Silicon Valley* (Oxford, U.K.: Oxford University Press, 2005), 214. I also gained useful perspective on industry work culture during this period from interviews with several former employees of Intel and other firms.
15. Ann Hardy, interview with the author, April 20, 2015, Stanford, Calif.
16. Noyce quoted in Berlin, *The Man Behind the Microchip*, 210.
17. Marty Goldberg and Curt Vendel, *Atari Inc.: Business Is Fun* (Carmel, N.Y.: Syzygy Press, 2012), 101–3.

18. William D. Smith, "Electronic Games Bringing a Different Way to Relax," *The New York Times*, December 25, 1975, 33; "Atari Sells Itself to Survive Success," *BusinessWeek*, November 15, 1976, 120–21; Leonard Herman, "Company Profile: Atari," in Mark J. P. Wolf, ed., *The Video Game Explosion: A History from PONG to Playstation and Beyond* (Westport, Conn.: Greenwood Press, 2008), 53–61. On this early period and its legacy, also see Michael Z. Newman, *Atari Age: The Emergence of Video Games in America* (Cambridge, Mass.: The MIT Press, 2017).
19. Tom McEnery, interview with the author, March 9, 2016, San Jose, Calif.; Glenna Matthews, *Silicon Valley, Women, and the California Dream: Gender, Class, and Opportunity in the Twentieth Century* (Standford, Calif.: Stanford University Press, 2002).
20. Kim-Mai Cutler, "East of Palo Alto's Eden: Race and the Formation of Silicon Valley," *TechCrunch*, January 10, 2015, https://techcrunch.com/2015/01/10/east-of-palo-altos-eden/, archived at https://perma.cc/7EMT-VSRD; Herbert G. Ruffin II, *Uninvited Neighbors: African Americans in Silicon Valley, 1769–1990* (Norman: University of Oklahoma Press, 2014).
21. Joan Didion, "Life at Court," *The New York Review of Books*, December 21, 1989, reprinted in *We Tell Ourselves Stories in Order to Live: Collected Nonfiction* (New York: Alfred A. Knopf, 2006); Margaret Pugh O'Mara, *Cities of Knowledge: Cold War Science and the Search for the Next Silicon Valley* (Princeton, N.J.: Princeton University Press, 2004), 132–39.
22. Bennett Harrison, "Regional Restructuring and 'Good Business Climates': The Economic Transformation of New England Since World War II," in *Sunbelt/Snowbelt: Urban Development and Regional Restructuring*, eds., Larry Sawers and William K. Tabb (Oxford, U.K.: Oxford, 1984), 49; David Lampe, ed., *The Massachusetts Miracle: High Technology and Regional Revitalization* (Cambridge, Mass.: The MIT Press, 1988), 4.
23. Lily Geismer, *Don't Blame Us: Suburban Liberals and the Transformation of the Democratic Party* (Princeton, N.J.: Princeton University Press, 2015), 22–23; AnnaLee Saxenian, *Regional Advantage: Culture and Competition in Silicon Valley and Route 128* (Cambridge, Mass.: Harvard University Press, 1994), 59. Geismer and Saxenian are both key texts for placing the growth and culture of the Boston-area high-tech industry in broader historical context.
24. Michael Widmer, "Basic Change Seen Solution to N.E. Economic Rebirth," *The Lowell Sun*, November 25, 1970.
25. Bank of Boston, "Look Out, Massachusetts!!!," reprinted in Lampe, *The Massachusetts Miracle*.
26. Fox Butterfield, "In Technology, Lowell, Mass., Finds New Life," *The New York Times*, August 10, 1982, 1.
27. Peter Krass, ed., *The Book of Entrepreneurs' Wisdom: Classic Writings by Legendary Entrepreneurs* (New York: John Wiley and Sons, 1999), 156.
28. Saxenian, *Regional Advantage*, 162.
29. Lee Wood, "It's convert or die on '128,'" *The Lowell Sun*, March 14, 1971, C10; Lampe, *The Massachusetts Miracle*, 11.
30. Victor K. McElheny, "High-Technology Jelly Bean Ace," *The New York Times*, June 5, 1977, F7.

CHAPTER 8: POWER TO THE PEOPLE

1. "Lee Felsenstein, 2016 Fellow," Computer History Museum, http://www.computerhistory.org/fellowawards/hall/lee-felsenstein/, archived at https://perma.cc/E26P-TXVV. On the prevalence of various forms of autism among notable scientists and technicians, including Felsenstein, see Steve Silberman, *NeuroTribes: The Legacy of Autism and the Future of*

Neurodiversity (New York: Avery / Penguin Random House, 2015), 223–60. Felsenstein's development of a better megaphone was a subversive act in and of itself, as amplification devices had been banned by university administrators (Jerry Gillam, "Sather Gate and All That," *The Los Angeles Times*, November 2, 1967, B4).

2. Lee Felsenstein, oral history interview by Kip Crosby, edited by Dag Spicer, May 7, 2008, CHM, 3–6.
3. Michael Swaine and Paul Freiberger, "Lee Felsenstein: Populist Engineer," *InfoWorld* 5, no. 45 (November 8, 1983): 105; Felsenstein, oral history interview, 6.
4. "Free Speech Movement: Do Not Fold, Bend, Mutilate, or Spindle," FSM Newsletter, c. 1964, The Sixties Project, Institute of Advanced Technology in the Humanities, University of Virginia, http://www2.iath.virginia.edu/sixties/HTML_docs/Resources/Primary/Manifestos/FSM_fold_bend.html, archived at https://perma.cc/BT7K-3Q7S.
5. Quoted in Swaine and Freiberger, "Lee Felsenstein"; Nan Robertson, "The Student Scene: Angry Militants," *The New York Times*, November 20, 1967, 1. As the focus of student protests shifted over the course of the 1960s, a multiracial civil rights coalition splintered into several movements—a largely white antiwar Left, and multiple racial identity and rights-based movements led by people of color.
6. Daryl E. Lembke, "Police Wield Clubs in Oakland to Quell War Demonstrators," *The Los Angeles Times*, October 18, 1967, 1; Felsenstein, oral history interview 9; John Markoff, *What the Dormouse Said: How the Sixties Counterculture Shaped the Personal Computer Industry* (New York: Penguin, 2005), 268–69.
7. "Alumnae," Helen Temple Cooke Library, Dana Hall School, Wellesley, Mass., http://library.danahall.org/archives/danapedia/alumnae/, archived at https://perma.cc/T69P-XZRS. Liza Loop, "Inside the 'Technical Loop,'" *Dana Bulletin* 58, no. 1 (Summer 1996); Loop, interview with Nick Demonte, July 19, 2013. Both reproduced at LO*OP Center, History of Computing in Learning and Education Virtual Museum (hcle.wikispaces.com), now offline and archived at https://perma.cc/X6RA-T5TN. Dana Hall, founded in the early 1880s as the feeder school for newly established Wellesley College, shared its sister institution's commitment to rigorous education using applied methods; see "The Woman's University," *The New-York Times*, January 4, 1880, 10.
8. B. F. Skinner, "Teaching Machines," *Science* 128, no. 1330 (October 24, 1958): 969–77; Ronald Gross, "Machines that Teach: Their Present Flaws, Their Future Potential," *The New York Times Book Review*, September 14, 1969, 36; Leah N. Gordon, *From Power to Prejudice: The Rise of Racial Individualism in Midcentury America* (Chicago: The University of Chicago Press, 2015). The question of education reform and technology was and is caught up in broader debates about behaviorist (or reinforcement and repetition) versus constructivist (or learning by doing) education; see Peter A. Cooper, "Paradigm Shifts in Designed Instruction: From Behaviorism to Cognitivism to Constructivism," *Educational Technology* 33, no. 5 (May 1993): 12–19. On the longer history of school reform, see Michael B. Katz, *Reconstructing American Education* (Cambridge, Mass.: Harvard University Press, 1989); David Tyack and Larry Cuban, *Tinkering Toward Utopia: A Century of Public School Reform* (Harvard, 1995).
9. Richard Martin, "Shape of the Future," *The Wall Street Journal*, February 13, 1967, 1; Gross, "Machines That Teach."
10. Dean Brown, "Learning Environments for Young Children," *ACM SIGCUE Outlook* 4, no. 4 (August 1970): 2.

11. Kevin Savetz, ANTIC Interview 38 – "Liza Loop, Technical Writer," *ANTIC: The Atari 8-bit Podcast*, April 27, 2015, http://ataripodcast.libsyn.com/antic-interview-38-liza-loop-technical-writer, archived at https://perma.cc/8C93-AZPP; Loop, interview with Nick Demonte, July 19, 2013.
12. Steven Levy, *Hackers: Heroes of the Computer Revolution* (New York: Anchor Press/Doubleday, 1984); Ron Rosenbaum, "Secrets of the Little Blue Box," *Esquire* 76, no. 4 (October 1971): 116. On the relationship between Vietnam-era countercultural politics and the emergence of the personal computer, as well as a much deeper dive into the lives and careers of the people discussed in this chapter, see Fred Turner, *From Counterculture to Cyberculture: Stewart Brand, the Whole Earth Network, and the Rise of Digital Utopianism* (Chicago: The University of Chicago Press, 2006); Markoff, *What the Dormouse Said*; and Michael Hiltzik, *Dealers of Lightning: Xerox PARC and the Dawn of the Computer Age* (New York: HarperBusiness, 1999).
13. Lee Felsenstein, "Resource One/Community Memory—1972–1973," http://www.leefelsenstein.com/?page_id=44 archived at https://perma.cc/4K8U-2BG3; Turner, *From Counterculture to Cyberculture*, 69–102; Claire L. Evans, *Broad Band: The Untold Story of the Women Who Made the Internet* (New York: Portfolio / Penguin, 2018), 95–108.
14. *People's Computer Company* 1, no. 1 (October 1972): 5, digitized online at the DigiBarn Computer Museum, http://www.digibarn.com/collections/newsletters/peoples-computer/peoples-1972-oct/index.html, archived at https://perma.cc/57DQ-L4FW.
15. Theodor H. (Ted) Nelson, *Computer Lib* (independently published, 1974; reprinted by Tempus Books of Microsoft Press, 1987), 30; Andreas Kitzmann, "Pioneer Spirits and the Lure of Technology: Vannevar Bush's Desk, Theodor Nelson's World," *Configurations* 9, no. 3 (September 2001): 452. "Nelson was the Tom Paine and his book was the *Common Sense* of the revolution," Michael Swaine and Paul Freiberger write in their definitive history, *Fire in the Valley: The Birth and Death of the Personal Computer,* 3rd ed. (Raleigh, N.C.: The Pragmatic Bookshelf, 2014), 103. Also see Robert Glenn Howard, "How Counterculture Helped Put the 'Vernacular' in Vernacular Webs," in *Folk Culture in the Digital Age: The Emergent Dynamics of Human Interaction*, ed. Trevor J. Blank (Logan: Utah State University Press, 2012), 25–46. Nelson's mother was the Hollywood actress Celeste Holm.
16. Turner, *From Counterculture to Cyberculture.*
17. Steven Lubar, "'Do Not Fold, Spindle or Mutilate': A Cultural History of the Punch Card," *Journal of American Culture* 15, no. 4 (Winter 1992): 43–55; Charles E. Silberman, "Is Technology Taking Over?," *Fortune*, February 1966, reprinted in *The Myths of Automation,* eds. Silberman and the editors of *Fortune* (New York: Harper & Row, 1966), 97.
18. Vance Packard, *The Naked Society* (New York: David McKay, 1964; repr., New York: Ig Publishing, 2014), 29–30.
19. Jacques Ellul, *The Technological Society,* trans. John Wilkinson (New York: Alfred A. Knopf, 1964).
20. Alvin Toffler, *Future Shock* (New York: Random House, 1970), 186.
21. Henry Raymont, "'Future Shock': The Stress of Great, Rapid Change," *The New York Times,* July 24, 1970, 28; Toffler, *Future Shock*, 155.
22. Sanford J. Ungar, Review of *Future Shock*, *The Washington Post*, August 7, 1970, B8; "Mom," *The Washington Post*, April 12, 1970, N2.
23. Toffler, *Future Shock*, 125.

24. *Congressional Record* 116, part 155, Sept. 8, 1970, 1662.
25. Neil Gallagher, "The Right to Privacy," speech delivered before the Institute of Management Sciences, Chicago Chapter, March 26, 1969, reprinted in *Vital Speeches of the Day* 35 (1969): 528–29; Gallagher, "The Computer as 'Rosemary's Baby,'" *Computers and Society* 1, no. 2 (April 1970): 1–12.
26. Berezin to Gallagher, January 10, 1967, encl. Berezin to Editors of *Datamation*, January 6, 1967, Box 21, FF 16, Cornelius Gallagher Papers, Carl Albert Center Archives, The University of Oklahoma; Baran quoted in John Lear, "Whither Personal Privacy?" *The Saturday Review*, July 23, 1966, 36.
27. *Privacy: The Collection, Use, and Computerization of Personal Data: Joint Hearings before the Ad Hoc Subcommittee on Privacy and Information Systems of the Committee on Government Operations and the Subcommittee on Constitutional Rights of the Committee on the Judiciary*, United States Senate, Ninety-Third Congress, Second Session, June 18, 19, and 20, 1974 (1974), 114–16.
28. Public Law 93-579, Ninety-third Congress, S. 3418, December 31, 1974.
29. Scott R. Schmedel, "Computer Convention Will Skip Esoterica and Focus on Layman," *The Wall Street Journal*, August 21, 1970, 4.
30. Charles Reich, excerpt from *The Greening of America*, *The New Yorker*, September 26, 1970, 42; E. F. Schumacher, *Small Is Beautiful: Economics as if People Mattered* (London: Blond & Briggs, 1973).
31. Philip A. Hart quoted in "The Industrial Reorganization Act: An Antitrust Proposal to Restructure the American Economy," *Columbia Law Review* 73, no. 3 (March 1973): 635; Michael O'Brien, *Philip Hart: The Conscience of the Senate* (East Lansing: Michigan State University Press, 1996). The IBM antitrust suit dragged on until 1982, when it was dropped by the Reagan Administration; for a comprehensive history, see Franklin M. Fisher, John J. McGowan, and Joen E. Greenwood, *Folded, Spindled and Mutilated: Economic Analysis and U.S. vs. IBM* (Cambridge, Mass.: The MIT Press, 1983).
32. Statement of Thomas R. Parkin, Vice President, Software, Control Data Corporation, in Senate Committee on the Judiciary, Subcommittee on Antitrust and Monopoly, *The Industrial Reorganization Act: Hearings*, Ninety-third Congress, S. 1167 (1974), 4868.

CHAPTER 9: THE PERSONAL MACHINE

1. Nancy L. Steffen, "King Calls for Further Action Before Crowd of Over 1800," *The Stanford Daily* 145, no. 43 (April 24, 1964): 1; Jon Roise, "Activists Come Home: Students Working in Strike," *The Stanford Daily* 148, no. 9 (October 6, 1965): 2; Bob Davis, "SNCC's Stokely Carmichael Will Lead Black Power Day," *The Stanford Daily* 150, no. 23 (October 25, 1966): 1; Nick Selby, "Crowd a Problem for Secret Service," *The Stanford Daily* 151, no. 17 (February 21, 1967): 1; "Peace Vigil," *The Stanford Daily* 152, no. 30 (November 2, 1967): 5; Gary Atkins, "Attacked for Politics, Policies; Critics Center on Hoover Boss," *The Stanford Daily* 160, no. 52 (January 7, 1972): 1.
2. Don Kazak, "Stanford University Under Siege," *Palo Alto Times*, April 13, 1994, https://www.paloaltoonline.com/news_features/centennial/1960SD.php archived at https://perma.cc/P6C8-K54R.
3. Douglas C. Engelbart and William K. English, "A Research Center for Augmenting Human Intellect," in American Federation of Information Processing Societies, *Proceedings of the*

1968 Fall Joint Computer Conference, San Francisco, Calif., December 9–11, 1968, 395–410; Jane Howard, "Inhibitions Thrown to the Gentle Winds," *Life* Magazine 65, no. 2 (July 12, 1968): 56; Paul Saffo, interview with the author, March 24, 2017, by phone.

4. Robert E. Kantor and Dean Brown, "On-Line Computer Augmentation of Bio-Feedback Processes," *International Journal of Bio-Medical Computing* 1, no. 4 (November 1970): 265–75; Saffo quoted in Michael Swaine and Paul Freiberger, *Fire in the Valley: The Birth and Death of the Personal Computer,* 3rd ed. (Raleigh, N.C.: The Pragmatic Bookshelf, 2014), 265.
5. "Xerox Plans Laboratory for Research in California," *The New York Times*, March 24, 1970, 89.
6. Lynn Conway, "Reminiscences of the VLSI Revolution," *IEEE Solid-State Circuits Magazine* 4, no. 4 (Fall 2012): 12. Immediately prior to PARC, Conway had worked briefly for Ed Zschau at System Industries.
7. Stewart Brand, "Spacewar: Fanatic Life and Symbolic Death Among the Computer Bums," *Rolling Stone*, December 7, 1972, 33–39; Fred Turner, *From Counterculture to Cyberculture: Stewart Brand, the Whole Earth Network, and the Rise of Digital Utopianism* (Chicago: The University of Chicago Press, 2006), 118.
8. A definitive profile of Taylor at PARC is found in Leslie Berlin, *Troublemakers: Silicon Valley's Coming of Age* (New York: Simon & Schuster, 2017), 89–106.
9. Swaine and Freiberger, *Fire in the Valley*, 102–06.
10. Ivan Illich, *Tools for Conviviality* (New York: Harper & Row, 1973), excerpted in "Ivan Illich: Inverting Politics, Retooling Society," *The American Poetry Review* 2, no. 3 (May /June 1973): 51–53.
11. Lee Felsenstein, "Tom Swift Lives!," *People's Computer Company*, c. 1974, 14–15; Felsenstein, "The Tom Swift Terminal or, A Convivial Cybernetic Device," c. 1975, http://www.leefelsenstein.com/?page_id=82, archived at https://perma.cc/Q3DM-DPFW.
12. "Ivan Illich: Inverting Politics," 52.
13. Lee Felsenstein quoted in Turner, *From Counterculture to Cyberculture*, 114.
14. Centers for Disease Control and Prevention, *Divorces and Divorce Rates: United States* (Washington, D.C.: Government Printing Office, April 1980).
15. Southern California Computer Society, *Interface* 1, no. 1 (September 1975), Box 1, Liza Loop Papers M1141, SU; John Markoff, *What the Dormouse Said: How the Sixties Counterculture Shaped the Personal Computer Industry* (New York: Penguin, 2005), 273–75.
16. Liza Loop, "Inside the 'Technical Loop,'" *Dana Bulletin* 58, no. 1 (Summer 1996); interview with Nick Demonte, July 19, 2013; both archived at https://perma.cc/X6RA-T5TN; Kevin Savetz, ANTIC Interview 38 – "Liza Loop, Technical Writer," *ANTIC: The Atari 8-bit Podcast*, April 27, 2015, http://ataripodcast.libsyn.com/antic-interview-38-liza-loop-technical-writer, archived at https://perma.cc/8C93-AZPP.
17. Sol Libes, "The S-100 Bus: Past, Present, and Future," Part I, *InfoWorld* 2, no. 3 (March 17, 1980): 7.

CHAPTER 10: HOMEBREWED

1. Homebrew Computer Club, *Newsletter*, no. 1 (March 15, 1975), reproduced in Len Shustek, "The Homebrew Computer Club 2013 Reunion," Computer History Museum, December 17, 2013, http://www.computerhistory.org/atchm/the-homebrew-computer-club-2013-reunion/, archived *at* https://perma.cc/RZ9J-M6ZN.

2. On the activism of Fred Moore, see John Markoff, *What the Dormouse Said:: How the Sixties Counterculture Shaped the Personal Computer Industry* (New York: Penguin, 2005), 31–40, 186–96; Lee Felsenstein, oral history interview by Kip Crosby, edited by Dag Spicer, May 7, 2008, CHM, 16.
3. Shustek, "The Homebrew Computer Club 2013 Reunion."
4. Markoff, *What the Dormouse Said*, 272, 274; Felsenstein, oral history interview, 24.
5. Felsenstein, oral history interview, 23.
6. Moore, "Amateur Computer Users Group," Homebrew Computer Club, *Newsletter,* no. 2 (April 12, 1975), Box 1, M1141, Liza Loop Papers, SU.
7. *Byte* 1, no. 1 (September 1975); Michael Swaine and Paul Freiberger, *Fire in the Valley: The Birth and Death of the Personal Computer,* 3rd ed. (Raleigh, N.C.: The Pragmatic Bookshelf, 2014), 184–86.
8. Swaine and Freiberger, *Fire in the Valley,* 194–95; Jim Warren, "We, the People, in the Information Age," January 1, 1991, *Dr. Dobb's Journal,* http://www.drdobbs.com/architecture-and-design/we-the-people-in-the-information-age/184408478, archived at https://perma.cc/KPN4-PSCW.
9. Homebrew Computer Club, *Newsletter,* no. 3 (May 10, 1975), 4, Liza Loop Papers, SU; John Doerr, "Low-cost microcomputing: The personal computer and single-board computer revolutions," *Proceedings of the IEEE* 66, no. 2 (February 1978): 129.
10. *Dr. Dobb's Journal* 2, no. 2 (February 1976), 2; Liza Loop Papers, M1141, FF 2, Box 1, SU; Warren, "We, the People"; Swaine and Freiberger, *Fire in the Valley,* 188–189.
11. Doerr, "Low-cost microcomputing."
12. Albert Yu, interview, September 15, 2005, Atherton, Calif., Silicon Genesis Project, SU.
13. Homebrew Computer Club, *Newsletter,* no. 3 (May 10, 1975), 4, Liza Loop Papers, SU; Doerr, "Low-cost microcomputing."
14. Lou Cannon, "The Puzzling Politics of Jerry Brown," *The Washington Post*, February 5, 1978, B1.
15. Duane Elgin and Arnold Mitchell, "Voluntary Simplicity," *Planning Review* 5, no. 6 (1977), 13–15; Joshua Clark Davis, *From Head Shops to Whole Foods: The Rise and Fall of Activist Entrepreneurs* (New York: Columbia University Press, 2017).

CHAPTER 11: UNFORGETTABLE

1. Sol Libes, "The S-100 Bus: Past, Present, and Future," *InfoWorld* 2, no. 3 (March 17, 1980,): 6.
2. Libes, "The S-100 Bus"; Michael Swaine and Paul Freiberger, *Fire in the Valley: The Birth and Death of the Personal Computer,* 3rd ed. (Raleigh, N.C.: The Pragmatic Bookshelf, 2014), 112–18.
3. Lee Felsenstein, oral history interview by Kip Crosby, edited by Dag Spicer, May 7, 2008, CHM, 24; Libes, "The S-100 Bus"; Swaine and Freiberger, *Fire in the Valley,* 119–21.
4. "Computer Coup," *Time* 119, no. 12 (March 22, 1962), 62; Adam Osborne, *Running Wild: The Next Industrial Revolution* (New York: Osborne/McGraw-Hill, 1979), 33–34; Vector Graphic, Inc., "Now. The Perfect Microcomputer," print advertisement, *Byte,* July 1977.
5. Steve Jobs, Speech, Cupertino, Calif., c. 1980, Computer History Museum, Gift of Regis McKenna.

6. Regis McKenna quoted in Swaine and Freiberger, *Fire in the Valley*, 217; Walter Isaacson, *Steve Jobs* (New York: Simon & Schuster, 2011).
7. Harry McCracken, "Regis McKenna's 1976 Notebook and the Invention of Apple Computer, Inc.," *Fast Company*, April 1, 2016.
8. "Apple Corporate Story," Lisa/Macintosh Positioning Memorandum, c. 1983, Apple Computer Inc. Records, 1977–1997, M1007, Series 7, Box 15, FF 3, SU.
9. McKenna, correspondence with the author, September 6, 2018; "CHM Revolutionaries: Regis McKenna in Conversation with John Markoff," video, The Computer History Museum, February 6, 2014; Memorandum, June 22, 1976, Regis McKenna Inc. Advertising, reproduced in McCracken, "Regis McKenna's 1976 Notebook."
10. McKenna, correspondence with the author, September 6, 2018; Donald T. Valentine, interview by Sally Smith Hughes, in "Early Bay Area Venture Capitalists: Shaping the Economic and Business Landscape," Regional Oral History Office, Bancroft Library, University of California, Berkeley, 2009, 33. On Markkula's pivotal role in launching and growing Apple, see Leslie Berlin, *Troublemakers: Silicon Valley's Coming of Age* (New York: Simon & Schuster, 2017), 146–58, 206–14, 292–307.
11. McKenna, correspondence with the author, September 6, 2018; Mike Cassidy, "Marketing Pioneer Recalls the Early Days of Apple," *The Seattle Times*, September 22, 2008, https://www.seattletimes.com/business/marketing-pioneer-recalls-early-days-of-intel-and-apple/, archived at https://perma.cc/JKN2-RVMA; McCracken, "Regis McKenna's 1976 Notebook."
12. McKenna, interview with the author, December 3, 2014; Meeting notes, in private possession of Regis McKenna and reproduced in McCracken, "Regis McKenna's 1976 Notebook"; "Introducing Apple II," print ad, *Scientific American*, September 1977; Luke Dormehl, "This day in tech history: The first Apple II ships," June 10, 2014, https://www.cultofmac.com/282972/day-tech-history-first-apple-ii-ships/, archived at https://perma.cc/W9K8-BE3J.
13. Harry McCracken, "Apple II Forever: A 35th-Anniversary Tribute to Apple's First Iconic Product," *Time*, April 16, 2012, http://techland.time.com/2012/04/16/apple-ii-forever-a-35th-anniversary-tribute-to-apples-first-iconic-product/, archived at https://perma.cc/CG5T-987V.
14. Jim C. Warren, *The First West Coast Computer Faire: Proceedings*, November 18, 1977, Silicon Valley Ephemera Collection, Series 1, Box 7, FF 2, SU.
15. Ted Nelson, "Those Unforgettable Next Two Years," in Warren, *The First West Coast Computer Faire: Proceedings*, 20–21.
16. Louise Cook, "Get Ready for Friendly Home Computers," *The Washington Post*, November 27, 1977, 166.
17. Quoted in Swaine and Freiberger, *Fire in the Valley*, 238.
18. Lee Dembart, "Computer Show's Message: 'Be the First on Your Block,'" *The New York Times*, August 26, 1977, A10.
19. Victor K. McElheny, "Computer Show: Preview of More Ingenious Models," *The New York Times*, June 16, 1977, D1.
20. Martin Campbell-Kelly, William Aspray, Nathan Ensmenger, and Jeffrey R. Yost, *Computer: A History of the Information Machine* (3rd ed.; Boulder, Colo.: Westview Press, 2013), 239.
21. Bill Gates, "An Open Letter to Hobbyists," Homebrew Computer Club, *Newsletter* 2, no. 1 (January 31, 1976), 2, Box 1, M1141, Liza Loop Papers, SU.
22. Stephen Manes and Paul Andrews, *Gates* (New York: Touchstone/Simon & Schuster, 1993), 58–60.

23. Bill Gates, *The Road Ahead* (New York: Random House, 1995), 44; Manes and Andrews, *Gates*, 63–71.
24. Manes and Andrews, *Gates*, 81.
25. Eric S. Raymond, *The Cathedral and the Bazaar: Musings on Linux and Open Source by an Accidental Revolutionary* (San Francisco: O'Reilly Media, 2001).
26. Christopher Evans, *The Micro Millennium* (New York: Viking, 1979), 67.

CHAPTER 12: RISKY BUSINESS

1. Ian Matthews, "Commodore PET History," Commodore.ca, February 2003, https://www.commodore.ca/commodore-products/commodore-pet-the-worlds-first-personal-computer/, archived at https://perma.cc/WY6J-UXT5.
2. "Has the Bear Market Killed Venture Capital?" *Forbes*, June 15, 1970, 28–37; Margaret A. Kilgore, "Public Urged to Invest in Technology," *The Los Angeles Times*, April 13, 1976, D7; Gene Bylinsky, *The Innovation Millionaires: How They Succeed* (New York: Charles Scribner's Sons, 1976), 25–46; Gary Klott, "Venture Capitalists Wary of Tax Plan," *The New York Times*, January 9, 1985, D1.
3. David Morgenthaler, interview with the author, November 3, 2015, by phone.
4. Burt McMurtry, interview with the author, January 15, 2015, Palo Alto, Calif.
5. Stewart Greenfield, interview with the author, May 19, 2015, by phone.
6. Reid Dennis, "Early Bay Area Venture Capitalists: Shaping the Economic and Business Landscape," interviews conducted by Sally Smith Hughes, Regional Oral History Office, The Bancroft Library, University of California, 2009, 43.
7. Ajay K. Mehrotra and Julia C. Ott, "The Curious Beginnings of the Capital Gains Tax Preference," *Fordham Law Review* 84, no. 6 (May 2016), 2517–36; "Capital Gains and Taxes Paid on Capital Gains, 1954–2009," Department of the Treasury, Office of Tax Analysis, 2012, https://www.taxpolicycenter.org/statistics/historical-capital-gains-and-taxes, archived at https://perma.cc/BTM4-57DV. This was part of a broader business push against the New Deal that included the formulation of the rhetoric of "free enterprise"; see Lawrence Glickman, "Free Enterprise versus the New Deal Order," paper presented at the "Beyond the New Deal Order" conference, Center for the Study of Work, Labor, and Democracy, University of California, Santa Barbara, September 24–26, 2015. Also see Kathryn S. Olmsted, *Right Out of California: The 1930s and the Big Business Roots of Modern Conservatism* (New York: The New Press, 2015); Kim Phillips-Fein, *Invisible Hands: The Businessmen's Crusade against the New Deal* (New York: W. W. Norton, 2010); and Julia C. Ott, *When Wall Street Met Main Street: The Quest for an Investors' Democracy* (Cambridge, Mass.: Harvard University Press, 2011).
8. "Curb Urged on Loans: Speculation Hit by Bankers," *The Los Angeles Times*, October 5, 1928, 2; "Capital Gains Tax," *The Wall Street Journal*, November 8, 1930, 1; "Whitney Attacks 'Excessive' Relief," *The New York Times*, February 27, 1935, 29. On Whitney, also see Mehrotra and Ott, "The Curious Beginnings."
9. "Tax Debate: Builders, Stock Brokers are Split," *The Wall Street Journal*, January 30, 1963, 1.
10. "Excerpts from Senator McGovern's Address Explaining His Economic Program," *The New York Times*, August 30, 1972, 22; James Reston, "The New Economic Philosophy," *The New York Times*, January 31, 1973, 41.
11. Quoted in Bylinsky, *The Innovation Millionaires*.

12. "Pension Fund Trustees Get Jitters Over Liability Laws," *The Los Angeles Times*, August 5, 1976, F15. On inflation during the decade, see Alan S. Blinder, "The Anatomy of Double-Digit Inflation in the 1970s," in *Inflation: Causes and Effects*, ed. Robert E. Hall (Chicago: University of Chicago Press, 1982), 261-82. On ERISA, see Christopher Howard, *The Hidden Welfare State: Tax Expenditures and Social Policy in the United States* (Princeton, N.J.: Princeton University Press, 1997), 130–34.
13. Pete Bancroft, "Reflections of an Early Venture Capitalist," March 28, 2000, unpublished manuscript in the author's possession.
14. David Morgenthaler, interview with the author, June 23, 2015, Palo Alto Calif; Pete Bancroft, interview with the author, November 3, 2015, San Francisco, Calif.
15. Jefferson Cowie, *Stayin' Alive: The 1970s and the Last Days of the Working Class* (New York: The New Press, 2010); Michael Reagan, "Capital City: New York in Fiscal Crisis, 1966–1978," PhD dissertation, University of Washington, 2017.
16. U.S. Department of Commerce, *The Role of Technical Enterprises in the United States Economy* (Washington, D.C.: U.S. Government Printing Office, January 1976); Robert Wolcott Johnson, "The Passage of the Investment Incentive Act of 1978: A Case Study of Business Influencing Public Policy," PhD dissertation, Harvard University Graduate School of Business Administration, 1980, 40–42.
17. Pete Bancroft and the National Venture Capital Association, *Emerging Innovative Companies—An Endangered Species*, November 29, 1976, unpublished manuscript in the author's possession, 1, 3.
18. Benjamin C. Waterhouse, *Lobbying America: The Politics of Business from Nixon to NAFTA* (Princeton, N.J.: Princeton University Press, 2013).
19. "Electronic firms seek broader political base," *The Los Angeles Times*, November 15, 1981, quoted in AnnaLee Saxenian, "In Search of Power: The Organization of Business Interests in Silicon Valley and Route 128," *Economy and Society* 18, no. 1 (February 1989): 40.
20. Patrick McNulty, "They Shrugged When Pete McCloskey Challenged the President," *The Los Angeles Times*, May 23, 1971, O24; Pete McCloskey, interview with the author, February 18, 2016, Rumsey, Calif.
21. McCloskey interview; Reid Dennis, interview with the author, May 26, 2015.
22. McCloskey interview.
23. Ed Zschau, interviews with the author, June 24, 2015, and January 19, 2016, Palo Alto, Calif.
24. Burt McMurtry, interview with the author, October 2, 2017, by phone; Kathie and Bob Maxfield, interview with the author, May 28, 2015, Los Gatos, Calif.
25. Greenfield interview; David Morgenthaler and Reid Dennis, interview with the author, May 26, 2015, Palo Alto, Calif.
26. Jimmy Carter, "Tax Reduction and Reform Message to the Congress," January 20, 1978, posted by Gerhard Peters and John T. Woolley, *The American Presidency Project*, http://www.presidency.ucsb.edu/ws/?pid=31055, archived at https://perma.cc/RXG7-F57W.
27. Saxenian, "In Search of Power."
28. Zschau interview, June 24, 2015; *Capital gains tax bills: Hearings before the Subcommittee on Taxation and Debt Management Generally of the Committee on Finance*, United States Senate, Ninety-fifth Congress, second session, June 28 and 29, 1978, 269.
29. McCloskey interview.
30. William A. Steiger, testimony, *Capital gains tax bills: Hearings.*

31. Barry Sussman, "Surprise: Public Backs Carter on Taxes: Roper Survey Shows Fairness Rated Above Tax Cut," *The Washington Post*, August 6, 1978, D5.
32. Clayton Fritchey, "Today's 'Forgotten Man': The Investor," *The Washington Post*, August 5, 1978, A15. Two years later, Massachusetts would follow suit, passing Proposition 2½—a measure that got a hefty financial push from the newly formed Massachusetts High Technology Council; see Saxenian, "In Search of Power."
33. Art Pine, "A Tax Break for the Rich in an Election Year?," *The Washington Post*, May 21, 1978, A16; "Rich, Poor, and Taxes," *The Washington Post*, June 2, 1978, A2; David Morgenthaler, testimony, H.R. 9549, *The Capital, investment, and business opportunity act: Hearing before the Subcommittee on Capital, Investment, and Business Opportunities of the Committee on Small Business*, House of Representatives, Ninety-fifth Congress, second session, February 22, 1978.
34. Art Pine, "Capital Gains Remarks by Carter Draw Hill Fire," *The Washington Post*, June 29, 1978, D12. For a comprehensive account of the bill's passage, see Johnson, "The Passage of the Investment Incentive Act of 1978."
35. Greenfield interview.
36. James M. Poterba, "Venture Capital and Capital Gains Taxation," in *Tax Policy and the Economy*, vol. 3, ed. Lawrence H. Summers (Cambridge, Mass.: The MIT Press, 1989), 47–67.
37. Ed Zschau, correspondence with the author, September 13, 2018.
38. *Memorial Services held in the House of Representatives and Senate of the United States, together with remarks presented in eulogy of William A. Steiger* (Washington, D.C.: USGPO, 1979).

ACT THREE

1. *The Man Who Shot Liberty Valance*, directed by John Ford, written by James Warner Bellah and Willis Goldbeck, based on the story by Dorothy M. Johnson (Paramount Pictures, 1962).

ARRIVALS

1. Trish Millines Dziko, interview with the author, April 2, 2018, by phone; correspondence with the author, September 7, 2018; Trish Millines Dziko, interview by Jessah Foulk, August 8, 2002, "Speaking of Seattle" oral history collection, Sophie Frye Bass Library, Museum of History and Industry, Seattle, Wash.
2. "Benjamin M. Rosen," *The Rosen Electronics Letter* 80, no. 10 (July 7, 1980), Catalog No. 102661121, Computer History Museum Archives, Mountain View, Calif.; John W. Wilson, *The New Venturers: Inside the High-Stakes World of Venture Capital* (Reading, Mass.: Addison-Wesley, 1985), 109–11.
3. Charles J. Elia, "Caution Increases on Semiconductor Issues Amid Signs of Slower Recovery by Industry," *The Wall Street Journal*, November 4, 1975, 43; Regis McKenna, interview with the author, December 3, 2014, Stanford, Calif.; Merrill Lynch, August 1978, report in the possession of Regis McKenna.
4. "From Little Apples Do Giant Orchards Grow," *The Rosen Electronics Letter* 80, no. 21 (December 31, 1980), 10, Catalog No. 102661121, Computer History Museum Archives, Mountain View, Calif.; Rosen, "Memories of Steve," *Through Rosen-Colored Glasses* blog, October 22, 2011, http://www.benrosen.com/2011/10/memories-of-steve.html, archived at https://perma.cc/9B2M-5462.

CHAPTER 13: STORYTELLERS

1. Display ad (Apple Computer Inc.), *Wall Street Journal*, August 13, 1980, 28.
2. Philip Shenon, "Investment Climate is Ripe for Offering by Apple Computer," *The Wall Street Journal*, August 20, 1980, 24.
3. Ben Rosen, "The Stock Market Looks Ahead—to the Golden Age of Electronics," *The Rosen Electronics Letter* 80, no. 15 (August 22, 1980), 1.
4. "High Technology: Wave of the future or market flash in the pan?" *BusinessWeek*, November 10, 1980, 86–97; Moore quoted in Wilson, *The New Venturers*, 189.
5. Carl E. Whitney, "Wall Street Discovers Microcomputers," *InfoWorld* 2, no. 18 (October 13, 1980), 4–5; James L. Rowe, Jr., "Speculation Fever Seeping Through Wall Street," *The Washington Post*, November 2, 1980, G1; Karen W. Arenson, "A 'Hot' Offering Retrospective," *The New York Times*, December 30, 1980, D1.
6. Ben Rosen, "Spectacular Year for Electronics Stocks," *The Rosen Electronics Letter* 80, no. 21 (December 31, 1980), 1. Also see Sally Smith Hughes, *Genentech: The Beginnings of Biotech* (Chicago: The University of Chicago Press, 2011).
7. Robert A. Swanson, oral history interviews by Sally Smith Hughes, 1996 and 1997, Regional Oral History Office, The Bancroft Library, University of California, Berkeley, 2001.
8. David Ahl, Interview with Gordon Bell, *Creative Computing* 6, no. 4 (April 1980), 88–89, via Garson O'Toole, "There is No Reason for Any Individual to Have a Computer in Their Home," Quote Investigator, https://quoteinvestigator.com/2017/09/14/home-computer/#return-note-16883-1 archived at https://perma.cc/5M5R-HQMA. At the time, the minicomputer market was $2.5 billion, and Digital controlled 40 percent of it. See Stanley Klein, "The Maxigrowth of Minicomputers," *The New York Times*, October 2, 1977, 3. Arthur Rock, interviews by Sally Smith Hughes, 2008 and 2009 "Early Silicon Valley Venture Capitalists," Regional Oral History Office, The Bancroft Library, University of California, Berkeley, California, 56.
9. David Morgenthaler, interviews with the author; David Morgenthaler, oral history interview, 41; Brent Larkin, "Cleveland's Quiet Business Visionary," *The Cleveland Plain Dealer*, January 15, 2012, G1.
10. William Bates, "Home Computers—So Near and Yet . . . ," *The New York Times*, February 26, 1978, F3; Wayne Green, "80 Remarks," column, *80 Microcomputing*, January 1980, quoted in Matthew Reed, "Was the TRS-80 affectionately known as the Trash-80?" TRS-80.org, undated, http://www.trs-80.org/trash-80/, archived at https://perma.cc/3J2G-7J9X.
11. Apple, "Personal Computer Market Fact Book" [c. 1983], 143, M1007, Series 7, Box 15, FF 1, Apple Computer Inc. Records, 1977–1997, SU; Regis McKenna, *The Regis Touch: Million-Dollar Advice from America's Top Marketing Consultant* (Reading, Mass.: Addison-Wesley, 1985), 28.
12. Regis McKenna, interview with the author, December 3, 2014.
13. Tom Hannaher, "Selling Apple Personal Computing with Advertising," Apple Computer Inc., 1983; "Personal Computer Market Fact Book," 160–62.
14. "Personal Computer Market Fact Book," 141, 146. As internal marketing documents like this one reveal, Apple marketed exclusively to men until its expansion into collegiate markets around the time of the introduction of the Macintosh (1984). On *Playboy* placement, see Michael Swaine and Paul Freiberger, *Fire in the Valley: The Birth and Death of the Personal Computer*, 3rd ed. (Raleigh, N.C.: The Pragmatic Bookshelf, 2014), 253.
15. McKenna, interview with the author, December 3, 2014.

16. Steve Jobs presentation, ca. 1980, gift of Regis McKenna, Catalog number 102746386, Lot number, X2903.2005, CHM. Also see Kay Mills, "The Third Wave: Whiz-Kids Make a Revolution in Computers," *The Los Angeles Times*, July 5, 1981, E3.
17. Mills, "The Third Wave."
18. Esther Dyson, "My iXperiences with Steve Jobs," August 26, 2011, *Reuters MediaFile*, http://blogs.re uters.com/mediafile/2011/08/26/my-ixperiences-with-steve-jobs/, archived at https://perma.cc/C9T3-L7WG.
19. "Osborne: From Brags to Riches," *BusinessWeek*, February 22, 1982, 82.
20. Schenker, "A Different Scenario: Personal Computers in the 80's," *InfoWorld* 2, no. 6 (April 14, 1980): 11, Box 1, Liza Loop Papers, M1141, SU; Peter J. Schuyten, "Subculture of Silicon Technology," *The New York Times*, May 10, 1979, D2.
21. McKenna, *The Regis Touch*, xi.
22. Alvin Toffler, *Future Shock* (New York: Random House, 1970), 29; Jobs quoted in Mills, "The Third Wave."
23. Christian Williams, "Future Shock Revisited: Alvin Toffler's 'Wave,'" *The Washington Post*, March 31, 1980, B1.
24. "Tandy Radio Shack Assaults the Small Computer Market," *The Rosen Electronics Letter* 80, no. 14 (August 8, 1980), Catalog No. 102661121, CHM; McKenna, *The Regis Touch*, 62.
25. Ben Rosen, "Memories of Steve," *Through Rosen-Colored Glasses* blog, October 22, 2011, http://www.benrosen.com/2011/10/memories-of-steve.html, archived at https://perma.cc/9B2M-5462.
26. C. Saltzman, "Apple for Ben Rosen: Use of Personal Computers by Securities Analysts," *Forbes* 124 (August 20, 1979), 54–55; Stratford P. Sherman, "Technology's Most Colorful Investor," *Fortune*, September 30, 1985, 156.
27. Adam Osborne, speech at West Coast Computer Faire, March 15, 1980, audio recording, Dan Bricklin's Web Site: www.bricklin.com, archived at https://perma.cc/EWM5-JVF7.
28. *The Rosen Electronics Letter*, August 8, 1980, 14, Catalog No. 102661121, CHM.
29. McKenna, interview with the author, December 3, 2014; Shenon, "Investment Climate is Ripe for Offering by Apple Computer"; Whitney, "Wall Street Discovers Microcomputers," 4–5.
30. Arthur Rock interview, "Early Bay Area Venture Capitalists"; "Making a Mint Overnight," *Time*, January 23, 1984, 44.
31. William M. Bulkeley, "In Venture Capitalism, Few Are As Successful as Benjamin Rosen," *The Wall Street Journal*, November 28, 1984, 1.
32. Frederick Golden, "Other Maestros of the Micro," *Time*, January 3, 1983; "The $1795 Personal Business Computer is changing the way people go to work," Osborne Computer Corp Ad, *Byte* 7, no. 9 (Sept 1982), 31.
33. *Inc.*, October 1981.

CHAPTER 14: **CALIFORNIA DREAMING**

1. Ronald Reagan, announcement of presidential candidacy, November 13, 1979, https://www.reaganlibrary.gov/11-13-79, archived at https://perma.cc/E7CL-GL47.
2. Quoted in Haynes Johnson, "The Perils of Paradise," *The Washington Post*, October 19, 1980, G1.

3. David Ignatius, "Political Evolution: Sen. Hart Seeks to Blur Left-Right Stereotypes in His Reelection Bid," *The Wall Street Journal*, August 20, 1980, 1.
4. Paul Tsongas, Testimony before the Senate Small Business Committee on the Elimination of the Capital Gains Differential, June 2, 1986, Ed Zschau Papers, Box 51, FF "Capital Gains II," Hoover Institution Archives, Stanford, Calif. (HH).
5. Bob Davis, "Future Gazers in the U.S. Congress," *The Wall Street Journal*, June 7, 2000, 3; David Shribman, "Now and Then, Congress Also Ponders the Future," *The New York Times*, March 14, 1982, E10.
6. Katie Zezima, "Ex-Gov. Edward J. King, 81, Who Defeated Dukakis, Dies," *The New York Times*, September 19, 2006, B8.
7. Elizabeth Drew, "The Democrats," *The New Yorker*, March 22, 1982, 130; William D. Marbach, Christopher Ma, et al., "High Hopes for High Tech," *Newsweek*, February 14, 1983, 61.
8. Editorial, "Jerry Brown on the 'Reindustrialization of America,'" *The Washington Post*, January, 14, 1980, A23; George Skelton, "Gaining Attention by Snubbing Tradition," *The Los Angeles Times*, October 17, 1978, A1; Skelton, "Waiting in Wings for 1980," *The Los Angeles Times*, November 8, 1978, B1.
9. Doug Moe, "35 Years On, Recalling 'Apocalypse Brown,'" *Wisconsin State Journal*, March 27, 2015, https://madison.com/wsj/news/local/columnists/doug-moe/doug-moe-years-on-recalling-apocalypse-brown/article_1b614603-1d07-51b7-a984-9b793fecf730.html, archived at https://perma.cc/C9KS-2594; Wayne King, "Gov. Brown, His Dream Ended, Returns to California," *The New York Times*, April 3, 1980, 34; Raymond Fielding, *The Technique of Special Effects Cinematography*, 4th ed. (Burlington, Mass.: Focal Press, 2013), 387–88.
10. Johnson, "The Perils of Paradise."
11. Rowland Evans and Robert Novak, "David Packard Gets on Board," *The Washington Post*, May 11, 1975, 39; Margot Hornblower, "Gold-Plated Panel Set to Raise, Spend Millions for Reagan," *The Washington Post*, July 10, 1980, A3; Debra Whitefield, "Business Leaders Jubilant; Wall Street Has Busiest Day," *The Wall Street Journal*, November 6, 1980, B1; Tom Redburn and Robert Magnuson, "Stung by Tax Bill, Electronics Firms Seek Broader Political Base," *The Los Angeles Times*, November 15, 1981, F1.
12. Tom Zito, "Steve Jobs: 1984 *Access* Magazine Interview," *Newsweek Access*, Fall 1984, reprinted at *The Daily Beast*, October 6, 2011, https://www.thedailybeast.com/steve-jobs-1984-access-magazine-interview, archived at http://perma.cc/A3W8-T4Q9; "InfoViews," *InfoWorld*, November 10, 1980, 12.
13. Whitefield, "Business Leaders Jubilant"; Ken Gepfert, "Defense Contractors Hail Reagan Win but Can They All Share in the Spoils?," *The Los Angeles Times*, November 30, 1980, F1; Nicholas Lemann, "New Tycoons Reshape Politics," *The New York Times*, June 8, 1986, Section M, 51. The Democrats' loss of the Senate came in the wake of a $700,000 ad blitz by National Conservative PAC (NCPAC), brainchild of former Nixon operative and lobbyist Roger Stone, whose later presidential campaigns included both George Bushes, Bob Dole, and Donald Trump. Warren Weaver Jr., "Conservatives Plan $700,000 Drive to Oust 5 Democrats From Senate," *The New York Times*, August 17, 1979, 1; "Attack PAC," *Time* 120, no. 17 (October 25, 1982), 28.
14. Both Tsongas and Hart quoted in Lawrence Martin, "Shift to Right in U.S. Begins to Hit Home," *The Globe and Mail*, November 8, 1980, 1. "Carter Told Major Threats Are Democrats," United Press International, May 4, 1977, wire service story.

15. Sidney Blumenthal, "Whose Side is Business On, Anyway?," *The New York Times*, October 25, 1981, 29.
16. Reagan, Proclamation 4829—Small Business Week, 1981, March 23, 1981; Reagan, "Remarks to the Students and Faculty at St. John's University," New York, March 28, 1985; Arthur Levitt Jr., "In Praise of Small Business," *The New York Times*, December 6, 1981, 136; Leslie Wayne, "The New Face of Business Leadership," *The New York Times*, May 22, 1983, B1; Don Oldenberg, "Entrepreneurs: The New Heroes?," *The Washington Post*, July 2, 1986, D5.
17. Levitt, "In Praise of Small Business"; William M. Bulkeley, "In Venture Capitalism, Few Are as Successful as Benjamin Rosen," *The Wall Street Journal*, November 28, 1984.
18. Ken Hagerty, "The Power of Grassroots Lobbying," *Association Management*, November 1979, collection of Ken Hagerty, in possession of the author; Bacon, "Lobbyists Say Options Tax Break is Needed to Spur Innovation," *The Wall Street Journal*, July 1, 1981, 27; Ken Hagerty, interview with the author, September 9, 2015, by phone.
19. Edward Cowan, "The Quiet Campaign to Cut Capital Gains Taxes," *The New York Times*, April 12, 1981, F8.
20. Otto Friedrich et al., "Machine of the Year: The Computer Moves In," *Time*, January 3, 1983; Jeanne Hayes, ed., *Microcomputer and VCR Usage in Schools, 1985–1986* (Denver, Colo.: Quality Education Data, 1986), 7.
21. Adam Smith, "Silicon Valley Spirit," *Esquire* 96, no. 11 (November 1981): 13–14; Reyner Banham, "Down in the Vale of Chips," *New Society* 56, no. 971 (June 25, 1981): 532.
22. Moira Johnston, "High Tech, High Risk, and High Life in Silicon Valley," *National Geographic* 162, no. 4 (October 1982): 459–77.
23. Quoted in Michael Moritz, *Return to the Little Kingdom: How Apple and Steve Jobs Changed the World* (New York: The Overlook Press, 2009), 142.
24. Smith, "Silicon Valley Spirit."
25. Mike Hogan, "Corporate Cultures Tell a Lot," *California Business*, November 1984, 92–96.
26. Margaret Comstock Tommervik and Craig Stinson, "Women at Work with Apples," *Softalk* 1, no. 7 (March 1981): 44–50; Jennifer Jones, interview with the author, November 14, 2014, Woodside, Calif.
27. Margaret Comstock Tommervik, "Exec Apple: Jean Richardson," *Softalk* 1, no. 7 (March 1981): 42–43.
28. C. W. Miranker, "What Makes Silicon Valley's Workforce Mostly Non-Union," Associated Press, December 24, 1983, Saturday AM cycle, retrieved from Nexis Uni.
29. "What Makes Tandem Run," *BusinessWeek*, July 14, 1980, 73–74; Smith, "Silicon Valley Spirit."
30. Fox Butterfield, "Two Areas Show Way to Success in High Technology," *The New York Times*, August 9, 1982, 1. Also see David Lampe, ed., *The Massachusetts Miracle: High Technology and Regional Revitalization* (Cambridge, Mass.: The MIT Press, 1988).
31. *Newsweek*, July 4, 1979; Carter, Speech to the Nation, July 15, 1979; McKenna, *The Regis Touch*, 28; Anthony J. Parisi, "Technology: Elixir for U.S. Industry," *The New York Times*, September 28, 1980, F1.
32. Moritz, *Return to the Little Kingdom*, 11; *Time* magazine, "Publisher's Letter," January 3, 1983.

CHAPTER 15: MADE IN JAPAN

1. Harry McCracken, "The Original Walkman vs. the iPod Touch," *Technologizer*, June 29, 2009, https://www.technologizer.com/2009/06/29/walkman-vs-ipod-touch/, archived at https://perma.cc/P92F-3WTL; "Ubiquitous Walkman Celebrates First Decade," *The Los Angeles Times*, June 21, 1989, C2.
2. Peter J. Brennan, "Advanced Technology Center: Santa Clara Valley, California," MO 443 Silicon Valley Ephemera Collection, Series 1, Box 1, FF 17, SU; Ben Rosen, "The Stock Market Looks Ahead—to the Golden Age of Electronics," *The Rosen Electronics Letter* 80, no. 15 (August 22, 1980); Maggie Canon, "Stanford and Japan Form Joint Industry Study," *InfoWorld*, November 24, 1980, 3.
3. Canon, "Stanford and Japan." For discussion of the Silicon Valley semiconductor industry's response to Japanese competition, and Noyce's leadership, see Leslie Berlin, *The Man Behind the Microchip: Robert Noyce and the Invention of Silicon Valley* (Oxford, U.K.: Oxford University Press, 2005), 257–80.
4. Marco Casale-Rossi, "The Heritage of Mead & Conway," *Proceedings of the IEEE* 102, no. 2 (February 2014): 114–19; Clair Brown and Greg Linden, "Offshoring in the Semiconductor Industry: Historical Perspectives," IRLE Working Paper No. 120-05, University of California, Berkeley, 2005.
5. Brennan, "Advanced Technology Center"; William Chapman, "High Stakes Race: Japanese Search for Breakthrough in Field of Giant Computers," *The Washington Post*, February 27, 1978.
6. Chalmers Johnson, *MITI and the Japanese Miracle: The Growth of Industrial Policy, 1925–1975* (Stanford, Calif.: Stanford University Press, 1982); Judith Stein, *Pivotal Decade: How the United States Traded Factories for Finance in the Seventies* (New Haven, Conn.: Yale University Press, 2010).
7. Thomas L. Friedman, "Silicon Valley's 'Underworld,'" *The New York Times*, December 3, 1981, B1; "Valley of Thefts," *Time*, December 14, 1981, 66; D. T. Friendly and Paul Abramson, "In Silicon Valley, Goodbye, Mr. Chips," *Newsweek*, May 12, 1980, 78.
8. Regis McKenna, interviews with the author, December 3, 2014, and May 31, 2016.
9. Hearings on H.R. 5805 "Chrysler Corporation Loan Guarantee Act of 1979," Subcommittee on Economic Stabilization, Committee on Banking, Finance, and Urban Affairs, House of Representatives, Ninety-sixth Congress, First Session, October 19, 1979; Charles K. Hyde, *Riding the Roller Coaster: A History of the Chrysler Corporation* (Detroit: Wayne State University Press, 2003); Stein, *Pivotal Decade*.
10. Johnson, "The Perils of Paradise"; Stone quoted in Susan Brown-Goebeler, "How Gray Is My Valley," *Time* 138, no. 20 (November 18, 1991): 90.
11. James Flanigan, "U.S., Japan Vie for Lead in Electronics," *Los Angeles Times*, October 12, 1980, 1; U.S. Department of Commerce, Industry and Trade Administration, *A Report on the U.S. Semiconductor Industry*, September 1979.
12. Hobart Rowen, "Entire Data Processing Industry Target of Japanese Companies," *The Washington Post*, March 23, 1980, E1.
13. McKenna, interview with the author, December 3, 2014.
14. Tom Redburn and Robert Magnuson, "Stung by Tax Bill, Electronics Firms Seek Broader Political Base," *The Los Angeles Times*, November 15, 1981, F1–4; "AeA Supports Two Bills Asking Tax Aid for R&D," *Computerworld*, June 1, 1981, 67; Ken Hagerty, interview with the author, September 9, 2015, by phone; Redburn and Magnuson, "Stung by Tax Bill."

15. David Harris, "Whatever Happened to Jerry Brown?," *The New York Times*, March 9, 1980, SM9. On Pat Brown's defeat and its implications, see Matthew Dallek, *The Right Moment: Ronald Reagan's First Victory and the Decisive Turning Point in American Politics* (New York: The Free Press, 2000).
16. McKenna, interview with the author, December 3, 2014.
17. Edmund G. Brown Jr., State of the State Address, January 8, 1981; "Governor Brown Boosts Microelectronics," *Science* 211, no. 4483 (February 13, 1981): 688–89.
18. William D. Marbach, "High Hopes for High Tech," *Newsweek*, February 14, 1983, 61.
19. California Commission on Industrial Innovation, "Winning Technologies: A New Industrial Strategy for California and the Nation," September 2, 1982, Silicon Valley Ephemera Collection, Series 1, Box 4, FF 21, SU.
20. Ronald Reagan, "Executive Order 12428—President's Commission on Industrial Competitiveness," June 28, 1983.
21. Ben Rosen, "Jerry Sanders' Humor," *The Rosen Electronics Letter* 82, no. 12 (August 25, 1982): 14–15.
22. House Democratic Caucus, *Rebuilding the Road to Opportunity: Turning Point for America's Economy* (Washington: USGPO, 1982).
23. "Steve Jobs and David Burnham," *Nightline*, ABC News, April 10, 1981, archived at https://perma.cc/4UER-Y3YV.
24. David Morrow, oral History interview with Steve Jobs, Palo Alto, Calif., April 20, 1995, Smithsonian Institution.
25. Quoted in Audrey Watters, "How Steve Jobs Brought the Apple II to the Classroom," *Hack Education.com*, February 25, 2015, http://hackeducation.com/2015/02/25/kids-cant-wait-apple, archived at https://perma.cc/3K62-ACW5.
26. Milton B. Stewart, "Polishing the Apple," *Inc.*, Feb. 1, 1983, https://www.inc.com/magazine/19830201/6207.html, archived at https://perma.cc/K7UQ-4ACC.
27. National Commission on Excellence in Education, *A Nation at Risk: The Imperative for Educational Reform* (April 1983).
28. Richard Severo, "Computer Makers Find Rich Market in Schools," *The New York Times*, December 10, 1984, B1.
29. Alan Maltun, "Students Beg to Stay After School to Use Computers"; David Einstein, "Bellflower Paces Area Schools in Computer Field"; Bob Williams, "Computer Parade Uneven," *The Los Angeles Times*, December 11, 1983, SB1.
30. Andrew Emil Gansky, "Myths and Legends of the Anti-Corporation: A History of Apple, Inc., 1976–1997," PhD dissertation, The University of Texas at Austin, 2017; Watters, "How Steve Jobs Brought the Apple II to the Classroom"; Harry McCracken, "The Apple Story is an Education Story: A Steve Jobs Triumph Missing from the Movie," *The 74*, October 15, 2015, https://www.the74million.org/article/the-apple-story-is-an-education-story-a-steve-jobs-triumph-missing-from-the-movie/, archived at https://perma.cc/EZV6-UGLT.
31. Natasha Singer, "How Google Took Over the Classroom," *The New York Times*, May 14, 2017, 1.
32. "'82 House Freshmen Eschew Partisanship and Posturing," *The Washington Post*, December 26, 1982, A1; Zschau, "Tax Policy Initiatives to Promote High Technology," May 13, 1983, Box 51, FF Capital Gains 1, Ed Zschau Papers, HH.
33. Mark Bloomfield, Memorandum to the Capital Gains Coalition, December 14, 1984, Box 51, Capital Gains II, Ed Zschau Papers, HH; "Testimony of Honorable Paul Tsongas (Foley,

Hoag, and Eliot, Boston, Mass.) before the Senate Small Business Committee on The Elimination of the Capital Gains Differential for Individuals and Its Impact on Small Business Capital Formation," June 2, 1986.

34. *Climate for Entrepreneurship and Innovation in the United States: Hearings Before the Joint Economic Committee*, August 27 and 28, 1984, 3.
35. Burt McMurtry, interview with the author, October 2, 2017.
36. Committee on Innovations in Computing and Communications, National Academy of Sciences, *Funding a Revolution: Government Support for Computing Research* (Washington, D.C.: National Academies Press, 1999), 52–61.
37. Michael Schrage, "Defense Budget Pushes Agenda in High Tech R&D," *The Washington Post*, August 12, 1984, F1; Schrage, "Computer Effort Falling Behind," *The Washington Post*, September 5, 1984, F1; Alex Roland with Philip Shiman, *Strategic Computing: DARPA and the Quest for Machine Intelligence, 1983–1993* (Cambridge, Mass.: The MIT Press, 2002).

CHAPTER 16: BIG BROTHER

1. Paul Andrews and Stephan Manes, "If Perot's So Smart, Why Did He Let Microsoft Slip Away?" *The Austin American-Statesman*, June 21, 1992, H1.
2. Stephen Manes and Paul Andrews, *Gates* (New York: Touchstone/Simon & Schuster, 1993), 120–21; Peter Rinearson, "Young Students Had Program to Make Millions," *The Seattle Times*, February 14, 1982, D3.
3. Manes and Andrews, *Gates*, 153.
4. Paul Andrews, "Mary Gates: She's Much More Than the Mother of Billionaire Bill," *The Seattle Times*, January 9, 1994, A1.
5. Ironically, given the fact that his story was one proof point Silicon Valley insiders gave for their animus toward Microsoft, Kildall was a third-generation Seattleite and a computer science graduate of the University of Washington. While never achieving the fame of Bill Gates, he continued to make and market CP/M, and became a familiar face on public television as the host of *The Computer Chronicles* before his untimely death at age 52, in 1994. Gary Kildall, *Computer Connections: People, Places, and Events in the History of the Personal Computer Industry*, unpublished manuscript in the possession of Scott and Kristen Kildall, reproduced online with permission by the Computer History Museum at http://www.computerhistory.org/atchm/in-his-own-words-gary-kildall/, archived at https://perma.cc/NU3B-M47B.
6. Rinearson, "Young Students."
7. Burt McMurtry, interview with the author, January 15, 2015; Leena Rao, "Sand Hill Road's Consiglieres: August Capital," *TechCrunch*, June 14, 2014, https://techcrunch.com/2014/06/14/sand-hill-roads-consiglieres-august-capital/, archived at https://perma.cc/6DN4-DERQ.
8. Charles Simonyi, interview with the author, October 4, 2017, Bellevue, Wash.; Michael Hiltzik, *Dealers of Lightning: Xerox PARC and the Dawn of the Computer Age* (New York: HarperBusiness, 1999), 194–210; Michael Swaine and Paul Freiberger, *Fire in the Valley: The Birth and Death of the Personal Computer*, 3rd ed. (Raleigh, N.C.: The Pragmatic Bookshelf, 2014), 271.
9. Simonyi interview; Charles Simonyi, oral history interview by Grady Booch, February 6, 2008, CHM, 30–34; Manes and Andrews, *Gates*, 167.
10. Intel Corporation, Annual Report, 1980 and 1984. The surge in growth came under the leadership of Andy Grove, who became CEO in 1987. Under Grove's tenure, Intel's 386

microprocessor became the industry standard and the company at last became a household name with its ubiquitous "Intel Inside" marketing campaign. See Richard S. Tedlow, *Andy Grove: The Life and Times of an American* (New York: Portfolio, 2006).

11. George Anders, "IBM Set to Announce Entry into Home-Computer Field," *The Wall Street Journal*, August 11, 1981, 35; "IBM to Announce More Small Computers," *InfoWorld*, August 17, 1981, 1.
12. Mike Markkula quoted in Paul Freiberger, "Apple Computer in News," *InfoWorld*, August 31, 1981, 1.
13. Display ad 25—no title, *The Wall Street Journal*, August 24, 1981, 7.
14. On the history of the Macintosh, see Steven Levy, *Insanely Great: The Life and Times of Macintosh, the Computer That Changed Everything* (New York: Viking, 1994); Andy Hertzfeld, *Revolution in the Valley: The Insanely Great Story of How the Mac Was Made* (Sebastopol, Calif.: O'Reilly Media, 2004); Swaine and Freiberger, *Fire in the Valley*, 262–75.
15. Margaret Comstock Tommervik, "The Women of Apple," *Softalk* 1, no. 7 (March 1981): 4–10, 38–39.
16. Floyd Kvamme, interview with the author, February 16, 2016, Stanford, Calif.; Guy Kawasaki, interview with the author, January 26, 2015, Menlo Park, Calif.; Andy Hertzfeld, "Pirate Flag, August 1983," Folklore.org, https://www.folklore.org/StoryView.py?story=Pirate_Flag.txt, archived at https://perma.cc/GET2-7LQN.
17. On Apple's internal analysis of the problem, see Clyde Folley, "Copy Strategy, Apple Computer Inc., MIS/OFFICE/EDP, Second Draft," January 5, 1983, Apple Computer Inc. Records, M1007, Series 7, Box 15, FF 1, SU.
18. Ben Rosen, "Evolutionary Computers Spawn Revolution: The Under-$10,000 Boom," *The Rosen Electronics Letter*, May 9, 1980, 1, 10.
19. *Fortune* (Dec 26, 1983, 142) quoted in Thomas & Company, "Competitive Dynamics in the Microcomputer Industry: IBM, Apple Computer, and Hewlett-Packard," 26, M1007, Series 7, Box 14, FF 5, Apple Computer Company Records, SU.
20. Martin Reynolds, "The Billionth PC Ships," Gartner Research Note, June 28, 2002; *Fortune* (Dec. 26, 1983, 142) and Jobs (*WSJ*, Oct, 4, 1983, 1) both quoted in Thomas & Company, "Competitive Dynamics in the Microcomputer Industry," 24, 26.
21. John Markoff, "Adam Osborne, Pioneer of the Portable PC, Dies at 64," *The New York Times*, March 26, 2003, C13; Daniel Akst, "The Rise and Decline of Vector Graphic," *The Los Angeles Times*, August 20, 1985, V_B5A; John Greenwald, Frederick Ungeheuer, and Michael Moritz, "D-Day for the Home Computer," *Time* 122, no. 20 (November 7, 1983): 74.
22. John Young, Paul Ely in *BusinessWeek*, October 3, 1983, quoted in Thomas & Company, "Competitive Dynamics in the Microcomputer Industry."
23. Thomas & Company, "Competitive Dynamics."
24. Smith, "Silicon Valley Spirit"; Robert Reinhold, "Life in High-Stress Silicon Valley Takes a Toll," *The New York Times*, January 13, 1984, 1.
25. Jean Hollands, *The Silicon Syndrome: A Survival Handbook for Couples* (Palo Alto, Calif.: Coastlight Press, 1983).
26. Reinhold, "Life in High-Stress Silicon Valley Takes a Toll"; Smith, "Silicon Valley Spirit."
27. Thomas & Company, "Competitive Dynamics"; Paul Freiberger, "IBM Counts its Chips, Invests $250 Million in Intel," *InfoWorld* 5, no. 5 (January 31, 1983): 30; Jean S. Bozman, "The IBM-Rolm Connection," *Information Week* 37 (October 21, 1985): 16; Katherine Maxfield,

Starting Up Silicon Valley: How ROLM became a Cultural Icon and Fortune 500 Company (Austin, Tex.: Emerald Book Co., 2014).

28. Mitch Kapor, interview with the author, October 19, 2017, Oakland, Calif.; Udayan Gupta, *Done Deals: Venture Capitalists Tell Their Stories* (Cambridge, Mass.: Harvard Business School Press, 2000), 83–88; John W. Wilson, *The New Venturers: Inside the High-Stakes World of Venture Capital* (Reading, Mass.: Addison-Wesley, 1985), 110–13.
29. Martin Campbell-Kelly, "Not Only Microsoft: The Maturing of the Personal Computer Software Industry, 1982–1995," *The Business History Review* 75, no. 1 (Spring 2001): 103–45.
30. Jeanne Hayes, ed., *Microcomputer and VCR Usage in Schools, 1985–1986* (Denver, Colo.: Quality Education Data, 1986), 4, 36, 38; U.S. Bureau of the Census, Robert Kominski, Current Population Reports, Special Studies, Series P-23, No. 155, *Computer Use in the United States: 1984* (Washington, D.C.: U.S. Government Printing Office, 1988).
31. Author interview with former associate of RMI, Inc., August 6, 2018.
32. Mike Hogan, "Fighting for the Heavyweight Title," *California Business*, November 1984, 78–93; *Computer Age*, December 12, 1983, quoted in Thomas & Company, "Competitive Dynamics"; Hogan, "Fighting for the Heavyweight Title."
33. SRI's Values and Lifestyles (VALS) program was relied upon heavily by Apple for its market research. See Macintosh Product Introduction Plan, October 7, 1983, M1007, Series 7, Box 13, FF 21, SU.
34. Chiat/Day, Macintosh Introductory Advertising Plan FY 1984, November 1983, M1007, Series 7, Box 14, FF 1, SU; Michael Moritz, *Return to the Little Kingdom: How Apple and Steve Jobs Changed the World* (New York: The Overlook Press, 2009), 123.
35. Haynes Johnson, "Election '84: Silicon Valley's Satisfied Society," *The Washington Post*, October 10, 1984, M3.
36. Patricia A. Bellew, "The Office Party is One Thing at Which Silicon Valley Excels," *The Wall Street Journal*, December 21, 1984, 1.

CHAPTER 17: **WAR GAMES**

1. Jim Treglio, "Briefing Paper for Paul Tsongas," July 28, 1983, Box 36B, FF 2, Paul E. Tsongas Collection, Center for Lowell History, University of Massachusetts Lowell (PT); John Lewis Gaddis, *The United States and the End of the Cold War: Implications, Reconsiderations, Provocations* (Oxford, U.K.: Oxford University Press, 1992); Frances FitzGerald, *Way Out There in the Blue: Reagan, Star Wars, and the End of the Cold War* (New York: Simon & Schuster, 2000).
2. Alex Roland with Philip Shiman, *Strategic Computing: DARPA and the Quest for Machine Intelligence, 1983–1993* (Cambridge, Mass.: The MIT Press, 2002), 83–95.
3. Jim Treglio, "Briefing Paper for Paul Tsongas," July 28, 1983, Box 36B, FF 2, PT; Reagan, "Address to the Nation on Defense and National Security," March 23, 1983; Union of Concerned Scientists, "The New Arms Race: Star Wars Weapons," Briefing Paper, October 1983, Cambridge Mass, PT. Useful discussion of pro- and anti-SDI campaigns and the role of the scientific community can be found in William M. Knoblauch, "Selling the Second Cold War: Antinuclear Cultural Activism and Reagan Era Foreign Policy," PhD Dissertation, History, Ohio University, 2012.
4. R. Jeffrey Smith, "New Doubts about Star Wars Feasibility," *Science* 229, no. 4711 (1985), 367–68; Gary Chapman, "Dear Colleague," undated (c. 1986), Silicon Valley Ephemera

Collection, Series 1, Box 7, FF 12, SU; Catherine Rambeau, "Badham's Movies Take Good Shots at Techno-Society," *Atlanta Journal-Constitution*, June 6, 1983, B11.

5. U.S. Bureau of the Census, Robert Kominski, Current Population Reports, Special Studies, Series P-23, No. 155, *Computer Use in the United States: 1984* (Washington D.C.: USGPO, 1988); Jerry Neumann, "Heat Death: Venture Capital in the 1980s," Reaction Wheel blog, January 8, 2015, http://reactionwheel.net/2015/01/80s-vc.html, archived at https://perma.cc/F5T2-GFS9; Steven Levy, *Hackers: Heroes of the Computer Revolution* (New York: Anchor Press/Doubleday, 1984); Fred Turner, *From Counterculture to Cyberculture: Stewart Brand, the Whole Earth Network, and the Rise of Digital Utopianism* (Chicago: The University of Chicago Press, 2006), 134–40.
6. Terry A. Winograd, "Strategic Computing Research and the Universities," Report no. STAN-CS-87-1160, Department of Computer Science, Stanford University, March 1987.
7. CPSR General Statement, Box 3, FF "Computer Professionals for Social Responsibility," Liza Loop Papers, Undated c. 1982, SU.
8. Winograd, "Some Thoughts on Military Funding," *CPSR Newsletter* 2, no. 2 (Spring 1984).
9. Winograd, "Strategic Computing Research and the Universities," Silicon Valley Research Group Working Paper No. 87-7, University of California, Santa Cruz, March 1987.
10. Roland, *Strategic Computing*, 86–91.
11. Zachary Wasserman, "Inventing Startup Capitalism: Silicon Valley and the Politics of Technology Entrepreneurship from the Microchip to Reagan," PhD dissertation, History, Yale University, 2015.
12. Mark Crawford, "In Defense of 'Star Wars,'" *Science* 228, no. 4699 (1985): 563; Cathy Werblin, "Lockheed, Silicon Valley's Mysterious Giant," *The Business Journal*, February 26, 1990, 23; Nicholas D. Kristof, "Star Wars Job Near at Lockheed," *The New York Times*, November 8, 1985, D2.
13. "March at Lockheed; 21 Star Wars Protesters Arrested in Sunnyvale," *The San Francisco Chronicle*, April 22, 1986, 4; Torri Minton, "50 Arrested at Star Wars Protest at Lockheed," *The San Francisco Chronicle*, October 21, 1986, 16.
14. Michael Schrage, "Defense Budget Pushes Agenda in High Tech R&D," *The Washington Post*, August 12, 1984, F1; "A Big Push for Pentagon Reform," Editorial, *The New York Times*, July 22, 1986, A24.
15. William Trombley, "Reagan Library Strains Link Between Stanford and Hoover Institution," *The Los Angeles Times*, March 8, 1987, A3.
16. Ron Lillejord and Seth Zuckerman, "The Hoover Institution: The Might of the Right?" *The Stanford Daily* 176, no. 29 (November 1, 1979): 3; Viewpoint, "Kennedy's Flawed 'Compromise,'" *The Stanford Daily* 184, no. 31 (November 7, 1983): 4.
17. Tom Bothell, "Totem and Taboo at Stanford," *National Review*, reprinted in *Stanford Review* 2, no. 1 (November 1987): 4; James Wetmore, "Former Hoover Director W. Glenn Campbell Discusses His Retirement," *Stanford Review* 4, no. 1 (October 8, 1989): 4–5.
18. Robert Marquand, "Stanford's core 'canon' debate ends in compromise," *The Christian Science Monitor*, April 8, 1988, https://www.csmonitor.com/1988/0408/dstan.html, archived at https://perma.cc/L5CV-XEXD; Andrew Hartman, *A War for the Soul of America: A History of the Culture Wars* (Chicago: The University of Chicago Press, 2015), esp. 222–52.
19. Goodwin Liu, "ASSU Urges Reforms," *The Stanford Daily* 192, no. 23 (October 28, 1987): 1; Josh Harkinson, "Masters of Their Domain," *Mother Jones*, June 20, 2007, https://www

.motherjones.com/politics/2007/06/masters-their-domain-2/, archived at https://perma.cc/FAC9-NV7L.

20. Jodi Kantor, "A Brand-New World in Which Men Ruled," *The New York Times*, December 23, 2014, 1; David O. Sacks and Peter A. Thiel, *The Diversity Myth: 'Multiculturalism' and the Politics of Intolerance at Stanford* (Oakland, Calif.: The Independent Institute, 1995).
21. Ann Hardy, phone conversation with the author, August 28, 2018. In contrast to a later generation of Silicon Valley executives who were bent on limiting screen time for their offspring, Hardy got her daughters in front of teletype machines as soon as they could sit upright; more than four decades on, she and they all agreed that the early exposure had worked to everyone's benefit.
22. Michael Weinstein, "Tymshare Puts McDonnell Douglas in Information Processing," *American Banker*, March 7, 1984, 15; Ann Hardy, interview with the author, April 20, 2015; Ann Hardy: An Interview Conducted by Janet Abbate, IEEE History Center, July 15, 2002, Interview #599 for the IEEE History Center, The Institute of Electrical and Electronic Engineers, Inc.
23. Michael A. Banks, *On the Way to the Web: The Secret History of the Internet and Its Founders* (Berkeley, Calif.: Apress, 2008), 38. Even more robust and publicly sponsored online networks emerged around the same time in countries that had maintained telecommunications service as a public utility, most notably France's Minitel system. See Julien Mailland and Kevin Driscoll, *Minitel: Welcome to the Internet* (Cambridge, Mass.: MIT, 2017).
24. Claire L. Evans, *Broad Band: The Untold Story of the Women Who Made the Internet* (New York: Portfolio / Penguin, 2018), 133; Turner, *From Counterculture to Cyberculture.*
25. Laura Smith, "In the early 1980s, white supremacist groups were early adopters (and masters) of the internet," *Medium*, October 11, 2017, https://timeline.com/white-supremacist-early-internet-5e91676eb847, archived at https://perma.cc/8UKG-UB8H; Kathleen Belew, *Bring the War Home: The White Power Movement and Paramilitary America* (Cambridge, Mass.: Harvard University Press, 2018). One early and leading participant in the Cypherpunk movement was Wikileaks founder Julian Assange, who made it the subject of a book-length treatise, *Cypherpunks: Freedom and the Future of the Internet* (New York: OR Books, 2016).
26. Peter H. Lewis, "Despite a New Plan for Cooling it Off, Cybersex Stays Hot," *The New York Times*, March 26, 1995, 1.
27. President's Blue Ribbon Commission on Defense Management, *A Quest for Excellence: Final Report to the President* (Washington, D.C.: USGPO, June 1986); William J. Broad, "What's Next for 'Star Wars'? 'Brilliant Pebbles,'" *The New York Times*, April 25, 1989, C1.
28. R. W. Apple Jr., "After the Summit," *The New York Times*, June 5, 1990, 1.

CHAPTER 18: BUILT ON SAND

1. Susan Brown-Goebeler, "How Gray Is My Valley," *Time* 138, no. 20 (November 18, 1991): 90.
2. Zschau, correspondence with the author, September 11, 2018; McKenna, correspondence with the author, September 6, 2018; "Cranston Rides into Zschau Country—Silicon Valley," *Los Angeles Times*, October 25, 1986, 37; Tom Campbell, interview with the author, February 17, 2016, by phone. Other dirty tricks marred the Zschau-Cranston race and possibly tipped the outcome, including a voter-payoff scheme in Orange County that sent its perpetrator to prison.
3. Tom Kalil, interview with the author, August 15, 2017, by phone; John Endean, "Let the 'Chips' Quote Fall on Whom It May" (Letter to the Editor), *The Wall Street Journal*, January 16, 1992, A13.

4. Mitchel Benson and David Kutzmann, "EPA Calls Valley Water Treatment, Air Pollution the Chief Cancer Risks," *San Jose Mercury News,* October 12, 1985, A1.
5. Judith E. Ayres, "Controlling the Dangers from High-Tech Pollution," *EPA Journal* 10, no. 10 (December 1984), 14–15; Judith Cummings, "Leaking Chemicals in California's 'Silicon Valley' Alarm Neighbors," *The New York Times,* May 20, 1982, A22; Chop Keenan, interview with the author, March 17, 2016, Palo Alto, Calif. On environmentally "clean" ideals and realities in the industry over time, see Margaret O'Mara, "The Environmental Contradictions of High-Tech Urbanism," in *Now Urbanism: The Future City is Here,* ed. Jeffrey Hou, Ben Spencer, Thaisa Way, and Ken Yocom (Abingdon, U.K.: Routledge, 2015), 26–42.
6. Lenny Siegel and John Markoff, *The High Cost of High Tech: The Dark Side of the Chip* (New York: Harper & Row, 1985); Glenna Matthews, *Silicon Valley, Women, and the California Dream: Gender, Class, and Opportunity in the Twentieth Century* (Redwood City, Calif.: Stanford University Press, 2002). The hazards of chip-making also could be deadly to the workers in the fabrication plants; see "Ailing Computer-Chip Workers Blame Chemicals, Not Chance," *The New York Times,* March 28, 1996, B1.
7. David Olmos, "Electronics Industry Resists Organized Labor," *Computerworld,* September 10, 1984, 113. As Timothy J. Sturgeon observes, in using contractors in this manner, electronics companies were on the front edge of what became a widely adapted form of industrial organization by U.S. manufacturers. Sturgeon, "Modular production networks: a new American model of industrial organization," *Industrial and Corporate Change* 11, no. 3 (2002). For more on the tech industry's workforce practices and its use of contractors, see Louis Hyman, *Temp: How American Work, American Business, and the American Dream Became Temporary* (New York: Viking, 2018).
8. Kenneth R. Sheets, "Silicon Valley Doesn't Hold All the Chips," *U.S. News & World Report,* August 26, 1985, 45.
9. Regis McKenna, "Marketing is Everything," *Harvard Business Review,* January-February 1991, https://hbr.org/1991/01/marketing-is-everything, archived at https://perma.cc/3RUZ-GVV5.
10. Scott Mace, "Apple Bets on the Macintosh," *InfoWorld,* February 13, 1984, 20; Dan'l Lewin, interview with the author, November 21, 2017, Seattle, Wash.; "Macintosh Product Introduction Plan," October 7, 1983, M1007, Series 7, Box 13, FF 12, SU.
11. Matthew Creamer, "Apple's First Marketing Guru on Why '1984' is Overrated," *Advertising Age,* March 1, 2012.
12. Barbara Rudolph, Robert Buderi, and Karen Horton, "Shaken to the Very Core: After Months of Anger and Anguish, Steve Jobs Resigns as Apple Chairman," *Time* 126, no. 13 (September 30, 1985): 64; Walter Isaacson, *Steve Jobs* (New York: Simon & Schuster, 2011), 192–211.
13. Phil Patton, "Steve Jobs: Out for Revenge," *The New York Times,* August 6, 1989, SM23.
14. Ron Wolf, "Amid Hoopla, 'Next' Computer is Unveiled by PC Pioneer Jobs," *The Washington Post,* October 13, 1988, C1; Mark Potts, "Computer Industry Wary of Jobs-Perot Alliance," *The Washington Post,* February 8, 1987, H2.
15. "NeXT," *Entrepreneurs,* dir. John Nathan, WETA-TV, Washington, D.C., 1986.
16. Doron P. Levin, *Irreconcilable Differences: Ross Perot versus General Motors* (New York: Little, Brown, 1989), 18; Ross Perot, "A Life of Adventure," The West Point Center for Oral History, 2010, archived at https://perma.cc/5T96-SLJU; Herbert W. Armstrong, "An Interview with H. Ross Perot," *The Plain Truth* Magazine 39, no. 3 (March 1974).

17. Robert Fitch, "H. Ross Perot: America's First Welfare Billionaire," *Ramparts Magazine*, November 1971, 42–51. Also see Fitch, "Welfare Billionaire," *The Nation* 254, no. 23 (June 15, 1992): 815–16; Eric O'Keefe, *A Unique One-Time Opportunity: The Story of How EDS Created Outsourcing*, self-published manuscript in the author's possession (2013); Stuart Auerbach, "Perot Medicare Bonanza Revealed," *The Washington Post*, September 29, 1971, A3.
18. Jon Nordheimer, "Billionaire Texan Fights Social Ills," *The New York Times*, November 28, 1969, 41; O'Keefe, *A Unique One-Time Opportunity*; Todd Mason, *Perot: An Unauthorized Biography* (Homewood, Ill.: Dow Jones–Irwin, 1990), 5.
19. Brenton R. Schlender, "Jobs, Perot Become Unlikely Partners in the Apple Founder's New Concern," *The Wall Street Journal*, February 2, 1987, 28.
20. Potts, "Computer Industry Wary of Jobs-Perot Alliance."
21. Lewin, interview with the author, November 21, 2017.
22. G. Pascal Zachary and Ken Yamada, "What's Next? Steve Jobs's Vision, So on Target at Apple, Now is Falling Short," *The Wall Street Journal*, May 25, 1993, A1.
23. Wes Smith, "Booming Seattle Tells Hip Californians Just to Stay Away," *The Chicago Tribune*, September 19, 1989, http://articles.chicagotribune.com/1989-09-19/features/8901140398_1_seattle-area-greater-seattle-californians, archived at https://perma.cc/G4K7-HBST.
24. Peter Huber, "Software's Cash Register," *Forbes*, October 18, 1993, 314.
25. Trish Millines Dziko, interview with the author, April 2, 2018, by phone.
26. Paul Andrews, "Inside Microsoft: A 'Velvet Sweatshop' or a High-Tech Heaven?," *The Seattle Times*, April 23, 1989, PM 8–17. On Microsoft stock price and employee wealth, see O. Casey Corr, "What's $1 Million Times 2,000?," *The Seattle Times*, February 27, 1992, A1.
27. Mark Leibovich, "Alter Egos: Two Sides of a High-Tech Brain Trust Make Up a Powerful Partnership," *The Washington Post*, December 31, 2000, A1; John Heilemann, *Pride Before the Fall: The Trials of Bill Gates and the End of the Microsoft Era* (New York: CollinsBusiness, 2001), 49.
28. Rachel Lerman, "Pam Edstrom Was Voice Behind Microsoft's Story, Dies at 71," *The Seattle Times*, March 30, 2017, https://www.seattletimes.com/seattle-news/obituaries/pam-edstrom-was-voice-behind-microsofts-story/, archived at https://perma.cc/3ZNJ-53HL.
29. Stephen Manes and Paul Andrews, *Gates* (New York: Touchstone/Simon & Schuster, 1993), 148, 244.
30. Kara Swisher, *AOL.com: How Steve Case Beat Bill Gates, Nailed the Netheads, and Made Millions in the War for the Web* (New York: Crown Business, 1998), xvii; Manes and Andrews, *Gates*, 403.
31. Brenton R. Schlender, "Computer Maker Aims to Transform Industry and Become a Giant," *The Wall Street Journal*, March 18, 1988, 1.
32. "April Fool Pranks in Sun Microsystems Over the Years," *Hacker News*, February 14, 2006, last updated January 26, 2014, https://news.ycombinator.com/item?id=7121224, archived at https://perma.cc/G5GH-FN6F.
33. Nancy Householder Hauge, "Misogyny in the Valley," and "Life in the Boy's Dorm: My Career at Sun Microsystems," *Consulting Adult*, January 29, 2010, http://consultingadultblog.blogspot.com/2010/01/life-in-boys-dorm-my-career-at-sun.html, archived at https://perma.cc/26WB-KTV9.
34. AnnaLee Saxenian, "Regional Networks and the Resurgence of Silicon Valley," *California Management Review* 33, no. 1 (Fall 1990): 89–113.

35. Mark Potts, "Rebellious Apple Finally Grows Up," *The Washington* Post, June 14, 1987, D1; Haynes Johnson, "Future Looks Precarious to Silicon Valley Voters," *The Washington Post*, October 24, 1988, A1; Brown-Goebeler, "How Gray Is My Valley."
36. "White House Won't Back Chip Subsidy," *The New York Times*, November 30, 1989, B1.
37. Constance L. Hays, "An Inventor of the Microchip, Robert N. Noyce, Dies at 62," *The New York Times*, June 4, 1990, A1.
38. "Companies with over 500 Employees," November 1986, Silicon Valley Ephemera Collection, Series 1, Box 5, FF 12, SU; Ken Siegmann, "Lockheed Cutting Thousands of Jobs," *San Francisco Chronicle*, August 4, 1993, B1; Michelle Quinn, "The Turbulence at Lockheed," *San Francisco Chronicle*, June 23, 1995, B1; Alan C. Miller, "Berman Feels the Heat Over Defense Cuts," *The Los Angeles Times*, June 23, 1991, A3.
39. Glenn Rifkin, "Light at the End of Digital's Tunnel," *The New York Times*, October 29, 1991, D1.
40. Fox Butterfield, "Chinese Immigrant Emerges as Boston's Top Benefactor," *The New York Times*, May 5, 1984, 1; Dennis Hevesi, "An Wang, 70, is Dead of Cancer; Inventor and Maker of Computers," *The New York Times*, March 25, 1990, 38.
41. David Morgenthaler, interviews with the author.

ACT FOUR

1. Christopher E. Martin, Khary Turner, "Ten Crack Commandments," Sony/ATV Music Publishing, 1997.

ARRIVALS

1. Brent Schlender, "How a Virtuoso Plays the Web," *Fortune* 141, no. 5 (March 6, 2000): 79–83.
2. United States Census, 1970, 1990. For more on the growth of the South Bay's Asian-American population and the social and political impacts of its growth, see Willow S. Lung-Amam, *Trespassers? Asian Americans and the Battle for Suburbia* (Berkeley: University of California Press, 2017), especially 19–52.
3. Lowell B. Lindsay, "A Long View of America's Immigration Policy and the Supply of Foreign-Born STEM Workers in the United States," *American Behavioral Scientist* 53, no. 7 (2010): 1029–44; AnnaLee Saxenian, "Silicon Valley's New Immigrant Entrepreneurs" (Public Policy Institute of California, 1999); Vivek Wadhwa, AnnaLee Saxenian, Ben Rissing, and Gary Gereffi, "America's New Immigrant Entrepreneurs," Master of Engineering Management Program, Duke University; School of Information, University of California, Berkeley, January 4, 2007. Another significant impetus for this immigration: international educational exchange and foreign student programs, which had their origins in Cold War diplomacy; see Margaret O'Mara, "The Uses of the Foreign Student," *Social Science History* 36, no. 4 (Winter 2012): 583–615.
4. Stanford School of Engineering, *Yahoo!: Jerry & Dave's Excellent Venture*, video recording (Mill Valley, Calif.: Kantola Productions, 1997), Stanford Libraries, Stanford, Calif.

CHAPTER 19: INFORMATION MEANS EMPOWERMENT

1. Rory J. O'Connor and Tom Schmitz, "U.S. Raids Hackers," *San Jose Mercury News*, May 9, 1990, A1.

2. Neil Steinberg, "Hacker Sting Nets Arrests in 14 Cities," *Chicago Sun-Times*, May 11, 1990, 16.
3. John Markoff, "Drive to Counter Computer Crime Aims at Invaders," *The New York Times*, June 3, 1990, 1.
4. Mitch Kapor, interview with the author, September 19, 2017, Oakland, Calif.
5. Fred Turner, *From Counterculture to Cyberculture: Stewart Brand, the Whole Earth Network, and the Rise of Digital Utopianism* (Chicago: The University of Chicago Press, 2006), 168–72; Alexei Oreskovic, "Who's Who in the Digital Revolution," *Upside* 6, no. 12 (December 1994): 52.
6. Rachel Parker, "Kapor Strives to Establish Rules for Living in a Computer Frontier," *InfoWorld*, July 23, 1990, 39; John Perry Barlow quoted in Turner, *From Counterculture to Cyberculture*, 172. Like the electronic frontier, the American West wasn't as unsettled or lawless as Kapor and Barlow understood it to be, but the historical comparison wasn't entirely off base. Rather than purely a realm of bootstrapping individualists, the West was a world made possible by government intervention—the drawing of boundary lines, the apportionment of land and resources, the removal of native peoples and replacement by American homesteaders, and the heavy subsidy of major infrastructure projects like the transcontinental railroad. See Richard White, *Railroaded: The Transcontinentals and the Making of Modern America* (New York: W. W. Norton, 2011).
7. Tim Berners-Lee, "Information Management: A Proposal," March 1989, May 1990, w3.org, https://www.w3.org/History/1989/proposal.html, archived at https://perma.cc/56D4-RJLE.
8. On the critical role of academic communication in shaping the NSFNET and the subsequent commercial Internet, see Juan D. Rogers, "Internetworking and the Politics of Science: NSFNET in Internet History," *The Information Society* 14, no. 3 (2006): 213–28. Also see John Markoff, "The Team That Put the Net in Orbit," *The New York Times*, December 9, 2007, B5.
9. Testimony of Mitchell Kapor, Management of NSFNET: Hearing before the Subcommittee on Science of the Committee on Science, Space, and Technology, U.S. House of Representatives, 102nd Congress, second session, March 12, 1992, 2; Katie Hafner and Matthew Lyon, *Where Wizards Stay Up Late: The Origins of the Internet* (New York: Simon & Schuster, 1996), 253–57.
10. Marty Tenenbaum, interview with the author, February 9, 2018, by phone.
11. Berners-Lee, "Longer Biography," https://www.w3.org/People/Berners-Lee/Longer.html, archived at https://perma.cc/VHJ4-C8GG. Also see Janet Abbate, *Inventing the Internet* (Cambridge, Mass.: The MIT Press, 1999), 214–18.
12. Berners-Lee quoted in Abbate, *Inventing the Internet*, 215. Also see Tim Berners-Lee with Mark Fischetti, *Weaving the Web: The Original Design and Ultimate Destiny of the World Wide Web* (San Francisco: HarperSanFrancisco, 1999).
13. National Research Council, *Toward a National Research Network* (Washington, D.C.: National Academies Press, 1988); Armand Mattelart, *The Information Society: An Introduction* (SAGE, 2003), 110–11.
14. Jane Bortnick, ed., Transcription of "Information and Communications," Congressional Clearinghouse on the Future, Chautauquas for Congress, March 1979, Congressional Research Service, Library of Congress, June 12, 1979; Cindy Skrzycki, "The Tekkie on the Ticket," *The Washington Post*, October 18, 1992, H1; Interview with W. Daniel Hillis, "Al

Gore, 'the Ozone Man,'" *Web of Stories*, https://www.webofstories.com/play/danny.hillis/173, archived at https://perma.cc/KGK5-NKWB.

15. High Performance Computing Act of 1991, P.L. 102-194. A decade later, as a sitting Vice President running against George W. Bush for the top job, Gore inelegantly declared that he "took the initiative in creating the Internet," precipitating widespread mockery by political opponents, pundits, and late-night comedians. The drubbing overlooks the fact that Gore did indeed play an important role in opening the Internet to commercialization. (Gore appearance on CNN *Late Edition*, March 9, 1999.)
16. John Heilemann, "The Making of the President 2000," *WIRED*, December 1, 1995, https://www.wired.com/1995/12/gorenewt/, archived at https://perma.cc/YE76-4JG4. On the propitious timing of NSF's dropping of its commercial restrictions, and the subsequent growth of Internet Service Providers (ISPs), see Shane Greenstein, "Commercialization of the Internet: The Interaction of Public Policy and Private Choices, or Why Introducing the Market Worked so Well," in *Innovation Policy and the Economy*, vol. 1, ed. Adam B. Jaffe, Josh Lerner and Scott Stern (Cambridge, Mass.: The MIT Press, 2001), 151–86.
17. Timothy C. May, "The Crypto Anarchist Manifesto," September 1992, https://www.activism.net/cypherpunk/crypto-anarchy.html, archived at https://perma.cc/F584-5SDY. May also delivered versions of this manifesto during at least two Hackers' Conferences.
18. "Names of 40 Who Gave Democrats Each $100,000 Disclosed," *The Washington Post*, November 3, 1988, N1; Testimony of Mitchell Kapor, *Management of NSFNET*, 2 [p. 6 of his prepared statement, p. 76 of hearing].
19. Kapor, interview with the author.
20. Jill Abramson, "Once Again, Clinton Has Met the Enemy, and He is Brown, Not Bush," *The Wall Street Journal*, March 27, 1992, A16.
21. Margaret O'Mara, *Pivotal Tuesdays: Four Elections That Shaped the Twentieth Century* (Philadelphia: University of Pennsylvania Press, 2015), 178–82.
22. Lawrence (Larry) Stone, interview with the author, April 7, 2015, San Jose, Calif. A full account of the Democrats' 1990s-era wooing of Silicon Valley is found in Sara Miles, *How to Hack a Party Line: The Democrats and Silicon Valley* (New York: Farrar, Straus and Giroux, 2001).
23. Regis McKenna, correspondence with the author, September 6, 2018.
24. Michael S. Malone, "Democrat Days in Silicon Valley," *The New York Times*, March 7, 1993, B27.
25. *Nominations of David J. Barram to Be Deputy Secretary of Commerce and Steven O. Palmer to Be Assistant Secretary for Governmental Affairs of the Department of Transportation; hearing before the Committee on Commerce, Science, and Transportation*, United States Senate, 103rd, First Session, September 15, 1993 (1995); Malone, "Democrat Days in Silicon Valley"; Stone interview; Regis McKenna, interviews with the author, December 3, 2014 and April 21, 2015.
26. Stone interview.
27. Skrzycki, "The Tekkie on the Ticket."
28. Calvin Sims, "Silicon Valley Takes a Partisan Leap of Faith," *The New York Times*, October 29, 1992, B1; Daniel Southerland, "The Executive With Clinton's Ear: Hewlett-Packard CEO John Young Finds Ally on Competitiveness," *The Washington Post*, October 20, 1992, C1.

29. Southerland, "The Executive with Clinton's Ear."
30. Sims, "Silicon Valley Takes a Partisan Leap of Faith."
31. Martha Groves and James Bates, "California Prospecting: State Business Executives Rumored as Possible Clinton Appointees," *The Los Angeles Times*, November 6, 1992, B5; "Excerpts from Clinton's Conference on the State of the Economy," *The New York Times*, December 15, 1992, B10.
32. Dan Pulcrano, "Guess Who's Coming to Dinner?," *Los Gatos Weekly-Times*, February 28, 1993, 1.
33. Lee Gomes, "Bridging the Culture Gap," *San Jose Mercury News*, January 24, 1994, D1.
34. Philip J. Trounstine, "Clinton's High-Tech Initiative," *San Jose Mercury News*, February 23, 1993, A1.
35. Lee Gomes, "Silicon Graphics Staff Impressed by Visitors," *San Jose Mercury News*, February 23, 1993, A1; John Markoff, "Conversations/T. J. Rodgers: Not Everyone in the Valley Loves Silicon-Friendly Government," *The New York Times*, March 7, 1993, E7; "William J. Clinton: Remarks and a Question-and-Answer Session With Silicon Graphics Employees in Mountain View, California," February 23, 1993.
36. The World Bank, Internet users (per 100 people), https://data.worldbank.org/indicator/IT.NET.USER.P2?view=map&year=1993, archived at https://perma.cc/YTL8-WSKD; Tom Kalil, interview with the author, August 8, 2017.
37. Significantly, the operation tagged with marketing and executing the audacious project wasn't the FCC (even though the man pegged to run it, Reed Hundt, had been a close friend of Gore's since they were schoolmates). It was the National Telecommunications and Information Administration, or NTIA, headed by a gregarious and K-Street-savvy Ed Markey aide named Larry Irving. Jube Shiver, Jr., "Agency Steps into the Telecomm Limelight," *The Los Angeles Times*, September 20, 1993, D1.
38. Thomas Kalil, "Public Policy and the National Information Infrastructure," *Business Economics* 30, no. 4 (October 1995): 15–20; National Telecommunications and Information Administration, U.S. Department of Commerce "20/20 Vision: The Development of a National Information Infrastructure," NTIA-Spub-94-28, March 1994; "NII Advisory Council Members," Domestic Policy Council, Carol Rasco, and Meetings, Trips, Events Series, "NII Advisory Meeting February 13, 1996," Clinton Digital Library, accessed August 3, 2017, https://clinton.presidentiallibraries.us/items/show/20743.
39. Computer Professionals for Social Responsibility, "Serving the Community: A Public-Interest Vision of the National Information Infrastructure," October 1993, http://cpsr.org/prevsite/cpsr/nii_policy.html/, archived at https://perma.cc/3VRD-Z9BU; Kapor, "Where is the Digital Highway Really Heading? The Case for a Jeffersonian Information Policy," *WIRED*, March 1, 1993, https://www.wired.com/1993/03/kapor-on-nii/, archived at https://perma.cc/VXZ6-NA56.
40. John Schwartz and John Mintz, "Gore: Federal Encryption Plan Flexible," *The Washington Post*, February 12, 1994, C1.
41. Domestic Policy Council, Carol Rasco, and Meetings, Trips, Events Series, "NII Advisory Meeting February 13, 1996," Clinton Digital Library, accessed August 3, 2017, https://clinton.presidentiallibraries.us/items/show/20743.
42. Heilemann, "The Making of the President 2000."

CHAPTER 20: SUITS IN THE VALLEY

1. John Doerr, "The Coach," interview by John Brockman, 1996, Edge.org, https://www.edge.org/digerati/doerr/, archived at https://perma.cc/9KWX-GLWK.
2. John Markoff, interview with Kara Swisher, *Recode: Decode* podcast, February 17, 2017, https://www.recode.net/2017/2/17/14652832/full-transcript-tech-reporter-john-markoff-silicon-valley-recode-decode-podcast, archived at https://perma.cc/XE3U-FCPC.
3. Michael Schrage, "Nation's High-Tech Engine Fueled by Venture Capital," *The Washington Post*, May 20, 1984, G1; Udayan Gupta, *Done Deals: Venture Capitalists Tell Their Stories* (Cambridge, Mass.: Harvard Business School Press, 2000), 374–5; Regis McKenna, interview with the author, May 31, 2016.
4. Michael Lewis, *The New New Thing: A Silicon Valley Story* (New York: W.W. Norton, 1999).
5. Gupta, *Done Deals*, 380.
6. Marc Andreessen interviewed by David K. Allison, *Computerworld* Honors Program Archives, June 1995, Mountain View, Calif.
7. David Bank, "Why Sun Thinks Hot Java Will Give You a Lift," *San Jose Mercury News*, March 23, 1995, 1A; Karen Southwick, *High Noon: The Inside Story of Scott McNealy and the Rise of Sun Microsystems* (New York: Wiley, 1999), 131.
8. Malia Wollan, "Before Sheryl Sandberg Was Kim Polese – the Original Silicon Valley Queen," *The Telegraph.co.uk*, November 11, 2013, https://www.telegraph.co.uk/technology/people-in-technology/10430933/Before-Sheryl-Sandberg-was-Kim-Polese-the-original-Silicon-Valley-queen.html, archived at https://perma.cc/Z7Y6-G2HM.
9. Elizabeth Corcoran, "Mother Hen to an Industry," *The Washington Post*, October 13, 1996, H1.
10. James Gibbons, interview with the author, November 4, 2015.
11. Brent Schlender, "How a Virtuoso Plays the Web," *Fortune* 141, no. 5 (March 6, 2000): 79–83.
12. Vindu Goel, "When Yahoo Ruled the Valley: Stories of the Original 'Surfers,'" *The New York Times*, July 16, 2016, B1.
13. "Don Valentine," in Gupta, *Done Deals,* 173; "History," Yahoo.com, October 1996, Archive.org, https://web.archive.org/web/19961017235908/http://www2.yahoo.com:80/.
14. Jared Sandberg, "Group of Major Companies is Expected to Offer Goods, Services on the Internet," *The Wall Street Journal*, April 8, 1994, B2; John Markoff, "Commerce Comes to the Internet," *The New York Times*, April 13, 1994, D5.
15. Elizabeth Perez, "Store on Internet is Open Book," *The Seattle Times*, September 19, 1995, E1; Brad Stone, *The Everything Store: Jeff Bezos and the Age of Amazon* (New York: Little, Brown, 2013); Randall E. Stross, *The eBoys: The True Story of the Six Tall Men who Backed eBay and Other Billion-Dollar Startups* (New York: Ballantine Books, 2000), 48–57.
16. Craig Torres, "Computer Powerhouse of D. E. Shaw & Co. May be Showing Wall Street's Direction," *The Wall Street Journal*, October 15, 1992, C1.
17. Robert Spector, *Amazon.com: Get Big Fast* (New York: HarperBusiness, 2000), 2–5.
18. Bezos, job posting for Cadabra.Inc, Usenet, c. 1994, reproduced in Kif Leswing, "Check out the first job listing Jeff Bezos ever posted for Amazon," *Business Insider*, August 22, 2018, https://www.businessinsider.com/amazon-first-job-listing-posted-by-jeff-bezos-24-years-ago-2018-8, archived at https://perma.cc/B3WS-PXS5.

19. "I did locate Amazon in Seattle because of Microsoft," Bezos told an interviewer in 2018. "I thought that that big pool of technical talent would provide a good place to recruit talented people from." Jeff Bezos, interview with David M. Rubenstein, The Economic Club of Washington, D.C., September 13, 2018.
20. "Jeff Bezos, Founder and CEO, Amazon.com." *Charlie Rose* (interview #12656), November 16, 2012.
21. Julia Kirby and Thomas A. Stewart, "The Institutional Yes," *Harvard Business Review*, October 2007, https://hbr.org/2007/10/the-institutional-yes, archived at https://perma.cc/XV5H-GULN.
22. United States Securities and Exchange Commission, Form 10-Q, Amazon.com, Inc., June 30, 1997; *60 Minutes*, "Amazon.com," January 1999.
23. Michael McCarthy, "Brand Innovators: Virtual Reality," *Adweek*, June 14, 1999, https://www.adweek.com/brand-marketing/brand-innovators-virtual-reality-31935/, archived at https://perma.cc/JC4A-6Z2W.
24. Bart Ziegler, "Internet Bulls Get On Line for Performance Systems," *The Wall Street Journal*, March 28, 1995, C1; Joseph E. Stiglitz, "The Roaring Nineties," *The Atlantic* 290, no. 3 (October 2002): 75–89; Sebastian Mallaby, *The Man Who Knew: The Life and Times of Alan Greenspan* (New York: Bloomsbury, 2016).
25. David Einstein, "Netscape Mania Sends Stock Soaring," *The San Francisco Chronicle*, August 10, 1995, D1; Lewis, *The New New Thing*, 85.
26. Rory J. O'Connor, "Microsoft Previews On-Line Service," *San Jose Mercury News*, November 15, 1994, D1.
27. Saul Hansell, "Flights of Fancy in Internet Stocks," *The New York Times*, November 22, 1998, B7; Patrick McGeehan, "Research Redux: Morgan Prints a Sleeper," *The Wall Street Journal*, March 20, 1996, C1.
28. Susanne Craig, "A Female Wall St. Financial Chief Avoids Pitfalls that Stymied Others," *The New York Times*, November 10, 2010, B1; John Cassidy, "The Woman in the Bubble," *The New Yorker*, April 26, 1999, 48. Ruth Porat's brother Marc was also the author of a first-of-its-kind 1977 Commerce Department study of the information economy (part of which originated as his Stanford PhD thesis): Marc Uri Porat and Michael Rogers Rubin, *The Information Economy*, U.S. Department of Commerce, Office of Telecommunications (1977). On General Magic, the company Marc Porat founded and many of whose employees went on to play seminal roles in the development of Apple's iPhone and Google's Android, see Sarah Kerruish, Matt Maude, and Michael Stern, *General Magic: The Movie* (Palo Alto, Calif.: Spellbound Productions, 2018).
29. Michael Siconolfi, "Under Pressure: At Morgan Stanley, Analysts Were Urged to Soften Harsh Views," *The Wall Street Journal*, July 14, 1992, A1.
30. Peter H. Lewis, "Once Again, Wall Street is Charmed by the Internet," *The New York Times*, April 3, 1996, D1.
31. Laurence Zuckerman, "With Internet Cachet, Not Profit, A New Stock is Wall Street's Darling," *The New York Times*, August 10, 1995, A1.
32. "New Accounting Rule Will Affect Employee Stock Options," *Morning Edition*, National Public Radio, April 11, 1994. Also see Steve Kaufman, "FASB Foes Make Last Stand," *San Jose Mercury News*, March 24, 1994, 1E.

33. Arthur Levitt, interviews with the author, May 7 and July 10, 2015, New York City and Westport, Conn.; Levitt, interview in "Bigger than Enron," PBS *Frontline*, 2002; Roger Lowenstein, "Coming Clean on Company Stock Options," *The Wall Street Journal*, June 26, 1997, C1; Max Walsh, "No Free Lunch but Lots of Options," *The Sydney Morning Herald*, July 8, 1997, 25; James J. Mitchell, "Stock Options Accounting Bill Already Panned," *San Jose Mercury News*, April 16, 1997, 1C.
34. Janelle Brown, "Start-up-cum-Goliath Works Hard to Get Help," *Wired*, August 22, 1997, https://www.wired.com/1997/08/start-up-cum-goliath-works-hard-to-get-help/, archived at https://perma.cc/U62L-PRRS.
35. Julia Angwin and Laura Castaneda, "The Digital Divide: High-tech boom a bust for blacks, Latinos," *San Francisco Chronicle*, May 4, 1998, A1.
36. Trish Millines Dziko, interview with the author, April 3, 2018; Millines Dziko, oral history interview by Jessah Foulk, Museum of History and Industry, "Speaking of Seattle," August 8, 2002, 28–29.

CHAPTER 21: MAGNA CARTA

1. Esther Dyson et al., "Cyberspace and the American Dream: A Magna Carta for the Knowledge Age" (Release 1.2, August 22, 1994), *The Information Society* 12, no. 3 (1996): 295–308; Boyce Rensberger, "White House Science Advisor is Cheerleader for Reagan," *The Washington Post*, November 12, 1985, A6; Philip M. Boffey, "Science Advisor Moves Beyond Rocky First Year," *The New York Times*, October 20, 1982, B8; Henry Allen, "The Word According to Gilder," *The Washington Post*, February 18, 1981, B1; Edward Rothstein, "The New Prophet of a Techno Faith Rich in Profits," *The New York Times*, September 23, 2000, B9; Fred Turner, *From Counterculture to Cyberculture: Stewart Brand, the Whole Earth Network, and the Rise of Digital Utopianism* (Chicago: The University of Chicago Press, 2006), 229. Also see Paulina Borsook, *Cyberselfish: A Critical Romp Through the Terribly Libertarian Culture of High Tech* (New York: PublicAffairs, 2000).
2. Paulina Borsook, "Release," *Wired* 1:5 (November 1993); "Esther Dyson," in *Internet: A Historical Encyclopedia*, vol. 2, ed. Laura Lambert, Chris Woodford, Hilary W. Poole, Christos J. P. Moschovitis (Santa Barbara, Calif.: ABC-CLIO, 2005), 88–92. Other important connectors in the Internet-age Valley salon were Tim O'Reilly and Stewart Alsop, each of whom built influential empires around annual conferences and publications aimed at the industry.
3. Quoted in Lambert et al., "Esther Dyson."
4. See, for example, Richard Barbrook and Andy Cameron's incendiary take on Silicon Valley mythmaking, "The Californian Ideology," *Science as Culture* 6, no. 1 (January 1996): 44–72. On the National Performance Review, see Al Gore, *The Gore Report on Reinventing Government: Creating a Government that Works Better and Costs Less* (New York: Three Rivers Press, 1993). Silicon Valley also played a role in this plank of the Clinton-Gore agenda; the Vice President hailed Sunnyvale (whose then mayor was Larry Stone) as a national example of and test bed for municipal reinvention, and several policies originating there became final recommendations of the performance review.
5. Claudia Dreifus, "Present Shock," *The New York Times*, June 11, 1995, SM46.
6. Dyson, "Friend and Foe," *Wired*, August 1, 1995, https://www.wired.com/1995/08/newt/, archived at https://perma.cc/NCP6-FHBP; Dyson et al., "Cyberspace and the American Dream."

7. Gingrich, remarks at the launch of Thomas.gov, January 5, 1995, Washington, D.C.
8. John Heilemann, "The Making of the President 2000," *WIRED*, December 1, 1995, https://www.wired.com/1995/12/gorenewt/.
9. Daniel Pearl, "Futurist Schlock," *The Wall Street Journal*, September 7, 1995, 1.
10. Brett D. Fromson and Jay Mathews, "Executives Wary But Hopeful About Prospects," *The Washington Post*, November 10, 1994, B13; David Hewson, "McNealy Trains His Sights on Computing's Big Guns," *The Sunday Times* (UK), January 28, 1996, via Nexis Uni (accessed August 30, 2018).
11. Mitch Betts, "The Politicizing of Cyberspace," *Computerworld* 29, no. 3 (January 16, 1995): 20.
12. John Heilemann, "The Making of the President 2000."
13. Philip Elmer-Dewitt, "Online Erotica: On a Screen Near You," *Time*, June 24, 2001, http://content.time.com/time/magazine/article/0,9171,134361,00.html, archived at https://perma.cc/DX42-A8JD.
14. Kara Swisher and Elizabeth Corcoran, "Gingrich Condemns On-Line Decency Act," *The Washington Post*, June 22, 1995, D8; Steve Lohr, "A Complex Medium That Will Be Hard to Regulate," *The New York Times*, June 13, 1996, B10; Nat Hentoff, "The Senate's Cybercensors," *The Washington Post*, July 1, 1995, A27; 47 U.S. Code, Section 230.
15. Daniel S. Greenberg, "Porn Does the Internet," *The Washington Post*, July 16, 1997, A19.
16. Elizabeth Darling, "Farewell to David Packard," *Palo Alto Times*, April 3, 1996, https://www.paloaltoonline.com/weekly/morgue/news/1996_Apr_3.PACKARD.html, archived at https://perma.cc/5B2A-HDPE.
17. Becky Morgan, interview with the author, May 13, 2016, by phone; Jim Cunneen, interview with the author, February 1, 2016, San Jose, Calif.; Tom Campbell interview; Ed Zschau interviews.
18. History of the National Economic Council and Clinton Administration History Project, "NEC—Education/Technology Initiative [2]," Clinton Digital Library, accessed August 7, 2017, https://clinton.presidentiallibraries.us/items/show/4837.
19. William J. Clinton: "Remarks on NetDay in Concord, California," March 9, 1996, posted by Gerhard Peters and John T. Woolley, *The American Presidency Project*, https://www.presidency.ucsb.edu/node/222473, archived at https://perma.cc/48LT-X5ZB.
20. Regis McKenna, interviews with the author; Don Bauder, "Out of Prison, Living in Luxury," *San Diego Reader*, May 26, 2010, https://www.sandiegoreader.com/news/2010/may/26/city-light-1/#, archived at https://perma.cc/3WAB-GD6P; Karen Donovan, "Bloodsucking Scumbag," *Wired*, November 1, 1996, https://www.wired.com/1996/11/es-larach/, archived at https://perma.cc/8TC6-JDRX.
21. Douglas Jehl, "Clinton to Fight Measure Revising Rules on Lawsuits," *The New York Times*, March 6, 1995, A1; Jerry Knight, "A Measure of Security on Securities Suits," *The Washington Post*, December 7, 1995, B11; Mark Simon, "Even Republicans Endorse Clinton," *San Francisco Chronicle*, August 21, 1996, C1.
22. John Markoff, "A Political Fight Marks a Coming of Age for a Silicon Valley Titan," *The New York Times*, October 21, 1996, D1.
23. John Doerr, "The Coach," interview by John Brockman, 1996, Edge.org, https://www.edge.org/digerati/doerr/, archived at https://perma.cc/9KWX-GLWK.
24. Lawrence (Larry) Stone, interview with the author, April 7, 2015, San Jose, Calif.; Sara Miles, *How to Hack a Party Line: The Democrats and Silicon Valley* (New York: Farrar,

Straus and Giroux, 2001); Philip Trounstine, "Clinton Opposes Lawsuit Measure," *San Jose Mercury News*, August 8, 1996, A1.

25. "Telephone Conversation with President Bill Clinton, Vice President Al Gore, and California Technology Executives," The White House, August 20, 1996, History of the Office of the Vice President and Clinton Administration History Project, "OVP—Gore Tech/Tech Outreach [1]," Clinton Digital Library, accessed August 10, 2017, https://clinton.presideniallibraries.us/items/show/5066.
26. Mark Simon, "GOP Voice in Silicon Valley," *The San Francisco Chronicle*, September 25, 1996, A13.
27. T. J. Rodgers, "Why Silicon Valley Should Not Normalize Relations with Washington, D.C.," Cato Institute, 1997.
28. Tom Campbell, interview with the author, February 17, 2016; Luis Buhler, interview with the author, February 8, 2016, by phone; Markoff, "A Political Fight Marks a Coming of Age."
29. Michelle Quinn, "Valley Execs Celebrate Decisive Ballot Victory," *The San Jose Mercury News*, November 6, 1996, EL1.
30. Brockman, "The Coach."
31. Lizette Alvarez, "High-Tech Industry, Long Shy of Politics, Is Now Belle of Ball," *The New York Times*, December 26, 1999, 1.

CHAPTER 22: DON'T BE EVIL

1. Michele Matassa Flores, "Gore Tells CEOs to Put Their Hearts Into It," *The Seattle Times*, May 9, 1997, A18; Howard Fineman, "The Microsoft Primary," *Newsweek*, May 19, 1997, 55; Alex Fryer, "Gates' Techno-Home Still a Work in Progress," *The Seattle Times*, May 7, 1997, A1.
2. "Microsoft Juggernaut Keeps on Rolling," *The Los Angeles Times*, April 20, 1994, 4; Clinton, Speech at Shoreline Community College, February 24, 1996, Office of Speechwriting; and James (Terry) Edmonds, "Seattle, WA (Shoreline Community College) 2/24/96 [1]," Clinton Digital Library, accessed August 15, 2017, https://clinton.presidentiallibraries.us/items/show/33816. Microsoft CTO Nathan Myhrvold was on the White House NII advisory council; the Silicon Valley-heavy Gore-Tech meetings of 1997 did not include Microsoft representatives.
3. *BusinessWeek*, February 24, 1992, quoted in Gary L. Reback, *Free the Market! Why Only Government Can Keep the Marketplace Competitive* (New York: Portfolio, 2009).
4. John Heilemann, *Pride Before the Fall: The Trials of Bill Gates and the End of the Microsoft Era* (New York: CollinsBusiness, 2001), 58, 91.
5. Karen Southwick, *High Noon: The Inside Story of Scott McNealy and the Rise of Sun Microsystems* (New York: Wiley, 1999), 45, 48.
6. Bill Gates, *The Road Ahead* (New York: Random House, 1995), x. Andreessen later said he was requoting 3Com's Bob Metcalfe when he made that famous Windows slam; see Chris Anderson, "The Man Who Makes the Future," *Wired*, April 24, 2014, https://www.wired.com/2012/04/ff-andreessen/, archived at https://perma.cc/6D5K-XGWJ.
7. Gates, Memorandum, "The Internet Tidal Wave," May 26, 1995, Exhibit 20, *United States v. Microsoft Corporation* 253 F.3d 34 (D.C. Cir. 2001).
8. Reback, *Free the Market!*; Heilemann, Pride *Before the Fall*, 64–67; Joel Brinkley and Steve Lohr, *The U.S. v. Microsoft: The Inside Story of the Landmark Case* (New York: McGraw-Hill Education, 2000), 4, 48–49.

9. *Time*, "The Golden Geeks," February 19, 1996, cover; "Whose Web Will It Be?," September 16, 1996, cover.
10. James Lardner, "Trying to Survive the Browser Wars," *U.S. News & World Report* 124, no. 13 (April 6, 1998); Heilemann, *Pride Before the Fall*, 91.
11. Brinkely and Lohr, *The U.S. v. Microsoft*, 38–40.
12. Heilemann, *Pride Before the Fall*, 42.
13. James Taranto, "Nader's Raiders Try to Storm Bill's Gates," *The Wall Street Journal*, November 18, 1997, A22; Nader, "The Microsoft Menace," *Slate*, October 30, 1997, http://www.slate.com/articles/briefing/articles/1997/10/the_microsoft_menace.html, archived at https://perma.cc/9ZEQ-8UWK.
14. Elizabeth Corcoran and Rajiv Chandrasekaran, "Nader Joins Chorus of Microsoft Critics," *The Washington Post*, November 14, 1997, G1; Gerald F. Seib, "Freedom Fighters: Antitrust Suits Expand and Libertarians Ask, Who's The Bad Guy?" *The Wall Street Journal*, June 9, 1998, A1.
15. Corcoran and Chandrasekaran, "Nader Joins Chorus."
16. Neukom quoted in David Lawsky, "Microsoft Urges Government to Drop Antitrust Case," Reuters, reprinted in *The Times of India*, November 26, 1998, 15; also see Steve Lohr, "Microsoft Presses Its View About Rivals' 3-Way Deal," *The New York Times*, January 7, 1999, C2.
17. Lizette Alvarez, "High-Tech Industry, Long Shy of Politics, Is Now Belle of Ball," *The New York Times*, December 26, 1999, 1.
18. Hiren Shah, "Y2K: The Bug of the Millennium," *The Times of India*, October 19, 1998, 14; Stephen Barr, "Social Security Killed Y2K Bug, President Says," *The Washington Post*, December 29, 1998, A2; Eric Lipton, "2-Digit Problem Means 9-Digit Bill for Local Governments," *The Washington Post*, August 4, 1998, A1.
19. Abhi Raghunathan, "Thanks for Coming. Now Go," *The New York Times*, July 15, 2001, NJ1.
20. Kathleen Kenna, "Commander in Geek," *Toronto Star*, May 24, 1999, 1; Paul A. Gigot, "Gore Slams Doerr on Silicon Valley," *The Wall Street Journal*, May 21, 1999, 21.
21. Jon Swartz, "Tech's Star Capitalist," *The San Francisco Chronicle*, November 13, 1997, D3; Marc Gunther and Adam Lashinsky, "Cleanup Crew," *Fortune* 156, no. 11 (November 26, 2007): 82–92.
22. Quoted in John Schwartz, "A Judge Overturned by an Appearance of Bias," *The New York Times*, June 29, 2001, p. C1.
23. Joel Brinkley, "U.S. vs. Microsoft: The Lobbying," *The New York Times*, September 7, 2001, C5.
24. Quoted in Alvarez, "High-Tech Industry."
25. James Gibbons, interview with the author, November 4, 2015.
26. "Bill Gates Stanford Dedication—Jan. 30, 1996," *Microsoft News*, https://news.microsoft.com/1996/01/30/bill-gates-stanford-dedication-jan-30-1996/, archived at https://perma.cc/XCK6-9SS6.
27. Ellen Ullman, *Life in Code: A Personal History of Technology* (New York: Farrar, Straus and Giroux, 2017), 100.
28. "Turning an Info-Glut into a Library," *Science* 266 (October 7, 1994), 20; Bruce Schatz and Hsinchun Chen, "Building Large-Scale Digital Libraries," *Computer* 29, no. 5 (May 1996): 22–26.
29. John Battelle, *The Search: How Google and Its Rivals Rewrote the Rules of Business and Transformed Our Culture* (New York: Portfolio, 2005), 65–75; Rich Scholes, "Uniquely

Google," *Stanford Technology Brainstorm*, Stanford Office of Technology Licensing, March 2000.

30. Scholes, "Uniquely Google."
31. Battelle, *The Search*, 90; "Sergey Brin's Home Page," http://infolab.stanford.edu/~sergey/, accessed May 20, 2018, archived at https://perma.cc/XH2S-RW58.

ARRIVALS

1. Chamath Palihapitiya, interview with the author, December 5, 2017.
2. Walter Mossberg, "Behind the Lawsuit: Napster Provides Model for Music Distribution," *The Wall Street Journal*, May 11, 2000, C1.
3. Cyrus Farivar, "Winamp's Woes: How the Greatest MP3 Player Undid Itself," *Ars Technica*, July 3, 2017; "The Biggest Media Merger Yet," *The New York Times*, January 11, 2000, A24. On the AOL Time Warner merger and its effects, see Kara Swisher with Lisa Dickey, *There Must Be a Pony in Here Somewhere: The AOL Time Warner Debacle* (New York: Three Rivers Press, 2003).

CHAPTER 23: THE INTERNET IS YOU

1. Joint Venture Silicon Valley, *2002 Index* (Palo Alto, Calif.: Joint Venture Silicon Valley, 2002); Gregory Zuckerman, "A Year After the Peak, How the Mighty Have Fallen," *The Wall Street Journal*, March 5, 2001, C1; Scott Berinato, "What When Wrong at Cisco in 2001," *CIO Magazine* 14, no. 20 (August 2001): 52–59.
2. Edward Helmore, "Lost Stock & Two Smoking Analysts," *The Guardian*, March 15, 2001, B12.
3. Zach Schiller, "Morgenthaler Scores in IPO," *Cleveland Plain Dealer*, March 4, 2000, C3; Alex Berenson, "Stocks End Gloomy First Quarter," *The New York Times*, March 31, 2001, C1; Burt McMurtry, interview with the author, October 2, 2017; Ann Hardy, interview with the author, September 19, 2017.
4. Mike Tarsala, "Pets.com Killed by Sock Puppet," *MarketWatch*, November 8, 2000, https://www.marketwatch.com/story/sock-puppet-kills-petscom, archived at https://perma.cc/T6WU-HKW5. Also see Jennifer Thornton and Sunny Marche, "Sorting through the dot bomb rubble: how did high-profile e-tailers fail?" *International Journal of Information Management* 23, no. 2 (April 2003): 121–38.
5. Joint Venture Silicon Valley, *2002 Index*, 6–7; Julekha Dash, "Former dot-com workers find slow start in new year," *Computerworld*, January 8, 2001, https://www.computerworld.com/article/2590192/it-careers/former-dot-com-workers-find-slow-start-in-new-year.html, archived at https://perma.cc/X4J2-JS7M.
6. Fred Vogelstein, "Google @ $165: Are These Guys for Real?," *Fortune*, December 13, 2004, http://archive.fortune.com/magazines/fortune/fortune_archive/2004/12/13/8214226/index.htm, archived at https://perma.cc/YQU9-QV94; "Liorean," comment thread "Google 1G Mail," *CodingForums.com*, June 2, 2004, https://www.codingforums.com/geek-news-and-humour/39589-google-1g-mail.html, archived at https://perma.cc/549J-5HNY; also see Kevin Marks, "Epeus' epigone," March 21, 2012, http://epeus.blogspot.com/2012/03/when-youre-merchandise-not-customer.html, archived at https://perma.cc/EP7P-ZBTN.
7. "From the Garage to the Googleplex," Alphabet, Inc., https://www.google.com/about/our-story/, archived at https://perma.cc/63XD-AZCA; "A Building Blessed with Tech Success,"

CNET, October 14, 2002, https://www.cnet.com/news/a-building-blessed-with-tech-success/, archived at https://perma.cc/H4W8-RSJE; Verne Kopytoff, "The Internet Kid is Growing Up Fast," *The San Francisco Chronicle*, September 11, 2000, A24.

8. Ken Auletta, *Googled: The End of the World as We Know It* (New York: Penguin Press, 2010), 20.
9. Stephanie Schorow, "Web heads go ga-ga for Google, for good reason," *Boston Herald*, December 4, 2001, 51.
10. Fred Turner, "Burning Man at Google: a cultural infrastructure for new media production," *New Media & Society* 11, nos. 1 & 2 (2009): 73–94; "Ten Principles of Burning Man," https://burningman.org/culture/philosophical-center/10-principles/, archived at https://perma.cc/KS28-9M36.
11. Auletta, *Googled*, 80.
12. John Battelle, *The Search: How Google and Its Rivals Rewrote the Rules of Business and Transformed Our Culture* (New York: Portfolio, 2005).
13. Google, "Ten things we know to be true," https://www.google.com/intl/en/about/philosophy.html, archived at https://perma.cc/G865-BALX.
14. Doerr quoted in Matt Marshall, "Is Google Like Microsoft? In Some Ways," *The San Jose Mercury News*, September 25, 2003, 1C.
15. Vogelstein, "Google @ $165."
16. Shirin Sharif, "Web Site Allows Students to Make Friends from Faces in the Crowd," *The Stanford Daily*, March 5, 2004, 1.
17. Sharif, "Web Site Allows"; U.S. Securities and Exchange Commission Form S-1, Facebook, Inc., February 1, 2012.
18. On the emergence of social networking and the struggles and triumphs of its pre-Facebook companies, see Julia Angwin, *Stealing MySpace: The Battle to Control the Most Popular Web Site in America* (New York: Random House, 2009).
19. The definitive history of the early years of Facebook (and the basis for the not altogether charitable portrayal of Zuckerberg and his company in the 2011 Hollywood film *The Social Network*) is David Kirkpatrick, *The Facebook Effect: The Inside Story of the Company that is Connecting the World* (New York: Simon & Schuster, 2010).
20. This data comes from, naturally, Wikipedia. "List of most popular websites," Wikipedia, March 2018, https://en.wikipedia.org/wiki/List_of_most_popular_websites, archived at https://perma.cc/9QBA-ABF6.
21. Lev Grossman, "You—Yes, You—Are TIME's Person of the Year," *Time*, December 25, 2006.
22. Esther Dyson et al., "Cyberspace and the American Dream: A Magna Carta for the Knowledge Age" (Release 1.2, August 22, 1994), *The Information Society* 12, no. 3 (1996): 295–308.
23. Ryan Singel, "Silicon Valley Lacks Vision? Facebook Begs to Differ," *Wired*, October 8, 2010, https://www.wired.com/2010/10/facebook-matters/, archived at https://perma.cc/VV4J-2JMS.
24. Chamath Palihapitiya, interview with the author; Palihapitiya interviewed by Kara Swisher, *Recode Decode* podcast, March 20, 2016, https://www.recode.net/2016/3/21/11587128/silicon-valleys-homogeneous-rich-douchebags-wont-win-forever-says, archived at https://perma.cc/PK2L-DDCR; Evelyn M. Rusli, "In Flip-Flops and Jeans, An Unconventional Venture Capitalist," DealBook blog, *The New York Times*, October 6, 2011, https://dealbook.nytimes.com/2011/10/06/in-flip-flops-and-jeans-the-unconventional-venture-capitalist/,

archived at https://perma.cc/C7X7-KWJ2; Eugene Kim, "Early Facebook Executive on Mark Zuckerberg," *Business Insider*, November 23, 2014, https://www.businessinsider.com.au/chamath-palihapitiya-on-mark-zuckerberg-2014-11, archived at https://perma.cc/9CLK-S8RS.

25. Caroline McCarthy, "Facebook f8: One Graph to Rule them All," CNET, April 21, 2010, https://www.cnet.com/news/facebook-f8-one-graph-to-rule-them-all/, archived at https://perma.cc/W4T5-49CM. Scholars began raising red flags about the ethics of such information-sharing practices as soon as they started. See for example Michael Zimmer, "'But the Data is Already Public': On the Ethics of Research in Facebook," *Ethics and Information Technology* 12, no. 4 (December 2010): 313–25; Rebecca McKee, "Ethical Issues in Using Social Media for Health and Health Care Research," *Health Policy* 110, nos. 2–3 (May 2013): 298–301. As Facebook's user base skyrocketed and "like" buttons metastasized around the Web, the company attracted the attention of the FTC, which required Facebook to adopt stricter and more transparent privacy standards (Federal Trade Commission, Decision and Order in the Matter of Facebook, Inc., Docket No. C-4365, August 10, 2012).
26. Nick Bilton, "A Walk in the Woods with Mark Zuckerberg," *The New York Times*, July 7, 2011, https://bits.blogs.nytimes.com/2011/07/07/a-walk-in-the-woods-with-mark-zuckerberg/, archived at https://perma.cc/86DU-LAWF.
27. Heather Brown, Emily Guskin, and Amy Mitchell, "The Role of Social Media in the Arab Uprisings," Pew Research Center, November 28, 2012; Benjamin Gleason, "#Occupy Wall Street: Exploring Informal Learning About a Social Movement on Twitter," *American Behavioral Scientist* 57, no. 7 (2013): 966–82; André Brock, "From the Blackhand Side: Twitter as a Cultural Conversation," *Journal of Broadcasting and Electronic Media* 56, no. 4 (2012): 529–49; Russell Rickford, "Black Lives Matter: Toward a Modern Practice of Mass Struggle," *New Labor Forum* 25, no. 1 (2016): 34–42.
28. Joshua Green, "The Amazing Money Machine," *The Atlantic*, June 1, 2008, https://www.theatlantic.com/magazine/archive/2008/06/the-amazing-money-machine/306809/, archived at https://perma.cc/V67S-PX4W; Brian Stelter, "The Facebooker Who Friended Obama," *The New York Times*, July 7, 2008, https://www.nytimes.com/2008/07/07/technology/07hughes.html, archived at https://perma.cc/U74U-XQ7Z.
29. Kristina Peterson, "Obama opening Silicon Valley office," *Palo Alto Daily News*, January 13, 2008, 1; Green, "The Amazing Money Machine"; Cecilia Kang and Perry Bacon Jr., "Obama Holds Silicon Valley Summit with Tech Tycoons," *The Washington Post*, February 18, 2011, C1.
30. "I am Barack Obama, President of the United States—AMA," Reddit, August 29, 2012, https://www.reddit.com/r/IAmA/comments/z1c9z/i_am_barack_obama_president_of_the_united_states/, archived at https://perma.cc/BB8Z-D7GZ; Brody Mullins, "Google Makes Most of Close Ties to the White House," *The Wall Street Journal*, March 24, 2015, https://www.wsj.com/articles/google-makes-most-of-close-ties-to-white-house-1427242076; David Dayen, "The Android Administration," *The Intercept*, April 22, 2016, https://theintercept.com/2016/04/22/googles-remarkably-close-relationship-with-the-obama-white-house-in-two-charts/, archived at https://perma.cc/NUP2-6XW6; Cecilia Kang and Juliet Eilperin, "A Clear Affinity Between White House, Silicon Valley," *The Washington Post*, February 28, 2015, http://www.pressreader.com/usa/the-washington-post/20150228/281784217548185. Also see Thomas Kalil, "Policy Entrepreneurship at the White House," *Innovations* 11, nos. 3/4 (2017): 4–22.

31. "The 'Anti-Business' President Who's Been Good for Business," *Bloomberg Businessweek*, June 27, 2016, https://www.bloomberg.com/features/2016-obama-anti-business-president/, archived at https://perma.cc/RG5N-VP2P.
32. Barack Obama, Speech at the White House Summit on Cybersecurity, Stanford, Calif., February 13, 2015. Obama's speechwriters had produced a knowing riff on the cathedral-vs.-bazaar software metaphors so familiar to Silicon Valley insiders, gloriously updated for a social-media age.

CHAPTER 24: SOFTWARE EATS THE WORLD

1. John C. Abell, "Aug. 6, 1997: Apple Rescued—By Microsoft," *Wired*, August 6, 2009, https://www.wired.com/2009/08/dayintech-0806/, archived at https://perma.cc/2RRH-FUBH.
2. Ken Siegmann, "Veteran Apple Exec Leaves for Top Job at Go," *The San Francisco Chronicle*, January 19, 1991, 1C; Walter Isaacson, *Steve Jobs* (New York: Simon & Schuster, 2011), 308.
3. Brian Merchant, *The One Device: The Secret History of the iPhone* (New York: Little, Brown, 2017), 148–62.
4. Morgenthaler Partners had been an investor in the company that made that voice recognition software, called Siri in homage to being developed at SRI. A DARPA grant had helped seed its early development. See SRI International, "Siri," https://www.sri.com/work/timeline-innovation/timeline.php?timeline=computing-digital#!&innovation=siri, archived at https://perma.cc/7SNR-V6MQ.
5. "For Apple Chief, Gadgets' Glitter Outshines Scandal," *The New York Times*, January 9, 2007, B1; Erica Sadun, "Macworld 2007 Keynote Liveblog," *Engadget*, January 9, 2007, https://www.engadget.com/2007/01/09/macworld-2007-keynote-liveblog/, archived at https://perma.cc/4394-QYDG.
6. Martyn Williams, "In his own words: The best quotes of Steve Ballmer," *PC World*, August 19, 2014.
7. Merchant, *The One Device*, 162–71; Doug Gross, "Apple trademarks 'There's an app for that,'" *CNN*, October 12, 2010.
8. Ken Auletta, *Googled: The End of the World as We Know It* (New York: Penguin Press, 2010), 204, 207–210.
9. "Mobile Fact Sheet," Pew Research Center, February 5, 2018, http://www.pewinternet.org/fact-sheet/mobile/, archived at https://perma.cc/44L8-W6EN.
10. Horace Dediu, "The iOS Economy, Updated," Asymco blog, January 8, 2018, http://www.asymco.com/2018/01/08/the-ios-economy-updated/, archived at https://perma.cc/W2Z5-MT6G.
11. Bruce Newman, "Steve Jobs, Apple Co-Founder," *San Jose Mercury News*, October 5, 2011.
12. "Remembering Steve," Apple.com, https://www.apple.com/stevejobs/, archived at https://perma.cc/7SES-3F5F; Maria L. LaGanga, "Steve Jobs' death saddens Apple workers and fans," *The Los Angeles Times*, October 6, 2011.
13. "Steve Jobs' Memorial Service: 6 Highlights," *The Week*, October 25, 2011.
14. "What Happened to the Future?" Founders Fund, http://foundersfund.com/the-future/, archived at https://perma.cc/82XW-VA2A.
15. Adam Lashinsky, "Amazon's Jeff Bezos: The Ultimate Disrupter," *Fortune* (December 2012); Jeff Bezos, "1997 Letter to Shareholders," Investor Relations, Amazon.com.

16. Jeff Bezos, "2005 Letter to Shareholders," Investor Relations, Amazon.com; Julia Kirby and Thomas A. Stewart, "The Institutional Yes," *Harvard Business Review*, October 2007, 8, https://hbr.org/2007/10/the-institutional-yes, archived at https://perma.cc/XV5H-GULN.
17. Jeff Bezos, "2011 Letter to Shareholders," Investor Relations, Amazon.com.
18. Ingrid Burrington, "Why Amazon's Data Centers are Hidden in Spy Country," *The Atlantic*, January 8, 2016.
19. Frank Konkel, "Daring Deal," *Government Executive*, July 9, 2014. An advantage for Amazon's securing national security cloud business was that it was not one of the American tech companies ensnared in PRISM, the intelligence-gathering program revealed in 2013 by NSA contractor Edward Snowden. Nearly every other boldface tech name appeared in the cache of classified documents, but 98 percent of the data came from only three: Yahoo!, Google, and Microsoft. The NSA had been in the electronic surveillance business since its 1947 inception, but involvement of consumer tech's biggest brands—including the "don't be evil" empire of Page and Brin—precipitated a major scandal. See Barton Gellman and Laura Poitras, "U.S., British intelligence mining data from nine U.S. Internet companies in broad secret program," *The Washington Post*, June 7, 2013, A1.
20. Nick Wingfield, "Amazon Reports Annual Net Profit for the First Time," *The Wall Street Journal*, January 28, 2004; Ron Miller, "How AWS Came to Be," *TechCrunch*, July 2, 2016; Jordan Novet, "Microsoft narrows Amazon's lead in cloud, but the gap remains large," CNBC, April 27, 2018.
21. Ashton B. Carter with Marcel Lettre and Shane Smith, "Keeping the Technological Edge," in *Keeping the Edge: Managing Defense for the Future*, ed. Ashton B. Carter, John Patrick White (Cambridge, Mass.: The MIT Press, 2001), 130–63.
22. Peter Thiel, "The Education of a Libertarian," The Cato Institute, April 13, 2009.
23. Rachel Riederer, "Libertarians Seek a Home on the High Seas," *The New Republic*, June 1, 2017; George Packer, "No Death, No Taxes," *The New Yorker*, November 28, 2011.
24. Andy Greenberg, "How a 'Deviant Philosopher' Built Palantir, A CIA-Funded Data-Mining Juggernaut," *Forbes*, September 2, 2013.
25. Rick E. Yannuzzi, "In-Q-Tel: A New Partnership between the CIA and the Private Sector," *Defense Intelligence Journal* (2000), Central Intelligence Agency, https://www.cia.gov/library/publications/intelligence-history/in-q-tel#copy, archived at https://perma.cc/AV9M-JTCA.
26. Greenberg, "How a 'Deviant Philosopher' Built Palantir"; Ellen Mitchell, "How Silicon Valley's Palantir wired Washington," *Politico*, August 14, 2016.
27. Anonymous, comment to "What is the interview process like at Palantir?," Quora, February 17, 2011, archived at https://perma.cc/R4FM-LPXL.
28. Julie Bort, "What It's Like to Work at the Valley's Most Secretive Startup," *Business Insider*, July 31, 2016; Ryan Singel, "Anonymous vs. EFF?" *Wired*, November 14, 2011.
29. Andrew Ruiz, Twitter, April 30, 2018, archived at https://perma.cc/FZ6X-VU84. Thiel also was unafraid to throw his weight around, most notably when he bankrolled a defamation suit brought by wrestler Hulk Hogan against the online newspaper Gawker, which had also outed Thiel against his wishes. After the suit went in Hogan's favor, a bankrupt Gawker had to shut down.

CHAPTER 25: MASTERS OF THE UNIVERSE

1. Adam Gorlick, "'I wanted to see with my own eyes the origin of success,' Russian president tells Stanford audience," *Stanford Report*, June 23, 2010; "Dmitry Medvedev visits Twitter HQ and tweets," *The Telegraph* (UK), June 24, 2010.
2. "Medvedev targeted with mock Twitter account," *The Telegraph* (UK), July 5, 2010; @KermlinRussia, Twitter, January 8, 2011, archived at https://perma.cc/K4V8-K7VK.
3. Vivek Wadhwa, AnnaLee Saxenian, and F. Daniel Siciliano, *Then and Now: America's New Immigrant Entrepreneurs*, part VII, Kauffman Foundation Research paper, 2012; "International Students," Stanford Engineering, accessed May 27, 2018, archived at https://perma.cc/EFS3-3X7N.
4. Marc Andreessen, "Why Software Is Eating the World," *The Wall Street Journal*, August 20, 2011.
5. Richard L. Florida and Martin Kenney, "Venture Capital, High Technology and Regional Development," *Regional Studies* 22, no. 1 (1988): 33–48; Florida, "America's Leading Metros for Venture Capital," *CityLab*, June 17, 2013; Chris DeVore, "The Venture Capital Stack + Regional Seed VC," *Crash/Dev*, June 15, 2017, archived at https://perma.cc/T493-FALD.
6. Tim Wu, tweet, 5/24/2018 8:14AM; Regis McKenna, interview with the author December 3, 2014. Peter Thiel believed in this market-definition-and-domination strategy so strongly that he co-wrote a book on the subject, *Zero to One: Notes on Startups, or How to Build the Future* (New York: Random House, 2014).
7. Zuckerberg, Facebook post, March 30, 2015, archived at https://perma.cc/S9DW-RVPW.
8. Margaret O'Mara, "The Other Tech Bubble," *The American Prospect*, Winter 2016.
9. Katie Hafner, "Google Options Make Masseuse a Multimillionaire," *The New York Times*, November 12, 2007; Kevin Maney, "Marc Andreessen puts his money where his mouth is," *Fortune*, July 10, 2009.
10. In September 2018, after reported infighting within Kleiner, Meeker abruptly quit, bringing along several other of its senior late-stage investors to start a new venture firm under her own leadership. Theodore Schleifer, "Mary Meeker, the Legendary Internet Analyst, Is Leaving Kleiner Perkins," *Recode*, September 14, 2018, https://www.recode.net/2018/9/14/17858582/kleiner-perkins-mary-meeker-split, archived at https://perma.cc/FJ8S-DVUM.
11. Jesse Drucker, "Kremlin Cash Behind Billionaire's Twitter and Facebook Investments," *The New York Times*, November 5, 2017; Michael Wolff, "How Russian Tycoon Yuri Milner Bought His Way into Silicon Valley," *Wired*, October 21, 2011.
12. Chris William Sanchirico, "As American as Apple Inc.: International Tax and Ownership Nationality," *Tax Law Review* 68, no. 2 (2015): 207–74; Rebecca Greenfield, "Senators Turn Tim Cook's Hearing into a Genius Bar Visit," *The Atlantic*, May 21, 2013.
13. David Kirkpatrick, "Inside Sean Parker's Wedding," *Vanity Fair*, August 1, 2013.
14. "Yammer Raises $17 Million in Financing Round Led by The Social+Capital Partnerhip," *Marketwire*, September 27, 2011.
15. Ben Horowitz, *The Hard Thing about Hard Things: Building a Business When There Are No Easy Answers* (New York: HarperBusiness, 2014), 62.
16. Ellen McGirt, "Al Gore's $100 Million Makeover," *Fast Company*, July 1, 2007.
17. John Doerr, "Salvation (and profit) in Greentech," TED2007, March 2007; Marc Gunther and Adam Lashinsky, "Cleanup Crew," *Fortune* 156, no. 11 (November 26, 2007).

18. Jon Gertner, "Capitalism to the Rescue," *The New York Times*, October 3, 2008.
19. Jerry Hirsch, "Elon Musk's growing empire is fueled by $4.9 billion in government subsidies," *The Los Angeles Times*, May 30, 2015; Sarah McBride and Nichola Groom, "Insight: How cleantech tarnished Kleiner and VC star John Doerr," *Reuters Business News*, January 15, 2013.
20. David Streitfeld, "Kleiner Perkins Denies Sex Bias in Response to a Lawsuit," *The New York Times*, June 14, 2012; Ellen Huet, "Kleiner Perkins' John Doerr and Ellen Pao: A Mentorship Sours," *Forbes*, March 4, 2015.
21. Gené Teare and Ned Desmond, "The first comprehensive study on women in venture capital and their impact on female founders," *TechCrunch*, April 19, 2016; "Despite More Women, VCs Still Mostly White Men," *The Information*, December 14, 2016.
22. Laszlo Bock, "Getting to work on diversity at Google," Google blog, May 28, 2014; Maxine Williams, "Building a More Diverse Facebook," Facebook Newsroom, June 25, 2014; Mallory Pickett, "The Dangers of Keeping Women out of Tech," *Wired*, January 26, 2018.
23. Emily Chang, *Brotopia: Breaking Up the Boys' Club of Silicon Valley* (New York: Portfolio, 2018), 145–46; John Doerr, interview with Emily Chang, Bloomberg TV, June 18, 2015.
24. Graham, "Why to Move to a Startup Hub," PaulGraham.com, October 2007, archived at https://perma.cc/TYF6-G3KT.
25. Kara Swisher, interview with Chamath Palihapitiya, *Recode/Decode*, March 20, 2016; Palihapitiya, interview with the author, December 5, 2017.
26. Ashley Carroll, "Capital-as-a-Service: A New Operating System for Early-Stage Investing," *Medium*, October 25, 2017, https://medium.com/social-capital/capital-as-a-service-a-new-operating-system-for-early-stage-investing-6d001416c0df, archived at https://perma.cc/G5QD-DUCF. Social Capital did not have much opportunity to test whether the new model would make a significant difference in the diversity of venture-funded entrepreneurs. The partnership imploded in the early fall of 2018 after an exodus of Palihapitiya's co-founders and other key executives, leaving the future of the firm and "CaaS" unclear.
27. Trish Millines Dziko, interview with the author, April 3, 2018.
28. Issie Lapowsky, "Clinton Owns Silicon Valley's Vote Now That Bloomberg's Out," *Wired*, March 8, 2016.
29. Thiel, "Trump Has Taught Us This Year's Most Valuable Political Lesson," *The Washington Post*, September 6, 2016.
30. Mitch Kapor, interview with the author, September 19, 2017.

DEPARTURE: INTO THE DRIVERLESS CAR

1. This also was a reminder of the Pentagon spending still lurking behind the Valley's entrepreneurial audacity, for a DARPA "Grand Challenge" competition a decade earlier had revved up the race to bring driverless vehicles to market. As ever, the Valley's next generation was helped along by the military's willingness to make far-out bets. See Alex Davies, "Inside the Races that Jump-Started the Self-Driving Car," *Wired*, November 10, 2017, https://www.wired.com/story/darpa-grand-urban-challenge-self-driving-car/, archived at https://perma.cc/EWN5-8XCD.
2. Tiernan Ray and Alex Eule, "John Doerr on Leadership, Education, Google, and AI," *Barron's*, May 5, 2018, https://www.barrons.com/articles/john-doerr-on-leadership-education-google-and-ai-1525478401, archived at https://perma.cc/S2W5-5GMY; James Morra, "Groq to reveal potent artificial intelligence chip next year," *ee News: Europe*, November 17,

2017, http://www.eenewseurope.com/news/groq-reveal-potent-artificial-intelligence-chip-next-year, archived at https://perma.cc/FQ3G-YAEK.

3. Maria di Mento, "Technology Investor Pledges $32 Million to Rice U," *Chronicle of Philanthropy* 18, no. 18 (June 29, 2006), via Nexis Uni, accessed August 30, 2018; Burt and Deedee McMurtry, "Remarks at the McMurtry Building Groundbreaking Ceremony," May 15, 2013, Stanford Arts, https://arts.stanford.edu/remarks-by-burt-and-deedee-mcmurtry/, archived at https://perma.cc/H59G-B2LW; McMurtry, interview with the author, October 2, 2017.
4. Gary Morgenthaler, e-mail correspondence with the author, August 17, 2016; "Startup developing new battery technology wins $12,000 in first MIT ACCELERATE Contest," *MIT News*, March 6, 2012, http://news.mit.edu/2012/battery-technology-startup-wins-accelerate-contest, archived at https://perma.cc/QAN9-ZHXB; Morgenthaler, interviews with the author, 2015 and 2016. David Morgenthaler died on June 16, 2016, at the age of 96, survived by his wife Lindsay, sons Gary and Todd, daughter Lissa, seven grandchildren, and four great-grandchildren. Katie Benner, "David T. Morgenthaler, Who Shaped Venture Capitalism, Dies at 96," *The New York Times*, June 21, 2016, A21.
5. Ann Hardy, interview with the author, September 19, 2017; phone conversation with the author, August 28, 2018. The continuing demographic imbalances in tech and the consequences for the products it builds and markets produced a wave of books during this period; see, for example: Emily Chang, *Brotopia: Breaking Up the Boys' Club of Silicon Valley* (New York: Portfolio, 2018); Safiya Umoja Noble, *Algorithms of Oppression: How Search Engines Reinforce Racism* (New York: New York University Press, 2018); Virginia Eubanks, *Automating Inequality: How High-Tech Tools Profile, Police, and Punish the Poor* (New York: St. Martin's Press, 2018); Sara Watcher-Boettcher, *Technically Wrong: Sexist Apps, Biased Algorithms, and Other Threats of Toxic Tech* (New York: W. W. Norton, 2017).
6. Yaw Anokwa and Hélène Martin, interview with the author, June 7, 2018, Seattle, Wash.

IMAGE CREDITS

1. Morgenthaler family photograph, Courtesy of Gary Morgenthaler.
2. Personal photograph, Courtesy of Ann Hardy.
3. Reproduced with Permission, Palo Alto Historical Association.
4. Reproduced with Permission, MIT Museum.
5. Reproduced with Permission, Courtesy of Agilent Technologies, Inc.
6. Reproduced with Permission, Palo Alto Historical Association.
7. Wayne Miller / Magnum Photos.
8. Schlesinger Library, Radcliffe Institute, Harvard University.
9. Computer History Museum, Gift of Lee Felsenstein.
10. Bruce Burgess / Lakeside School, Courtesy of Paul Allen and Bill Gates.
11. Computer History Museum, Gift of Jim Warren.
12. Personal photograph, Courtesy of Ed Zschau.
13. Personal photograph, Courtesy of Regis McKenna.
14. Personal photograph, Courtesy of Regis McKenna.
15. AP Photo / Neal Ulevich.
16. Computer History Museum.
17. Ann E. Yow-Dyson / Getty Images.
18. Ed Kashi / VII / Redux.
19. Ann E. Yow-Dyson / Getty Images.
20. AP Photo / Paul Sakuma.
21. AP Photo / Denis Paquin.
22. AP Photo / Namas Bhojani.
23. AP Photo / Andy Rogers.
24. AP Photo / Justin Sullivan.
25. AP Photo / Ben Margot.
26. Kimberly White / Getty Images.
27. AP Photo / Marcio Jose Sanchez.
28. Brian Ach / Getty Images.
29. Morgenthaler family photograph, Courtesy of Gary Morgenthaler.
30. Personal photograph, Courtesy of Ann Hardy.
31. Stanford News Service / Steve Castillo.
32. Gilles Mingasson / Getty Images.

INDEX